Energy Balance Climate Models

Wiley Series in Atmospheric Physics and Remote Sensing
Series Editor: Alexander Kokhanovsky

Wendisch, M. / Brenguier, J.-L. (eds.)
Airborne Measurements for Environmental Research
Methods and Instruments

2013

Coakley Jr., J. A. / Yang, P.
Atmospheric Radiation
A Primer with Illustrative Solutions

2014

Stamnes, K. / Stamnes, J. J.
Radiative Transfer in Coupled Environmental Systems
An Introduction to Forward and Inverse Modeling

2015

Tomasi, C. / Fuzzi, S. / Kokhanovsky, A.
Atmospheric Aerosols
Life Cycles and Effects on Air Quality and Climate

2016

Weng, F.
Passive Microwave Remote Sensing of the Earth
for Meteorological Applications

2017

North, G. R. / Kim, K.-Y.
Energy Balance Climate Models

Forthcoming:

Kokhanovsky, A. / Natraj, V.
Analytical Methods in Atmospheric Radiative Transfer

Huang, X. / Yang, P.
Radiative Transfer Processes in Weather and Climate Models

Davis, A. B. / Marshak, A.
Multi-dimensional Radiative Transfer
Theory, Observation, and Computation

Minnis, P. *et al.*
Satellite Remote Sensing of Clouds

Zhang, Z. *et al.*
Polarimetric Remote Sensing
Aerosols and Clouds

Energy Balance Climate Models

Gerald R. North and Kwang-Yul Kim

Authors

Prof. Gerald R. North
Texas A&M University
Department of Atmospheric Sciences
College Station
TX 77843-3150
United States

Prof. Kwang-Yul Kim
Seoul National University
College of Natural Sciences
1 Gwanak-ro, Gwanak-gu
Seoul 08826
Republic of Korea

A book of the Wiley Series in Atmospheric Physics and Remote Sensing

The Series Editor
Dr. Alexander Kokhanovsky
EUMETSAT
EUMETSAT-Allee 1
64295 Darmstadt
Germany

All books published by **Wiley-VCH** are carefully produced. Nevertheless, authors, editors, and publisher do not warrant the information contained in these books, including this book, to be free of errors. Readers are advised to keep in mind that statements, data, illustrations, procedural details or other items may inadvertently be inaccurate.

Library of Congress Card No.: applied for

British Library Cataloguing-in-Publication Data
A catalogue record for this book is available from the British Library.

Bibliographic information published by the Deutsche Nationalbibliothek
The Deutsche Nationalbibliothek lists this publication in the Deutsche Nationalbibliografie; detailed bibliographic data are available on the Internet at <http://dnb.d-nb.de>.

© 2017 Wiley-VCH Verlag GmbH & Co. KGaA, Boschstr. 12, 69469 Weinheim, Germany

All rights reserved (including those of translation into other languages). No part of this book may be reproduced in any form – by photoprinting, microfilm, or any other means – nor transmitted or translated into a machine language without written permission from the publishers. Registered names, trademarks, etc. used in this book, even when not specifically marked as such, are not to be considered unprotected by law.

Print ISBN: 978-3-527-41132-0
ePDF ISBN: 978-3-527-68383-3
ePub ISBN: 978-3-527-68381-9
Mobi ISBN: 978-3-527-68384-0
oBook ISBN: 978-3-527-69884-4

Cover Design Grafik-Design Schulz
Typesetting SPi Global Private Limited, Chennai, India
Printing and Binding CPI Group (UK) Ltd, Croydon, CR0 4YY

Printed on acid-free paper

We wish to dedicate this book to the memory of those who worked with us on EBMs but were taken away much too soon, Thomas J. Crowley, John Mengel, and Wan-Ho Lee.

Contents

Preface *xiii*

1	**Climate and Climate Models** *1*	
1.1	Defining Climate *3*	
1.2	Elementary Climate System Anatomy *7*	
1.3	Radiation and Climate *9*	
1.3.1	Solar Radiation *9*	
1.3.2	Albedo of the Earth–Atmosphere System *13*	
1.3.3	Terrestrial Infrared Radiation into Space (The IR or Longwave Radiation) *14*	
1.4	Hierarchy of Climate Models *15*	
1.4.1	General Circulation Models (GCMs) *16*	
1.4.2	Energy Balance Climate Models *17*	
1.4.3	Adjustable Parameters in Phenomenological Models *19*	
1.5	Greenhouse Effect and Modern Climate Change *20*	
1.6	Reading This Book *20*	
1.7	Cautionary Note and Disclaimer *22*	
	Notes on Further Reading *23*	
	Exercises *23*	
2	**Global Average Models** *27*	
2.1	Temperature and Heat Balance *27*	
2.1.1	Blackbody Earth *28*	
2.1.2	Budyko's Empirical IR Formula *29*	
2.1.3	Climate Sensitivity *30*	
2.1.4	Climate Sensitivity and Carbon Dioxide *31*	
2.2	Time Dependence *31*	
2.2.1	Frequency Response of Global Climate *32*	
2.2.2	Forcing with Noise *35*	
2.2.3	Predictability from Initial Conditions *37*	
2.2.4	Probability Density of the Temperature *39*	
2.3	Spectral Analysis *40*	
2.3.1	White Noise Spectral Density *41*	
2.3.2	Spectral Density and Lagged Correlation *41*	
2.3.3	AR1 Climate Model Spectral Density *42*

2.3.4	Continuous Time Case	42
2.4	Nonlinear Global Model	44
2.4.1	Ice-Albedo Feedback	44
2.4.2	Linear Stability Analysis: A Slope/Stability Theorem	46
2.4.3	Relaxation Time and Sensitivity	47
2.4.4	Finite Amplitude Stability Analysis	48
2.4.5	Potential Function and Noise Forcing	49
2.4.6	Relation to Critical Opalescence	52
2.5	Summary	52
	Suggestions for Further Reading	53
	Exercises	53
3	**Radiation and Vertical Structure**	**57**
3.1	Radiance and Radiation Flux Density	58
3.2	Equation of Transfer	61
3.2.1	Extinction and Emission	61
3.2.2	Terrestrial Radiation	62
3.3	Gray Atmosphere	63
3.4	Plane-Parallel Atmosphere	64
3.5	Radiative Equilibrium	65
3.6	Simplified Model for Water Vapor Absorber	68
3.7	Cooling Rates	72
3.8	Solutions for Uniform-Slab Absorbers	73
3.9	Vertical Heat Conduction	75
3.9.1	$K > 0$	77
3.10	Convective Adjustment Models	77
3.11	Lessons from Simple Radiation Models	79
3.12	Criticism of the Gray Spectrum	80
3.13	Aerosol Particles	82
	Notes for Further Reading	83
	Exercises	83
4	**Greenhouse Effect and Climate Feedbacks**	**85**
4.1	Greenhouse Effect without Feedbacks	85
4.2	Infrared Spectra of Outgoing Radiation	85
4.2.1	Greenhouse Gases and the Record	92
4.2.2	Greenhouse Gas Computer Experiments	92
4.3	Summary of Assumptions and Simplifications	99
4.4	Log Dependence of the CO_2 Forcing	101
4.5	Runaway Greenhouse Effect	102
4.6	Climate Feedbacks and Climate Sensitivity	105
4.6.1	Equilibrium Feedback Formalism	107
4.7	Water Vapor Feedback	108
4.8	Ice Feedback for the Global Model	109
4.9	Probability Density of Climate Sensitivity	110
4.10	Middle Atmosphere Temperature Profile	112
4.10.1	Middle Atmosphere Responses to Forcings	113

4.11	Conclusion *115*
	Notes for Further Reading *116*
	Exercises *116*

5	**Latitude Dependence** *119*
5.1	Spherical Coordinates *120*
5.2	Incoming Solar Radiation *121*
5.3	Extreme Heat Transport Cases *122*
5.4	Heat Transport Across Latitude Circles *122*
5.5	Diffusive Heat Transport *123*
5.6	Deriving the Legendre Polynomials *125*
5.6.1	Properties of Legendre Polynomials *127*
5.6.2	Fourier–Legendre Series *128*
5.6.3	Irregular Solutions *128*
5.7	Solution of the Linear Model with Constant Coefficients *129*
5.8	The Two-Mode Approximation *129*
5.9	Poleward Transport of Heat *133*
5.10	Budyko's Transport Model *134*
5.11	Ring Heat Source *136*
5.12	Advanced Topic: Formal Solution for More General Transports *137*
5.13	Ice Feedback in the Two-Mode Model *138*
5.14	Polar Amplification through Ice Cap Feedback *140*
5.15	Chapter Summary *141*
5.15.1	Parameter Count *142*
	Notes for Further Reading *142*
	Exercises *142*

6	**Time Dependence in the 1-D Models** *145*
6.1	Differential Equation for Time Dependence *146*
6.2	Decay of Anomalies *146*
6.2.1	Decay of an Arbitrary Anomaly *147*
6.3	Seasonal Cycle on a Homogeneous Planet *148*
6.4	Spread of Diffused Heat *153*
6.4.1	Evolution on a Plane *155*
6.5	Random Winds and Diffusion *157*
6.6	Numerical Methods *159*
6.6.1	Explicit Finite Difference Method *159*
6.6.2	Semi-Implicit Method *162*
6.7	Spectral Methods *163*
6.7.1	Galerkin or Spectral Method *163*
6.7.2	Pseudospectral Method *164*
6.8	Summary *166*
6.8.1	Parameter Count *166*
	Notes for Further Reading *167*
	Exercises *167*
6.9	Appendix to Chapter 6: Solar Heating Distribution *169*
6.9.1	The Elliptical Orbit of the Earth *171*

6.9.2	Relation Between Declination and Obliquity	*172*
6.9.3	Expansion of $S(\mu, t)$	*172*

7   **Nonlinear Phenomena in EBMs**   *175*
7.1   Formulation of the Nonlinear Feedback Model   *176*
7.2   Stürm–Liouville Modes   *178*
7.2.1   Orthogonality of SL Modes   *179*
7.3   Linear Stability Analysis   *180*
7.4   Finite Perturbation Analysis and Potential Function   *184*
7.4.1   Neighborhood of an Extremum   *185*
7.4.2   Relation to Gibbs Energy or Entropy   *187*
7.4.3   Attractor Basins—Numerical Example   *187*
7.5   Small Ice Cap Instability   *187*
7.5.1   Perturbation of an Exact Ice-Free Solution   *190*
7.5.2   Frequency Dependence of the Length Scale   *191*
7.6   Snow Caps and the Seasonal Cycle   *193*
7.7   Mengel's Land-Cap Model   *193*
7.8   Chapter Summary   *196*
       Notes for Further Reading   *199*
       Exercises   *199*

8   **Two Horizontal Dimensions and Seasonality**   *203*
8.1   Beach Ball Seasonal Cycle   *203*
8.2   Eigenfunctions in the Bounded Plane   *205*
8.3   Eigenfunctions on the Sphere   *208*
8.3.1   Laplacian Operator on the Sphere   *208*
8.3.2   Longitude Functions   *209*
8.3.3   Latitude Functions   *209*
8.4   Spherical Harmonics   *211*
8.4.1   Orthogonality   *211*
8.4.2   Truncation   *212*
8.5   Solution of the EBM with Constant Coefficients   *212*
8.6   Introducing Geography   *214*
8.7   Global Sinusoidal Forcing   *216*
8.8   Two-Dimensional Linear Seasonal Model   *217*
8.8.1   Adjustment of Free Parameters   *219*
8.9   Present Seasonal Cycle Comparison   *220*
8.9.1   Annual Cycle   *220*
8.9.2   Semiannual Cycle   *220*
8.10   Chapter Summary   *220*
       Notes for Further Reading   *224*
       Exercises   *224*

9   **Perturbation by Noise**   *229*
9.1   Time-Independent Case for a Uniform Planet   *230*
9.2   Time-Dependent Noise Forcing for a Uniform Planet   *234*
9.3   Green's Function on the Sphere: $f = 0$ *235*

9.4	Apportionment of Variance at a Point 237
9.5	Stochastic Model with Realistic Geography 238
9.6	Thermal Decay Modes with Geography 243
9.6.1	Statistical Properties of TDMs 246
	Notes for Further Reading 248
	Exercises 249

10	**Time-Dependent Response and the Ocean** 253
10.1	Single-Slab Ocean 254
10.1.1	Examples with a Single Slab 255
10.1.2	Eventual Leveling of the Forcing 258
10.2	Penetration of a Periodic Heating at the Surface 259
10.3	Two-Slab Ocean 262
10.3.1	Decay of an Anomaly with Two Slabs 266
10.3.2	Response to Ramp Forcing with Two Slabs 268
10.4	Box-Diffusion Ocean Model 269
10.5	Steady State of Upwelling-Diffusion Ocean 271
10.5.1	All-Ocean Planetary Responses 273
10.5.2	Ramp Forcing 274
10.6	Upwelling Diffusion with (and without) Geography 274
10.7	Influence of Initial Conditions 276
10.8	Response to Periodic Forcing with Upwelling Diffusion Ocean 277
10.9	Summary and Conclusions 280
	Exercises 282

11	**Applications of EBMs: Optimal Estimation** 287
11.1	Introduction 287
11.2	Independent Estimators 288
11.3	Estimating Global Average Temperature 290
11.3.1	Karhunen–Loève Functions and Empirical Orthogonal Functions 292
11.3.2	Relationship with EBMs 296
11.4	Deterministic Signals in the Climate System 298
11.4.1	Signal and Noise 299
11.4.2	Fingerprint Estimator of Signal Amplitude 299
11.4.3	Optimal Weighting 299
11.4.4	Interfering Signals 302
11.4.5	All Four Signals Simultaneously 303
11.4.6	EBM-Generated Signals 306
11.4.7	Characterizing Natural Variability 310
11.4.8	Detection Results 311
11.4.9	Discussion of the Detection Results 314
	Notes for Further Reading 317
	Exercises 317

12	**Applications of EBMs: Paleoclimate** 321
12.1	Paleoclimatology 321
12.1.1	Interesting Problems for EBMs 322

12.2	Precambrian Earth *325*
12.3	Glaciations in the Permian *327*
12.3.1	Modeling Permian Glacials *327*
12.4	Glacial Inception on Antarctica *331*
12.5	Glacial Inception on Greenland *333*
12.6	Pleistocene Glaciations and Milankovitch *335*
12.6.1	EBMs in the Pleistocene: Short's Filter *338*
12.6.2	Last Interglacial *346*
12.6.3	EBMs and Ice Volume *348*
12.6.4	What Can Be Done without Ice Volume *350*
	Notes for Further Reading *350*
	Exercises *351*

References *353*

Index *365*

Preface

This book is the result of many decades of formulating, solving, and teaching about energy balance climate models (we call them *EBMs*). Both authors have used earlier versions of the material in graduate courses in the atmospheric and neighboring sciences. The book is designed to appeal to many types of students or readers, including atmospheric scientists, oceanographers, geologists (especially paleoclimatologists and paleooceanographers), mathematicians, physicists, engineers, environmental scientists, and those interested in impacts, such as hydrologists and economists. While the book delves deeply into the mathematical details of EBMs, many readers can skip the technical details of model solutions and proofs and jump to the results that are usually presented graphically. One of our aims is to provide a rigorous study of the subject, while teaching students some of the methods of classical mathematical physics that are often neglected in the traditional curriculum. Problems are included at the ends of most chapters to hone the skills of students in these methods. Another aim is to engage the mathematics and physics (and other) communities in climate modeling, especially the side of it that might appeal to those readers. There are many mathematical problems remaining to be solved in this subfield of climate science, and we hope those inclined will think of taking a few of them on.

The popularity of EBMs has been going up and down over the decades since they were introduced by Budyko and Sellers in the late 1960s. In the 1970s, they were in favor, but as general circulation models (GCMs) began to improve significantly, EBMs were often dismissed as too crude. But over time EBMs were recognized as important tools in the hierarchy of models. As GCMs include more and more processes and components (biology, carbon cycle, cryosphere, etc.) the output of their simulations becomes even more difficult to comprehend. EBMs and other simplified schemes can help to sort out some key processes in the system.

In introducing the subject to many readers with little climate science experience, we have endeavored to be conservative in our presentation of EBMs and their applications. Similar to climate models at all levels of complexity, EBMs are pretty blunt instruments. EBMs can provide insight by clarifying some of the cause-and-effect issues in the climate system. In some cases such as perturbations of the surface temperature field due to small changes in greenhouse gases (GHGs) or the Earth's orbital elements, they can be surprisingly helpful even to a quantitative extent. As with any model, the most important step is to ask the appropriate question. The EBMs provide a unique learning environment for seeing how simplified models can be solved by using applied mathematics. Many EBM formulations are found to be linear systems with all their general properties available to us. We can not only solve the problem in terms of well-known functions but we can also probe the structures with modes and frequency components that strictly numerical approaches do not readily provide. Valuable stores of intuition can be accumulated from linear thinking on the way to the nonlinear world.

We approach the hierarchy of models from a global perspective, then turn to increasing resolution in space and time, also in the complexity of the geography. This is done in a stepwise fashion, starting first with global averages. Next comes zonal symmetry but with horizontal dependence to the solar heating in steady state, then time dependence. Next, we consider a zonally symmetric planet with a homogeneous global surface upon which is imposed seasonal heating. After developing two-dimensional solutions on the bare planet, we break the homogeneous spherical symmetry with the land–sea geography of today or of ancient times. Temporally, we begin with steady state and later introduce the decay of anomalies, the periodic seasons, and eventually random fluctuations of the surface temperature field. In cases of high symmetry, analytical solutions and analyses using familiar functions give insight into how things work when the symmetries are relaxed one at a time. The style of the book is that of a physicist rather than a mathematician, although we hope to retain the mathematician's attention with minimal offence. We do not hesitate to use shortcuts such as the Dirac delta function. Our treatment of stochastic matters is likewise in physics style.

An interesting aspect of the climate system is how timescales of radiative relaxation (a month over land, a few years over a mixed-layer ocean) contrast with those of the weather anomalies (few days) that occur especially in the mid-latitudes. This wide separation of temporal scales allows us to make stochastic EBMs, that is, linear models with sluggish response to white noise forcing. Stochastically driven EBMs provide a whole family of models that can be solved completely and examined in detail. Often, the EBM framework can be used to illustrate a technique such as in estimation of faint deterministic signals in a noisy climate system. We devote a whole chapter to this and related problems.

We have both moved on long ago from the study of EBMs in our primary research interests, but we return to them again and again because of their intrinsic aesthetic appeal.

A book like this one could not have been completed without the help of many students and collaborators. We list here many who collaborated with us in this endeavor over the years to whom we are grateful: Steve Baum, Thomas L. Bell, Robert Cahalan, Petr Chyèk, James Coakley, Robert Chervin, Thomas Crowley, Marc Genton, Charles Graves, Gabi Hegerl, Louis Howard, Jian-Ping Huang, William Hyde, Philip B. James, Lai-Yung (Ruby) Leung, Wan-Ho Lee, Rai-Qing Lin, John Mengel, David Pollard, Haydee Salmun, Stephen H. Schneider, Sam (S. S. P.) Shen, David Short, Bruce Wielicki, Qigang Wu, Wei (Julia) Wu. Kelin Zhuang.

Others who perhaps did not publish with us on EBMs, but who have influenced us through private conversations or encouragements include Al Arking, David Atlas, Eric Barron, Kenneth P. Bowman, M. I. Budyko, Alex Dessler, Andrew Dessler, Robert Dickinson, Tsvi Gal-Chen, Lev S. Gandin, Michael Ghil, Georgi Golitsyn, L. D. Danny Harvey, Klaus Hasselmann, Isaac Held, John Imbrie, Igor Karol, C. E. Leith, Richard S. Lindzen, Kuo-Nan Liou, John Nielsen-Gammon, Richard Peltier, Alan Robock,

Joe Smagorinsky, William D. Sellers, Richard Somerville, Peter H. Stone, Max Suarez, Warren Washington, Manfred Wendisch, Robert Watts, Ping Yang, and many others.

We also thank Texas A&M University, College Station, Texas, USA, and Seoul National University, Seoul, Korea, for allowing us the time and resources to write this book. We are especially grateful to our families who endured the long process of getting this book together.

Seoul, Korea, 2017 *Kwang-Yul Kim*

1
Climate and Climate Models

The global climate system consists of a large number of interacting parts. The material components and their sub-members include the following:

1) the atmosphere and its constituents such as free molecules and radicals of different chemical species, aerosol particles, and clouds;
2) the ocean waters and their members such as floating ice, dissolved species including electrolytes and gases as well as undissolved matter such as of biological origin and dust;
3) the land components with characteristics such as snow and ice cover, permafrost, moisture, topographical features and vegetation with all its ramifications.

The space–time configuration of abstract fields that are used to characterize properties of interest (such as temperature, density, and momentum) attributed to these components and their sub-members vary with time and position and each exhibits its own spectrum of time and length scales. Heat (or more formally, enthalpy) fluxes, moisture, and momentum fluxes pass from one of these material components to another, sometimes through subtle mechanisms. Determination of whether and how these constituent parts combine to establish a statistical equilibrium may seem challenge enough, but the climate dynamicist also seeks to understand how the system responds to time-dependent changes in certain *control parameters* such as the Sun's brightness, or the chemical composition of the atmosphere. Although we have been at it for many decades now, the grand problem is still far too complicated to solve at the desired level of accuracy (no bias) and precision (error variance) even though preliminary engineering-like calculations are being used routinely in scenario/impact studies because policymakers must (should!) make use of even tentative information in their deliberations (IPCC, 2007, 2013).

Serious attempts at quantitative climate theories can be said to have begun in the late 1960s, although some very clever attempts predate that by decades (see Weart, 2008). The theory of global climate is emerging from its infancy but it hardly constitutes a set of principles that can be converted into reliable numerical forecasts of climate decades ahead or that can be unequivocally used in explaining the paleoclimatic record. However, some valuable insights have been gained and many problems can be cast into the form of conceptual frameworks that can be understood. We now have an idea of which of the components are important for solving certain idealized problems, and indeed, in some cases, it appears that the problems can be made comprehensible (but not strictly quantitative) with models employing only a few variables.

The field of climate dynamics is vast, embracing virtually every subfield of the geosciences (even "pure" physics, chemistry, and biology) from the quantum mechanics of photons being scattered, absorbed, and emitted by/from atmospheric molecules in radiative transfer processes to the study of proxies such as tree-ring widths and isotopic evidence based on fossilized species deposited and buried long ago in sediments deep below the ocean's floor. The in-depth coverage of these subfields is generally presented in the traditionally separate course offerings of curricula in the geosciences. This book is concerned with the integration of this array of material into a composite picture of the global climate system through simplified phenomenological models. The approach will be to pose and examine some problems that can be solved or analyzed with the classical techniques of mathematical physics. Throughout we attempt to use these analytical methods, but will introduce and use numerical methods and simulations when necessary. However, our main strategy will be to idealize the physical problem in such a way as to render it solvable or at least approachable, then compare or draw analogies either to the real world or to the results of solving a more believable model – hardly a foolproof procedure but likely to be instructive. In short, we hope to get at the heart of some climate problems in such a way that the reader's intuition for the composite system can be developed and more informed approaches can be taken toward the solution of specific problems.

The energy balance climate models (EBCMs) generally deal with an equation or a set of coupled equations whose solution yields a space–time average of the surface temperature field. Unfortunately, the solutions cannot usually describe the temperature field above the boundary layer of the atmosphere except in rare circumstances. This is a severe limitation, leaving us with only partial answers to many questions we would like to pose. On the other hand, we are blessed with many reasons supporting the importance of the surface temperature field:

1) Space–time averages of surface temperature are easily estimated and many instrumental records provide good data, not just contemporarily, but over the last century.
2) Space–time averages of surface temperature data are close to being normally distributed, making them easy to understand and treat. This is not so for precipitation and some other variables. Moreover, the larger the space–time scale, the more information from point sources can be combined into the average, resulting in a reduction in the random measurement errors on the mean estimates.
3) The time series of space–time averages of surface temperature is particularly simple, resulting in applicability of autoregressive behavior of order unity in many cases.
4) Nearly all paleoclimate indicators provide information about the surface temperature, extending the data base that can be used in testing. There are never enough data to check and adjust models, especially complex numerical models. Paleoclimatology can potentially provide more data that can be used to understand climate models.
5) As we will show, the surface temperature is also the easiest variable to model, especially for large area and time averages. It becomes more difficult as the space–time scales in the problem decrease. In this book, we will start with the largest space and time scales and find that there is a natural progression of estimates from the largest to the smallest space–time scales. Moreover, averaging over large scales reduces some errors in models as well as in measurements.
6) Most of the externally applied perturbations to the climate system that are of interest are directed at large spatial and temporal scales. This happens to be the case for the

four best known perturbations: greenhouse gases, volcanic dust veils, anthropogenic aerosols, and solar brightness. It is intuitively appealing (as well as motivated by physics, as we shall see) that the large space–time scale perturbations result primarily in the same large space–time scales of thermal response patterns in the climate system.
7) The study of energy balance models is cheap. This can be a factor when questions are posed from paleoclimatology, for example. Big models are simply too expensive to experiment with in the first trials. With the speed up of modern computers, many paleoclimate problems can be examined with general circulation models (GCMs), but not every one of them.
8) The study of exoplanets has become important in recent years. The habitable zone of a planet's orbital and atmospheric/oceanic dynamical/chemical parameters may fall into the purview of energy balance models.
9) Finally, the surface temperature is important for societal well-being and it is easily grasped, although the idea of large space–time scales is less easily identifiable and appreciated by the average person.

Unfortunately, as soon as we go above the near-surface environment, the mathematical difficulties of solving the climate problem even for the temperature becomes orders of magnitude more difficult. Also, for all its importance, precipitation cannot be solved by simple models because it depends too sensitively on the circulation of the atmosphere (and the ocean).

1.1 Defining Climate

Before proceeding, we must define what we mean by climate. As an illustrative example, we restrict ourselves at first to the global average surface temperature. Our definition is abstract and not strictly an operational one unless certain (reasonable but, unfortunately, unverifiable) conditions are fulfilled. When we examine records of globally averaged temperature at the Earth's surface we find that it fluctuates in time. Figure 1.1 shows a century-long record of both annual and global averages (*estimates* of these, to be more precise) and, except for a possible upward slope, we find departures from the mean linear trend that persist over a few years or even decades.

Consider an abstraction of the real system. We borrow from the discipline of time series analysis (which may have originated in the subdiscipline of theoretical physics called *statistical mechanics*) the concept of an *ensemble*.[1] By this, we mean to consider a segment of a record of some quantity versus time (e.g., the record of estimates of annual-mean and global-average temperatures in Figure 1.1) as a single *realization* drawn from an infinite number of statistically equivalent (imagined or hypothetical) manifestations of the record. It is presumed that all the realizations are generated from the same physical process (imagine a large number of Earths rattling along but each with slightly different initial conditions set long before the beginning of our "observation" record) for temperature distribution, winds, and so on, but otherwise all the externally imposed conditions such as the Sun's brightness and atmospheric composition are the

---

1 Ensemble: a group of items viewed as a whole rather than individually.

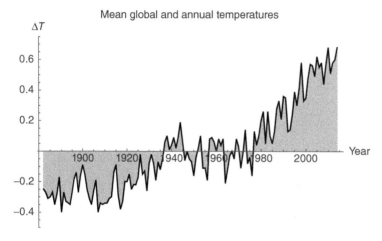

Figure 1.1 Time series of thermometer-based global average temperatures from the website of Goddard Institute for Space Studies: www.giss.nasa.gov. The units are in Kelvin and the temperature values are "anomalies" or deviations from a long-term mean (1951–1980). (Goddard Institute for Space Studies (NASA) (2017).)

same for all the "toy" Earths). The construction of an ensemble is strictly a mathematical convenience, as it allows for ease of computation of moments of the statistical distributions. It is our belief that the realizations form time series that are stationary (defined more precisely below) and that long-term averages of quantities are equivalent to averages across the ensemble members. The advantage of this scheme is that the ensemble provides us with a framework that makes thinking about the problem easier and it facilitates computation of statistical quantities. Also, from a practical point of view this is exactly the way we generate the model climate from a series of simulations from a big climate model (GCM).

The idea of studying a fluctuating system by examining the statistics of individual realizations is called the *frequentist* approach. One can perform statistical tests similarly by asking the probability of occurrence of an event by looking at many realizations of the process. In this book, we will assume the frequentist method is sufficient.

To illustrate the idea, consider a climate that is characterized by a single variable, its temperature[2] $T(t)$. The ensemble average of the temperature at time t is

$$\langle T(t) \rangle \equiv \lim_{N \to \infty} \frac{1}{N} \sum_{i=1}^{N} T^{(i)}(t), \tag{1.1}$$

where the superscript i is an index labeling the ith realization. We imagine calling upon some kind of algorithm that generates realizations for us on command. In the following, we will see how this can be done in a simplified statistical model driven by uncorrelated random numbers. The ensemble of realizations is to have the same statistics (moments) as the real process. We then average over a large number (N) of these to form $\langle T(t) \rangle$. The brackets are defined by the above averaging operation $\langle \cdot \rangle$ (taken in the limit of $N \to \infty$,

---

2 As always in this book, unless otherwise indicated, we refer to the air temperature a few meters above the land surface, ground or over water, the temperature of the surface level water itself. We will normally use units of Kelvin, but occasionally we will use °C.

but practical experience with GCM simulations suggests that 5–10 is enough for many purposes).

Often in geophysical problems, a long-term temporal mean is equivalent to the ensemble mean:

$$\langle T(t) \rangle \approx \lim_{T_A \to \infty} \frac{1}{T_A} \int_{t-T_A/2}^{t+T_A/2} T^{(i)}(t') \, dt'. \tag{1.2}$$

A relation like this holds for so-called *ergodic* systems. Roughly speaking, an ergodic system is one for which the physical timescales are bounded (more precision on this shortly). Now that we have the concept of ensemble averaging in mind, we can compute the second moment of $T(t)$:

$$\langle T(t)^2 \rangle \equiv \lim_{N \to \infty} \frac{1}{N} \sum_{i=1}^{N} (T^{(i)}(t))^2. \tag{1.3}$$

We could also define the probability density function (*pdf*) of the temperature at time t as $p(T(t))$. We would have

$$\langle T(t)^2 \rangle \equiv \int_{-\infty}^{\infty} T(t)^2 p(T(t)) \, dT(t), \tag{1.4}$$

and, for the nth moment,

$$\langle T(t)^n \rangle \equiv \int_{-\infty}^{\infty} T(t)^n p(T(t)) \, dT(t). \tag{1.5}$$

The *variance* is a *centered moment*

$$\text{var } T = \sigma_T^2 = \langle (T - \langle T \rangle)^2 \rangle. \tag{1.6}$$

We can also consider the *covariance* between the temperature at time t and at another time t'. This is defined as

$$\text{covar}(T(t), T(t')) = \langle (T(t) - \langle T(t) \rangle)(T(t') - \langle T(t') \rangle) \rangle. \tag{1.7}$$

We can think of a *bivariate pdf*, $p(T(t), T(t'))$, in this case. A time series is said to be *stationary* if the mean and variance are independent of t and if $\text{covar}(T(t), T(t'))$ depends only on the time difference $|t - t'|$. These statements mean effectively that there is no preferred origin along the t-axis (at least up to the second moments). In this case, we can write

$$\text{covar}(T(t), T(t')) = \sigma_T^2 \rho(\tau), \tag{1.8}$$

where $\rho(\tau)$ is called the *lagged autocorrelation function*[3] at lag $\tau \equiv |t - t'|$. Note that, by this definition, $\rho(0) = 1$ and $\rho(\tau) = \rho(-\tau)$. Figure 1.2 shows the lagged autocorrelation function for the data in Figure 1.1 (actually, it is the autocorrelation function of the residuals after detrending the data from Figure 1.1 with a straight regression line). The *autocorrelation time* is the integral of $\rho(\tau)$ over all lags (≈ 3.5 years).

$$\mathcal{T} \equiv \int_0^{\infty} \rho(\tau) \, d\tau. \tag{1.9}$$

---

3 Strictly speaking, this is called *wide-sense stationarity* because it only considers moments of up to the second order. If the time series elements are normally distributed, this is enough.

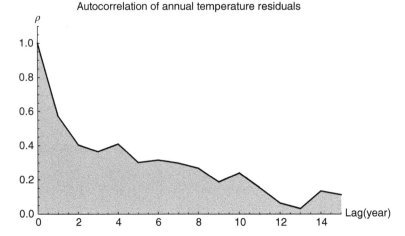

Figure 1.2 Autocorrelation function for the residuals about a linear regression line of the GISS global average surface temperatures. The abscissa is in lagged years. The decay of the autocorrelation indicates the lack of correlation over time. In this case, the autocorrelation function falls to 1/e in about 3.5 years. (Goddard Institute for Space Studies (NASA) (2017).)

Later, we will see how simple models of the system can reproduce curves very similar to that in Figure 1.1 with the added possibility of interpretation of the underlying processes. Here we can get a better idea of what an ergodic system is: the autocorrelation time should be finite. From a practical point of view, it must be short compared to the total length of the time series under consideration.

Instead of real data, we can also generate a time series that resembles a real climate variable. This illustration based on a simple time series algorithm should help in understanding the meaning of some of the above definitions. Figure 1.3 shows five realizations of a time series generated from the *stochastic process* defined by

$$T_n = \lambda T_{n-1} + \gamma Z_{n-1}, \tag{1.10}$$

where T_n is the temperature at time n, a discrete time index (such as an annual average temperature), and Z_n is a random number (variate) that at each time (drawing or *innovation*) takes on a value from a normal distribution with mean zero and standard deviation unity (statisticians indicate this by $Z_n \sim N(0,1)$); each drawing is statistically independent of the previous one. The constants λ and γ have values (arbitrarily chosen here for cosmetic purposes) 0.8 and 0.05, respectively. In Figure 1.3, the heavy line is the average across the five realizations. If there were a large number of realizations in this process, we would find the average approaching the x-axis (i.e., $\langle T_n \rangle_N \to 0$ as $N \to$ Large). More formally, the standard deviation of the individual points along the heavy curve approaches zero as $\sigma_T/\sqrt{N}$, where σ_T is the standard deviation of the variate T_n, and N is the number of realizations in the ensemble. The process (1.10) that relates the nth value of T_n as proportional to the $(n-1)$th plus a random normally distributed disturbance is called an *autoregressive process of order one* or *AR1*. This particular type of stochastic process is common in geophysical processes such as temperature field evolution. Higher-order autoregressive processes give weights to more distant values in the

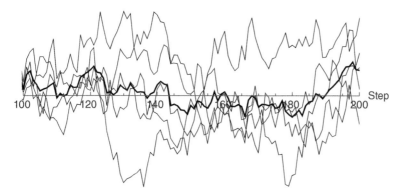

Figure 1.3 Five realizations from a time series generated from the AR1 algorithm $T_n = 0.8T_{n-1} + 0.05Z_{n-1}$, where T_n is a "model temperature" at the nth time step, and Z_n is a random number taken from a normal distribution with mean zero and standard deviation unity. The heavy line is the ensemble average across the five realizations. In this example, the realizations are started at $n = 1$ with $T_1 = 0$. The values from $n = 100$ to 200 are shown. The graphic shows how averaging over only five realizations smooths the time series, diminishing excursions from the mean (=0 here). An observed temperature time series is similar to a single realization.

past than just the last one. Although we will deal with a number of stochastic models in this book, we will find it unnecessary to go beyond the first-order autoregressive process.

We now understand what is meant by a stationary *univariate* climate. By *climate change*, we mean that some moment, typically, the ensemble mean, is subject to a temporal or secular change. For example, the time series in Figure 1.1 appears to have a secular drift upward of about 0.6–0.8 K per century, and over the last half of the twentieth century, even steeper. Another possibility is that the system might experience a step function shift in *forcing*, leading to a climate change from one statistical steady state to a different one after a suitable waiting period (more on this idea in later chapters). As we will see later, the term *forcing* implies an externally imposed imbalance of the planetary energy budget. Such forcings might be time dependent, for example, linear secular increase, abrupt increase (step function), and pulse (delta function).

1.2 Elementary Climate System Anatomy

The vertical structure of the Earth's atmosphere divides nicely into layers each having distinct properties. The layers are conveniently separated according to the slope of temperature profile with respect to the altitude. The *troposphere* lies between the surface and the *tropopause* which in the US Standard Atmosphere is at about 10 km, see Figure 1.4. The layer between 10 km and about 32 km is called the *lower stratosphere* and the part between 32 km and the next slope discontinuity above is the *stratosphere*, which is bounded above by the *stratopause*. In this book, we will confine our attention mainly to the troposphere.

The air above the ground flows with horizontal scales ranging from roughly 1000 km to even larger scales. As it rubs against the surface, turbulence occurs, resulting in a

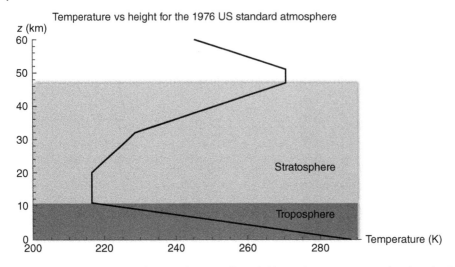

Figure 1.4 The US Standard Atmospheric Profile (solid line). The tropopause is the altitude of the temperature slope discontinuity at 10 km on this graph. The US Standard Atmosphere is an average around the globe. The level of the tropopause here is characteristic of the mid-latitudes. In the tropics, it lies at about 18 km.

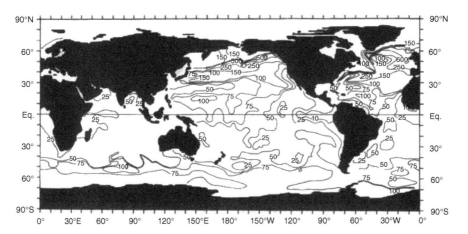

Figure 1.5 March mixed-layer depths (meters) based on a temperature criterion of 0.5 °C (difference from the temperature at the surface). The ocean's upper layer is well mixed by the action of wind stirring the water. The mixed-layer depth varies by location. It tends to be deeper at higher latitudes and shallow in the tropics. (Taken from Levitus (1982): NOAA Professional Paper, Figure 95a.)

boundary layer near the surface of depth 1–2 km, consisting of well-mixed air. At night, the boundary layer shrinks to a fraction of its daytime depth; then, as the Sun rises, it swells back to its maximum depth of 1–2 km, depending on season and location (see Figure 1.5).

The ocean is complicated, but for our simple-model considerations we will take the top 50 m (sometimes up to 100 m) to be the wind-driven mixed layer that is expected to be vertically homogenized over a period of days (see Figure 1.6) and note the tendency

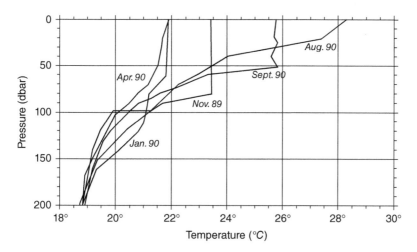

Figure 1.6 Growth and decay of the *mixed layer* and seasonal *thermocline* from November 1989 to September 1990 at the Bermuda Atlantic Time-series Station (bats) at 31.8°N 64.1°W. Data were collected by the Bermuda Biological Station for Research, Inc. Note that pressure in decibars is nearly the same as depth in meters. (Reproduced with kind permission of Robert Stewart.)

for shallow, mixed layers in the tropics and deeper ones in the higher latitudes. The temperature of the ocean is then nearly constant in that upper layer. The temperature falls off from the bottom of the mixed layer to its value (usually around 4 °C, the temperature at which sea water has its maximum density) at very deep levels (approximately several kilometers) approximately exponentially. The e-folding depth of the temperature profile is called the *thermocline*, typically around 500–800 m, depending on season and location (Figure 1.6). The ocean below the mixed layer becomes important when time-dependent perturbations are imposed on timescales longer than a few years. We consider that problem in Chapter 10.

1.3 Radiation and Climate

1.3.1 Solar Radiation

The climate of the Earth is ultimately controlled by the energy output of the Sun and the Earth's orbital elements (see Figure 1.7 for calculations of past values or impacts of the orbital elements, based on Berger, 1978b):

1) The mean annual Earth–Sun distance, currently 149 597 870 700 m. This defines 1 astronomical unit (AU) in planetary astronomy.
2) The eccentricity of the orbit, presently 0.0167 varying between nearly circular value of 0.005 up to 0.06 with a period of roughly 100 ky. See Figure 1.7, which is based on calculations by Berger (1978b). There is little or no effect of eccentricity on the annual, global mean because of Kepler's second law of equal areas of the orbital sweep being equal for equal time intervals. In other words, when the orbiting Earth is closest to the Sun, it is moving faster around its cycle than when it is near aphelion.

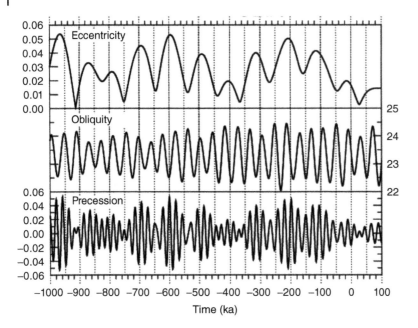

Figure 1.7 The temporal variations of the orbital elements. Upper: Eccentricity (dimensionless); the present value is 0.016. Middle: Obliquity (°) is the tilt angle of the Earth's spin axis from a perpendicular to the orbital plane (the *ecliptic plane*); the present value of the obliquity is 23.5°. Lower: Precession (solar radiation flux density at 65°N in W m^{-2}). This latter shows the variation of this radiation flux density at a latitude thought to be important in forming an ice sheet in North America. (Berger (1978b). © American Meteorological Society. Used with permission.)

3) The angle of the spin axis to a perpendicular to the plain of the orbit, called the *obliquity*, which varies from its present value of 23.5° to between 22.1° and 24.5°. The obliquity varies roughly sinusoidally with a period of about 41 ky. Larger obliquity leads to a larger swing of the seasonal surface temperatures. Zero obliquity leads to a perpetual equinox. Obliquity also has a small influence on the annual mean insolation; large obliquity leads to a slight warming of the annual mean with a north–south symmetrical, hemispherical minimum in lower latitudes.

4) The seasonal phase of *perihelion*, which is the point or calendar time of year on the orbit closest to the Sun along the Earth's elliptical path. The time of equinox shifts slowly through the calendar year owing to two effects: the precession of the spin axis like a top, with period about 26 ky (today the star Polaris sits above the North Pole, but it moves from that position over time, and this has been documented from comparing with ancient astronomers). The equinoxes also precess because of an actual rotation of the major axis of the elliptical figure around the Sun. The two effects cause the calendar date of perihelion to cycle over a 22 ky period. Today, perihelion occurs on December 21, but 11 ky ago it occurred on June 21. The result is a 6% difference in summertime *insolation*[4] at latitude 65°N between summers today and those 11 ky ago. Northern Hemisphere summers would have been warmer (and winters cooler) over continental interiors back then.

---

4 Insolation is the amount of radiation flux per unit surface area impinging on the Earth at a particular latitude and time of year. Units: W m^{-2}.

5) The chemical composition of the atmosphere. As is by now well accepted, the amount of CO_2 and other greenhouse gases in the atmosphere controls the mean annual temperature of the planet, whereas the aforementioned three effects tend mostly to control the seasonality and/or the latitudinal distribution of the insolation. The changes in CO_2 over the last 800 ky are correlated strongly with the time series of temperatures in Antarctica (Lüthi *et al.*, 2008).

The Sun's luminosity affects the global climate system through the so-called *total solar irradiance* or *TSI* (the amount of radiant energy passing through a unit area perpendicular to the line joining the Earth and the Sun averaged through the year to eliminate the small (~3%) variation of the Earth–Sun distance due to its slightly elliptical orbit). A number of artificial satellites have been launched over the last four decades for delivering estimates of the TSI. Techniques for analyzing these data have now been perfected sufficiently to provide unbiased estimates of the TSI's average value and its variability for the purposes of climate research. Figure 1.8 shows a graph of measurements from a combination of radiometers aboard a sequence of satellites since the mid-1970s. Much of the high-frequency part of the variation can be attributed to the passage of sunspots across the face of the Sun as it rotates with a period of about 25 days. The longer term trends are consistent with a weak quasi-periodicity of 11 years commensurate with the solar cycle of the frequency (number per year) of appearance of sunspots. The amplitude of the oscillation (at least over the few cycles observed during the *satellite era*) is about 0.1%. The most modern estimate of the absolute magnitude of the TSI is $\sigma_\odot \approx 1360.45 \mathrm{W\ m^{-2}}$. This value is somewhat smaller than the average shown in Figure 1.8 as the value quoted here is based on a recent highly reliable calibration (Kopp and Lean, 2011; Coddington *et al.*, 2016[5]). It is not yet clear whether the Sun's output varies on longer time scales.

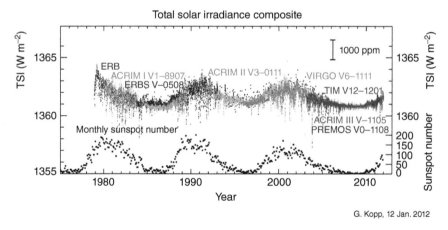

Figure 1.8 Time series of the TSI (total solar irradiance) over the last three solar cycles. The TSI is the radiation energy flux density (in W m^{-2}) reaching the top of the atmosphere. It is averaged through the year as the Earth passes around its elliptical orbit. The upper curve (points) are newly calibrated and reconciled data for the TSI from a variety of instruments over the satellite era (W m^{-2}). The lower curve is the monthly count of sunspots. (Reproduced with kind permission of Kopp (2017).)

---

5 Coddington *et al.* (2016) present explanations and graphics clarifying how the sunspots and facula (bright zones around the dark sunspots) compete in modulating the TSI.

A longstanding conjecture is that the Sun was less bright during the *Maunder Minimum*, a period of about a century starting in 1650 AD, in which there were essentially no sunspots (Eddy, 1976). So far no compelling evidence for the validity of this attractive conjecture has been published. Over ultralong time scales it is strongly suggested by astrophysical theory that the solar constant should have steadily increased by about 30% over the last 4.7×10^9 years, Gough (1981).

The Sun radiates approximately as a blackbody whose temperature is about 5770 K over most of the emission spectrum as seen in Figure 1.9. The distribution of the radiation by wavelength is important in determining how much of the radiation penetrates to various depths in the atmosphere before being absorbed or scattered. Very short wavelength radiation (X-rays, extreme UV, etc.) are absorbed in the upper atmosphere, while UV of shorter wavelength than 273 nm (nm = nanometers) is absorbed by ozone (O_3) in the stratosphere (Pierrehumbert, 2011).

Figure 1.9 (modified from Gray *et al.*, 2010) shows the distribution of solar flux as a function of wavelength. Most of the radiation power is in the visible part of the spectrum, but a large part is also in the near infrared ($\lambda > 800$ nm $= 800 \times 10^{-9}$ m). Disposition of the solar radiation entering the top of the atmosphere is as follows: 23% is reflected back to space by clouds and particles suspended in the air, 4% is reflected back to space by the

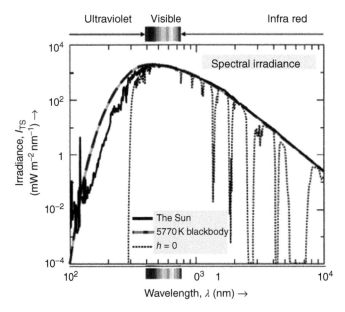

Figure 1.9 Spectral irradiance from the Sun (black line). Spectral irradiance means the Sun's rays are decomposed into wavelengths in order to reveal how different wavelength bands are disposed of by the Earth's atmosphere. The black and gray long-dashed smooth curve is the distribution of incoming radiation from an imaginary black body whose temperature is 5770 K, which is a rather good model for the Sun's radiation. The solid black curve indicates the actual radiation from the Sun. It differs some from the blackbody especially in the ultraviolet and shorter wavelengths. The dotted lines indicate the absorption of sunlight as a function of wavelength by the clear atmosphere as seen from the ground ($h = 0$). The attenuation is cut off by atmospheric gases for wavelengths below about 270 nm. Other absorption due to molecular interactions occurs in the infrared. (Gray *et al.* (2010). Reproduced with permission of American Geophysical Union.)

surface; 23% is absorbed by water vapor, clouds, dust, and ozone, and 47% is absorbed at the surface. About 30% of the total is reflected back to space (planetary albedo) and the rest goes into heating the system. Recent research suggests that the UV radiation varies appreciably more over the 11-year cycle than the TSI and that this variation can lead to a faint 11-year cycle in some climate variables (Haigh, 2010).

Figure 1.9 shows the total column absorption by the atmosphere as a function of wavelength as the gray-dotted line. The spectrum of absorptivity is very complicated to say the least. The interplay between the incoming and outgoing radiation and this spectrum along with the atmosphere's dynamic reaction to it is a key ingredient in determining the vertical structure of the atmospheric column of air at a point at the surface and ultimately in determining the horizontal movements in the system components as well.

1.3.2 Albedo of the Earth–Atmosphere System

The climate system only makes use of the solar radiation that is absorbed by the Earth–atmosphere combination. The unused fraction of the solar radiation flux reflected by the system back to space (referred to as the *planetary albedo*, averaged over the globe, through the diurnal and annual cycles) is governed by a number of factors, several of which are dynamically determined within the system. Artificial satellite-based instruments providing estimates of the albedo of the Earth–atmosphere system have been conducted since the mid-seventies. Trenberth *et al.* (2009, always subject to updates) summarize the current status of the Earth Radiation Budget estimates and the associated errors (see also Loeb *et al.*, 2009; Loeb and Wielicki, 2014). Figure 1.10 shows the flows of energy entering from space in the visible part of the spectrum and its

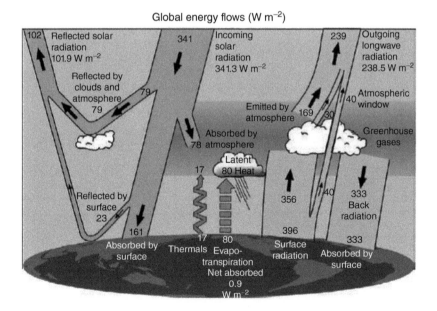

Figure 1.10 Schematic of energy flows in the globally averaged Earth–atmosphere system. Arrows indicate flux densities. The incoming solar flux is divided by 4 to take into account the ratio of the Earth's total area to the area of its silhouette. (Trenberth *et al.* (2009). © American Meteorological Society. Used with permission.)

apportionment into different flows in the Earth–atmosphere system. We can estimate the planetary albedo from the data in the figure ($102/341 \approx 0.30$).

1.3.3 Terrestrial Infrared Radiation into Space (The IR or Longwave Radiation)

Besides measuring the albedo[6] of the planet in small latitude–longitude boxes, satellite observatories also provide estimates of the outgoing infrared radiation fluxes at the top of the atmosphere. These are typically also for month-long averages but over $10° \times 10°$ boxes. It is possible to find a relationship of the outgoing radiation with the surface temperature by a comparison of the two data sets as shown in Figure 1.11. This suggests that, for many rough calculations, the outgoing infrared flux leaving the top of the atmosphere can be approximated by a linear relationship[7] with slope $1.90 \text{ W m}^{-2} \text{ °C}^{-1}$. Analysis of 10 years of data from the Nimbus 6 and Nimbus 7 satellites in mid-latitudes yields essentially the same relationship as with the Earth Radiation Budget Experiment

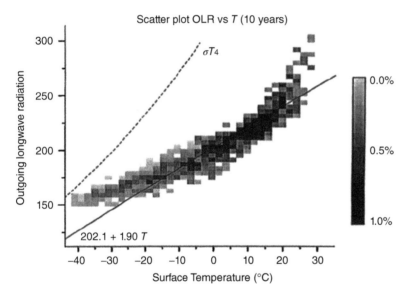

Figure 1.11 Density plot of outgoing infrared radiation flux versus surface temperature taken from satellite data. The radiation data from different locations and seasons against the local month-averaged temperature at the surface for the same actual month when the satellite data were collected. Darker shading indicates greater frequency of occurrence. These data are for the entire sky, that is, cloudy portions are not omitted. Note that the curve and its slope are lower than that of a blackbody curve (dashed curve). The linear radiation law due to Budyko is subject to many shortcomings, but as a tool it has proven useful in energy balance models. (Graves *et al.* (1993). Reproduced with permission of Wiley.)

---

6 Strictly speaking the satellite-based radiometer measures solar-reflected radiation as a function of the angles involved and these data are converted to values on a grid.

7 The tropical latitudes (not shown here) actually exhibit a negative correlation with local surface temperature because clouds (convection) tend to migrate to the hottest points on the surface. This clustering of clouds at surface hot spots leads to a decrease of outgoing radiation in these regions because the intense convection leads to high cloud tops that are cool and radiate less than a clear area would. In energy balance models, this is somewhat compensated by the fact that the albedo is increased where the clouds are more concentrated.

(ERBE) data (see Graves *et al.*, 1993). Furthermore, the low-frequency filtered part of the terrestrial radiation to space (periods between 1 and 10 years) also yield essentially the same regression slopes. This presumably means that the relationship between IR and surface temperature holds for slow climate changes as well as the faster seasonal cycle (see the intercomparison of GCM results for sensitivity by Cess *et al.*, 1990). It is important to realize that the relationship between outgoing IR and surface temperature is not a result of simple radiative-transfer calculations but is an empirical relationship between the equilibrium ground temperature and the outgoing radiation flux. The atmospheric column undergoes convective overturning adjustments in establishing the relationship (see Chapters 3 and 4). Despite the encouraging results just mentioned, this technique of obtaining the slope of the linear regression line is subject to a variety of errors. For example, the outgoing radiation from latitude to latitude and/or from season to season at a point might be very different from that occurring during a secular change in forcing. Hence, use of the linear infrared radiation rule (attributed to Budyko, 1968; who, having no satellite data, came upon the rule using radiative-transfer calculations) may be very convenient in our calculations, but the strict numerical values are not to be taken literally.

Convective adjustment happens automatically in the atmospheric column because the shape of the atmospheric profile derived without convection is unstable to overturning if the vertical profile of the temperature is determined solely by the radiative heating and cooling (so-called *radiative equilibrium*; see Chapter 3). Incidentally, convective adjustment leads to the global average constant lapse rate of about $6.5\,\text{K}\,\text{km}^{-1}$ in the troposphere. Referring to Figure 1.10, if we try to account for the unreflected $239\,\text{W}\,\text{m}^{-2}$ fraction of radiation energy out of $341\,\text{W}\,\text{m}^{-2}$ initially delivered to the top of the atmosphere (per unit area of the spherical Earth) by the Sun, we find that the energy flux density bifurcates taking a variety of paths through the Earth–atmosphere system before the $239\,\text{W}\,\text{m}^{-2}$ is finally returned to space. For example, much of the solar radiation absorbed by the surface ($161\,\text{W}\,\text{m}^{-2}$) is released from and cools the surface by *thermals* (or dry convection) and *evapotranspiration*, which includes the flux of water vapor from the biosphere. Both processes deposit the heat energy in sensible form in the atmosphere above, warming it. Huge amounts of energy are radiated from the surface ($396\,\text{W}\,\text{m}^{-2}$), most of which is absorbed in the atmosphere and radiated back down as well as upward ($333\,\text{W}\,\text{m}^{-2}$), rewarming the surface. The sky actually warms our faces as we stand outside on a hot day. Finally, the heat-induced radiation trickles out the top of the atmosphere to space. The various components thus come to a statistical equilibrium (fluctuating, but statistically stationary in time). For the seasonal cycle, the equilibrium is cyclic and the statistical term for it is *cyclostationary*. For an EBM example of its use, see Kim and North (1997).

1.4 Hierarchy of Climate Models

Climate models fit into a hierarchy[8] that we believe is helpful to understanding the complete system. At the low end of the model hierarchy are the global average planetary

---

[8] One of the earliest and best review papers discussing the hierarchy of climate models is that of Schneider and Dickinson (1974). Isaac Held makes a strong case for a hierarchical approach in Held (2005).

models to be discussed in the next chapter. In these global models, the climate consists of a single variable, the surface temperature. As we will see in the next few chapters, some global average models will include a vertical dependence, so that the climate consists of a single function of altitude. The hierarchy is topped off by the GCMs. Each long run can be considered a detailed *realization* of an artificial climate system including all the weather-scale fluctuations. In keeping with this view, we might consider a string of actual data about the Earth's global average temperature as a single realization taken from an imaginary ensemble of such realizations. The currently most sophisticated system models couple the circulation of the atmosphere with that of the ocean and other components such as the biosphere and the cryosphere (ice parts such as glaciers, sea ice, and permafrost). Part way along are the models that omit the slower components and concentrate on the circulation and thermodynamic indices of the atmosphere. We will have cause to discuss these models frequently, as they provide artificial realizations of the faster part of the climate system and they are especially helpful in understanding the relation of simpler model ideas to the greater system.

1.4.1 General Circulation Models (GCMs)

General Circulation Models (GCMs) have evolved from the numerical weather forecast models of the 1950s. Those short-range forecasting models have stood the test of time day in and day out over this period. The upgrading of numerical methods and the implementation of improved representation of the physical components and mechanisms have led to steady improvement of their forecasting performance and their ability to simulate, with reasonable certainty, the evolution of most middle-latitude storm systems that are so important in transporting heat and other quantities across latitude circles. In addition, much of the variability of the climate system originates in these disturbances.

A short-term weather forecast does not depend much on tiny trends of imbalance in the overall energy balance such as might occur at the top of the atmosphere owing to solar brightness changes or a trend of CO_2 amounting to 0.0014% per day. Not much happens owing to such an imbalance in a day or two. To make a climate simulation model, one has to go back to the fundamentals and include accurate radiative transfer modules in the computer code to take these seemingly tiny effects into account. Over decades, they matter. Today's models still struggle to properly include aerosols and clouds in their radiation budgets. These are known problems. There may very well exist problems we do not yet know about.

Following the pioneering numerical experiment by Phillips (1956) on modeling the atmosphere's general circulation with one of the original digital computers, modeling groups began to respond. The leaders included Kasahara, Washington, Smagorinsky, Manabe, Arakawa, and Leith (see the book by Donner *et al.*, 2011). The first climate models in the 1960s were run at mean annual solar distribution over the planet (no seasons) and a surface that was composed of dry land partially covered with moist surface and an ocean with wet surface. Among many pioneering studies, a particularly influential one was conducted by Manabe and Wetherald (1975). The model included very simple geography (land alternating with ocean at 60° longitude segments). There were no seasons and the ocean was a simple wet surface. The vertical structure consisted of nine layers and the horizontal resolution (grid-box size) was approximately 500 km. Water substance in the atmosphere was computed as vapor and was allowed to evaporate from the surface when conditions were right, and then carried by the simulated

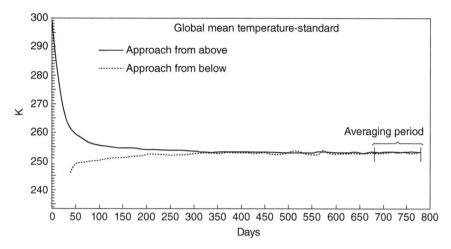

Figure 1.12 Time evolution of two runs from a very early general circulation model (GCM) with different initial conditions (one cold the other warm) by Manabe and Wetherald (1975). The globally averaged (mass-weighted) temperature of the atmosphere in the two runs eventually settles down to approximately the same statistical steady state. (Manabe and Wetherald (1975). © American Meteorological Society. Used with permission.)

winds; the model produced its own precipitation. Relative humidity was computed as fluxes of water into and out of grid boxes as warranted. Interestingly, the relative humidity near the surface remained rather steady through the integrations of climate change. The model was initialized in a cold state and a separate run was made with an initial warm state (Figure 1.12) and both solutions evolved to the same statistical steady state fluctuating endlessly but essentially randomly about a constant mean; that is, it continued to fluctuate but the statistics of the fluctuations formed a stationary time series. Moreover, the latitudinal dependence of the temperature distribution looked qualitatively similar to that of the annual average for the planet Earth. Surely, the authors were pleased with this very remarkable result, and the age of climate simulation was thus launched.

Emboldened, Manabe and Wetherald, and other fledgling GCM groups, proceeded to double the CO_2 in the model atmosphere from 300 to 600 ppm. The resulting globally averaged surface temperature increased by 2.93°, a result eerily close to the value of the most modern simulations nearly 40 years later. The latitudinal and vertical dependences of the temperature change are given in Figure 1.13. Note the cooling in the stratosphere, a finding that still holds in simulations from the most recent high-resolution models. The cooling of the lower stratosphere during CO_2-forced warming at the surface is also a rather simple consideration of energy balances in the vertical layers of the atmosphere that we will discuss in Chapter 4.

1.4.2 Energy Balance Climate Models

This book focuses on the low end of the climate model hierarchy because it is a good entry point for those wanting to learn about climate models, but also because the nomenclature and many concepts introduced at this level apply all the way up the hierarchy. Our primary focus is the class of so-called EBCMs (sometimes they are

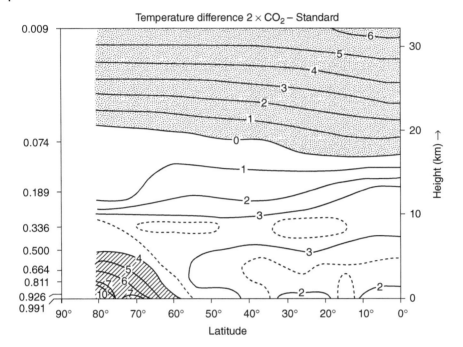

Figure 1.13 Change of temperature (°C) after reaching equilibrium from a change due to doubling CO_2, based on the 1975 GCM (same as in the previous figure). The stippled areas indicate negative changes. Note the appearance of a very interesting phenomenon: the cooling of the lower stratosphere. (Manabe and Wetherald (1975). © American Meteorological Society. Used with permission.)

referred to simply as *energy balance models* or *EBMs*). EBMs were introduced by Budyko (1968) and Sellers (1969) independently and they are often justly referred to as the *Budyko–Sellers* models. There was also an earlier paper by the astrophysicist (Öpik, 1965) that used an EBM with ice cover. Their first papers brought to light the possibility that a modest lowering (a small percentage) of the TSI would lead to an expansion of the polar ice caps from their present area of a about 5% of the Earth's surface area to a complete covering of the planet by an ice sheet. This alarming finding called attention to the potential fragility of the Earth's climate and it sparked the explosion of climate modeling research that followed. At first, the interest was not so much in the effects of increasing CO_2, but with solar brightness changes, the effect of screening of sunlight by volcanic dust veils, and anthropogenic aerosols. Also, there was the possibility of explaining the ice ages.

As GCMs began their remarkable ascent, in parallel, EBMs were subjects of experimentation by many groups and individuals. Held and Suarez (1974) studied the Budyko model and improved on it. Using a method inspired by Chýlek and Coakley (1975), North (1975a; 1975b) solved analytically the latitude-dependent ice cap model similar to the Budyko and Sellers versions but with a constant thermal diffusion coefficient. The stability of the solutions was investigated numerically by Ghil (1976) and Gal-Chen and Schneider (1976); also, analytically by Cahalan and North (1979) and North et al. (1979),

Golitsyn and Mokhov (1978). Many other references can be found in the early review article by North *et al.* (1981).

The essence of the EBMs is their conceptual simplicity, despite some of the rather involved mathematical methods required to solve them. We favor the analytical solution, not only because it is more elegant but also because such methods often lead to a deeper understanding. Often in this book and elsewhere in the literature, some less mathematically inclined readers can grasp the conceptual basis of a model and skip directly to the graphs that capture the solutions, with the comfort that the curves are backed up by an analytical or an uncontroversial straightforward numerical solution. There are many shades to the word "skip" here.

1.4.3 Adjustable Parameters in Phenomenological Models

All EBMs contain empirical parameters such as the slope and intercept of Budyko's infrared radiation rule or the thermal diffusion coefficient in latitudinally dependent models. As more independent variables are included, the number of such empirical coefficients increases. The beauty of the approach is that the model's use of phenomenological coefficients keeps the approach close to the large-scale observations. The downside is that we have departed from first principles, and we cannot know how such a coefficient might change as the climate changes. In the case of GCMs, the analog is the grid resolution: as the grid is made finer, or as more physical processes are included, there is more physical realism, but inevitably more parameters are needed. The values of the coefficients have to be "tuned" to fit what little data there are. It is often instructive to "back off" and view the system with less resolution (in its broadest sense).

With EBMs, there are two approaches:

1) Use as few empirical or phenomenological coefficients as possible to see if the main features of the model are robust. This would be important in early tests of a hypothesis where one is not interested in quantitative results over qualitative features. The advantage of this approach is to keep the number of free parameters as small as possible: the idea of *parsimony*, often discussed in fitting models in the statistical literature.[9] As we proceed in the story, we will try at each stage to tell what value is added (or lost) by the addition of a new phenomenological parameter.

2) One is interested in quantitative results in, for example, the detection of faint signals (such as the response of the surface temperature of the Earth to the solar cycle of brightness). Here one might introduce more than the minimal number of phenomenological parameters in order to establish a base state from which a perturbed solution is required. In this case, one might sacrifice parsimony of the number of freely adjustable coefficients in order to obtain the best value of the perturbed climate as possible within the EBM framework.

Cautionary Note: It is extremely tempting to add to a simple EBM. Do it at extreme risk. Depending on the problem you are addressing, you might be overfitting, that is, adding a new parameter which can be adjusted to get whatever result you wish. Such an addition could "explain" a certain phenomenon, but it is not likely to be unique.

---

9 For example, see p. 223 of Montgomery *et al.* (2001).

1.5 Greenhouse Effect and Modern Climate Change

The greenhouse effect has become a prototype of climate change studies. Solar radiation passes essentially unaltered through the atmosphere to the surface where about half of it is absorbed. Some absorption also occurs in the atmosphere, warming the local surroundings; in the troposphere, convection spreads this heat energy (enthalpy) from its top to its bottom. Heat (enthalpy) leaves the surface as infrared radiation, sensible heat flux (e.g., thermals or dry convection), and latent heat flux (heat that is removed from the surface by evaporation, then released aloft as sensible heat in cloud formation), all heating the troposphere. The heated troposphere radiates toward the surface as well as upward toward outer space. The eventual radiation to space is from colder material than that at the surface, the rate being that corresponding to a 255 K blackbody. The result is a surface temperature some 30 °C above what it would have been for a non-absorbing atmosphere. The ever-lurking question of what happens as the concentrations of certain trace gases such as carbon dioxide and methane increase exponentially over the next century will be discussed in Chapter 4.

1.6 Reading This Book

We expect readers from many disciplines to read or at least peruse this book. The primary audience is likely to be from the climate science community, which covers many subfields from meteorology, oceanography, atmospheric chemistry, paleoclimatology, and other geosciences. But in addition we hope to interest readers from physics, applied mathematics, statistical sciences, economics, and engineering. The models are rich in interesting problems, many remaining unsolved. We have chosen mostly to exploit problems in which standard low-level theoretical physics is employed, but there are forays into stochastic processes and more modern methods. We generally take the method of old-fashioned mathematical physics and some mathematically inclined readers may occasionally shudder at our glossing over mathematical technicalities. We hope our sins of omission lead others to be inspired to clean up after us. In our opinion, the field of climate science could benefit from the entry of mathematically, statistically, and physically talented innovators. As you read these chapters, you should find many intriguing problems to work on.

The following is a summary of the topics with help on what can be skipped on first reading.

- The book really begins in Chapter 2 in which most of the methods of the rest of the book are presented in the context of the model for the globally averaged temperature. In Chapter 2, the mean annual temperature of the planet is determined from the most elementary principles of radiation balance at the top of the atmosphere. We find how the planetary temperature returns to its equilibrium state if perturbed and we proceed to see how it responds to periodic disturbances. We can also see how the variations of planetary climate can be modeled by taking the fluctuating weather as a driving force, tickling the more sluggish response of the temperature field. This allows us to see how a noisy system like climate can be predictable. We will also see how the system responds to external forcings such as imposed imbalances such as changes in the solar brightness or changes in greenhouse gases.

- Chapters 3 and 4 can be skipped by many readers as, to some extent, they are a diversion from Chapter 2 and the later chapters. But they explain in some detail a few idealized models of the vertical structure of the atmosphere (radiative equilibrium and radiative-convective equilibrium in a gray atmosphere), and how that structure changes with perturbations. Chapter 4 gives a detailed look at the spectra of infrared radiation, the heart and soul of the greenhouse effect, without much mathematics. An online calculator is used to compute the effect of doubling CO_2 when no feedback mechanisms are in play. Then a detailed discussion of feedbacks is presented. Both of these chapters stand by themselves. Chapter 4 should interest many readers who might not be interested in other parts of the book, as it purports to show exactly how the rather subtle greenhouse effect actually works.
- Chapter 5 resumes the study of the surface temperature, but now it considers the latitudinal dependence of the surface temperature as modeled by diffusion of thermal energy (macroturbulent heat conduction) across latitude belts. As with Chapter 2, it delves into the aspects of steady-state models. The upshot is a model solution developed into a series of Legendre polynomials whose coefficients are temperature mode amplitudes corresponding to decreasing latitudinal space scales. It is shown that a satisfactory solution involves only the first two Legendre modes, indexed 0 and 2. A derivation of the poleward transport of heat as a function of latitude is provided with a comparison with data.
- Chapter 6 extends the previous chapter to incorporate time dependence, including the seasonal cycle. A derivation of the insolation function is presented, revealing its dependence on the orbital elements: eccentricity, obliquity, and precession. The insolation and model solutions again involve Legendre polynomial modes. Each thermal mode has a characteristic decay time scale. Smaller spatial scales lead to shorter time scales. There is a discussion of a heuristic connection between the random fluctuations of the mid-latitude storm systems and the transport of thermal energy as being diffusive in ensemble average. Also brief attention is given to numerical techniques.
- Chapter 7 introduces the nonlinear ice-cap feedback mechanism and finds an analytical solution to it for the one-dimensional models. The examination reveals multiple solutions for a particular set of external controls (such as solar brightness or CO_2 concentration). The nonlinear ice-cap model is a marvelous example of how bifurcations (in the popular literature: tipping points) can appear in these simple model structures. Mathematicians and theoretical physicists might find ways of extending some of these results.
- Chapter 8 begins a new class of models with two horizontal dimensions. This leap in dimension is achieved by allowing the local heat capacity of the air–land–ocean column to have a position dependence over the planet. The effective heat capacity over ocean (mixed layer) is about two orders of magnitude larger than that over land. Sea ice cover is somewhere in between. The spherical harmonic basis set is introduced to span the globe. With essentially no further changes, the model delivers the seasonal cycle surprisingly accurately over the globe. Because the Earth's land–sea geography is so complex (shorelines), one must resort to numerical solutions on the sphere. Readers with less interest in the mathematical details can read the first few pages and skip to the graphics.
- Chapter 9 extends the previous chapter by introducing white noise forcing to simulate weather fluctuations. Without any new parameters from the previous chapter,

the model shows the geographical dependence of the temperature variance. Moreover, even the correlation lengths and their frequency dependence come out of the solutions. Again, less mathematically inclined readers can read the first pages and skip to the figures.
- Chapter 10 returns to the problem of how the ocean delays and suppresses the response to time-dependent forcings, starting with a single mixed-layer slab. More complicated systems are treated, including slabs below the mixed layer. The problem of the response to periodic forcing at the surface is solved for vertically diffused heat.
- The book ends (Chapters 11 and 12) with a few applications of EBMs. Chapter 11 covers some estimation problems in climate science, such as the uncertainty of estimating global averages of surface temperature drawn from a finite number of dispersed point sources of information and the detection of faint signals in the climate system. In both cases, the EBMs are used to help understand the procedures involved in the estimation processes. In fact, they fare rather well against their bigger cousins.
- Chapter 12 surveys the use of EBMs in the pursuit of solutions of a variety of paleoclimate problems. Paleoclimatologists can see how simple EBMs can be used to treat some of these problems after they peruse the relevant earlier chapters and Chapters 8 and 9. First the faint Sun paradox is considered, then glaciations in deep time (the late Paleozoic) and the initiation of glaciation on Antarctica and Greenland. Finally, progress in understanding the glaciations of the Pleistocene is discussed briefly.

1.7 Cautionary Note and Disclaimer

Everyone reading this book should recognize that climate models are pretty blunt instruments and this especially holds for EBMs. We should think of the EBMs (others, too) as *analogies* to the climate system. They give us insight into the dominant features of this incredibly complex field of interacting components and help guide us to implement better models and/or more-relevant observing systems. It should be no surprise that after 40 years of this endeavor, the sensitivity to climate forcing is not known to better than about $\pm 50\%$. All models, no matter how complicated, have adjustable parameters that are used to fit the climate data we have in our hands. Different models use different parameterizations to do this fitting. Yet when the different models are advanced into the next century, their solutions for changes diverge from one another alarmingly ($\sim \pm 50\%$). Aside from our lack of knowledge of the effects of human intervention (or lack thereof) in changing the radiation budget by altering CO_2 or aerosol concentrations, the problems seem to be centered on our lack of understanding of the climate feedback mechanisms, which are discussed in many parts of this book. Often the blame is placed on the lack of resolution of the numerical grids or lack of inclusion of enough physical processes. But as soon as a finer grid or more physical processes are introduced, even more parameters have to be inserted and adjusted to fit almost the same amount of data. The process of improving the models often leads to a phenomenon known by statisticians as *overfitting*, wherein there are too many adjustable parameters for the number of available uncorrelated observations. The different GCM simulation groups naturally use different parameterizations to arrive at their final candidate model to be entered in the beauty contest. There is a multiplicity of ways to achieve a better goodness of fit. Different groups achieve their best results in different ways. It is an oversimplification to say the whole problem lies in the phenomenological coefficients. For example, there is freedom to choose exactly which and how many physical processes to include (or remove) to achieve a better match with available data. When conditions change into the future, the solutions diverge from one another. The great economist John Maynard Keynes once said something to the effect that "it is better to have a rough idea of the truth than a very precise estimate of an untruth." Another sage (C. E. Leith, a pioneer in climate modeling and theory) once said something to the effect of "1.1 or 1.0?" "Nonsense, this is climate science, they are equal to one another." It is not an excuse to delay action. As in medical research, decisions have to be made based on incomplete data sets (sampling errors). We press on sometimes a bit too hurriedly, but the process is surely self-correcting over time.

A final caution about EBMs. EBMs appear to work for the surface-temperature field. Some simple versions can be applied for a layered atmosphere or ocean, but the real value is at the surface where the radiation budget and pretty simple statistical and thermodynamical considerations dominate. As the focus lifts above the surface (or boundary layer), a host of new mechanisms are invoked. For example, at the surface, the response to a stimulating heating imbalance decays away in space a finite distance from the source. But above the boundary layer, such a disturbance can result in changes of local buoyancy and wavelike anomalies in density will radiate away from the source. We have been removed from the EBM world. We end these introductory remarks by cautioning the reader that our simple models are always highly idealized and might be best thought of as *analogs* to the climate system. "They are to be taken seriously, but not literally."[10]

Notes on Further Reading

Excellent descriptions of the climate system can be found in such books as Hartmann (2016) and Neelin (2011). The stratosphere and above are described in Andrews *et al.* (1987). Elementary accounts of the oceans are given in Picard and Emery (1990). The role of the Sun in the Earth's climate is nicely described at a beginners level in the book by Haigh and Cargill (2015). The articles in the volume edited by Archer and Pierrehumbert (2010) provide further historical material.

Exercises

1.1 Compute the emitted radiation of a black body whose temperature is 300 K in W m^{-2}. What is the total emitted radiation for a spherical black body the size of the Earth (radius 6000 km)?

1.2 Compute the same based on the graph in Figure 1.11.

1.3 In Figure 1.10, the energy rates are balanced at the top of the atmosphere. Show that a similar balance occurs at the surface.

---

10 This statement was made to GRN by the late Stephen Schneider in the 1980s.

1.4 Find the heat capacity at constant pressure for a column of air at sea level. Take the air pressure to be 10^5 Pa, leading to a mass of $P/g \approx 10^4$ kg. Now use the specific heat of dry air (≈ 1000 J kg^{-1} K^{-1}). Finally, if the whole column of air responds "rigidly" (i.e., the change is independent of altitude). Using the value of the radiation damping coefficient B, compute the relaxation time in days (and months) for this case. Would it be reasonable to use a mass less than that of the whole column for the diurnal cycle or the seasonal cycle?

1.5 Using the same approach as in the previous problem, compute the effective heat capacity and radiative relaxation time in months for a column of mixed layer of ocean water that has a depth of 50 m. How might this contrast in heat capacities for a square meter over land versus over ocean affect the seasonal cycle of the surface temperature field?

1.6 A certain random process has an autocorrelation function, $\rho(\tau) = e^{-\alpha\tau}$. What is the autocorrelation time of this process? How does your answer compare with the so-called e-folding scale of the autocorrelation function, that is, $\alpha\tau_\text{e-folding} = 1$?

1.7 A very simple climate model is defined by the energy balance

$$C\frac{dT(t)}{dt} = -BT(t),$$

where C and B are constants, t is time, and $T(t)$ is the temperature departure from equilibrium. Find the time-dependent solution for a given initial condition, $T(0) = T_0$.

1.8 In the presence of some noise, the simple climate model in Problem 1.7 can be written as

$$C\frac{dT(t)}{dt} + BT(t) = f(t),$$

where $f(t)$ is assumed to be a normally distributed white noise time series with mean zero and variance σ^2. Write the equation above as an AR1 process, that is,

$$T_n = \lambda T_{n-1} + \gamma Z_{n-1}, \quad T_n = T(n\Delta t), \quad Z_n \sim N(0, 1)$$

by determining the coefficients λ and γ.

1.9 Let $X_i \sim N(\mu, \sigma^2), i = 1, 2, \ldots, N$, be random variables with an identical normal distribution with mean μ and variance σ^2. Show that $Y = (X_1 + X_2 + \cdots + X_N)/N$ has a normal distribution with mean μ and standard deviation $\sigma/\sqrt{N}$.

1.10 According to Planck's law, radiation is determined by

$$B_\nu(T) = \frac{2h\nu^3}{c^2(e^{h\nu/kT} - 1)}, \quad \nu = c/\lambda,$$

where λ is wavelength, v is frequency, and the constant values are defined by

$h = 6.626 \times 10^{-34}$ J s : Planck's constant,
$k = 1.381 \times 10^{-23}$ J K^{-1} : Boltzmann's constant,
$c = 2.990 \times 10^8$ m s^{-1} : speed of light.

(The program "planck.f" can be found at the authors' (KYK) website.)

(a) Compute the radiation function for the wavelength range of (0, 2.0 μm) using temperature, 5770, 6000, and 7000 K. Plot the radiation functions in one plot.

(b) According to Wien's law, the wavelength at which the maximum radiation is reached is given by

$$\lambda_{max} T = \text{const.}$$

This constant is approximately 2900. Plot, the location of maximum radiation for the four temperature in part (a).

1.11 For this exercise, use the following files: t2m.data (2 m air temperature), insol.data (solar irradiance at TOA), nswt.data (net shortwave radiation at TOA), nsws.data (net shortwave radiation at surface), nlwt.data (net longwave radiation at TOA), and nlws.data (net longwave radiation at surface). These are the global averaged values for the period 1979–2015 (total of 813 points) derived from the NCEP/NCAR reanalysis product. (These files can be found at the author's (KYK) website.)

(a) Calculate the average albedo of the Earth.
(b) How much of the solar irradiance reaches the surface?
(c) What is the linear relationship between 2 m air temperature and net longwave radiation at the top of atmosphere (TOA)?
(d) What is the linear relationship between 2 m air temperature and net longwave radiation at the surface? Explain your result.
(e) How does the mean magnitude of net longwave radiation at TOA compare with the mean magnitude of net shortwave radiation at TOA? How do you interpret the result?

2

Global Average Models

2.1 Temperature and Heat Balance

Most textbooks on elementary physics and astronomy introduce estimates of the temperatures of the planets using the equality of the planet's solar absorption and its emitted radiation flux densities.[1] Usually, the textbook starts with the planet radiating as a blackbody according to the Stefan–Boltzmann T^4 law.

This is a convenient way to introduce this remarkable law. Stefan was the first to find the T^4 dependence by observations in the laboratory and Boltzmann was the first to show that it follows theoretically from the second law of thermodynamics. Many, especially older, modern physics textbooks tell this story (e.g., Richtmyer *et al.*, 1955). It is noted in the planetary climate problem that the temperature is not that of the planet's surface, but some "radiation" or "emission" temperature associated with an emission level usually located high in the planet's atmosphere (for the Earth, this altitude is of the order of 5–10 km). One can use a simple model to explain why this is necessary. Some gases in the atmosphere are good absorbers/emitters of infrared radiation, but essentially transparent in the visible. Since the air substance is cooler aloft, the radiation temperature is lower than the surface temperature.

For the Earth, a linear rule devised by Budyko (1968) and his colleagues, based on detailed radiative transfer calculations (of that era), provides a reasonably good approximation relating the outgoing infrared radiation to the surface temperature at least over the thermal range of interest. This rule is the basis of much of the modeling in this book. These models are called *energy balance climate models* (*EBCMs*), or simply *energy balance models* (*EBMs*, which we will adopt). Some authors have called them *heat balance models*, and they are also routinely referred to as "toy climate models." This class of climate models milks as much as possible about the climate solely on the basis of the thermodynamics in the problem, pretty much ignoring atmospheric dynamics except in very gross parameterizations. The success of the EBMs is remarkable if we do not ask too much of them. The beauty of the models is the few phenomenological (adjustable, based on data) parameters utilized in their construction.

---

1 A flux density is the rate of passage of a vector quantity through an infinitesimal window whose area is projected onto the direction perpendicular to the flow. For example, $\rho \mathbf{v}$ is mass flux density, where ρ is density and $\mathbf{v}$ is the vector wind. The units are those of the quantity (e.g., kg m^{-2} s^{-1}). Other examples include moisture flux density, $q\rho\mathbf{v}$, where q is the water vapor per kilogram of air; momentum flux density, $\rho v_i v_j$, where i and j are Cartesian components; and radiation energy flux density, $Ec\hat{\mathbf{r}}$, where $\hat{\mathbf{r}}$ is a unit vector along the direction of the beam, c is the velocity of light, and E is the radiation energy per unit volume.

Energy Balance Climate Models, First Edition. Gerald R. North and Kwang-Yul Kim.
© 2017 Wiley-VCH Verlag GmbH & Co. KGaA. Published 2017 by Wiley-VCH Verlag GmbH & Co. KGaA.

Hence, the models are kept very close to observational information. They fall into a class known as *semiempirical models* or *phenomenological models*. They are just a peg above "conceptual" models in the model hierarchy. As with regression analysis in statistics, one must be careful not to introduce more parameters than absolutely necessary for the problem at hand – we must be "parsimonious" with our fudge factors.[2]

The crudest of these models are the globally and seasonally averaged models that are to be treated in this chapter. We will refer to them as *global models*. In this case, the climate of our toy planet consists of only one variable, the global average temperature. After a couple of chapters on vertical structure,[3] we will rejoin in Chapter 5 the study of the surface temperature with each chapter, then include more detail as we explore just how far we can go with such a simplified picture of our climate system. For example, Chapter 5 includes the extension to latitude dependence. Chapter 6 allows for time dependence, but only for very simplified geographies. Later chapters introduce more complications such as two horizontal dimensions and simplified models of fluctuations. Each chapter then includes more geographical, geometrical, or temporal detail that will require the addition of more mathematical apparatus. It is hoped that readers can learn about mathematical methods in physics while enjoying the exploration of toy climate models.

2.1.1 Blackbody Earth

In this chapter, we examine globally and seasonally averaged climate models. We consider the globally averaged temperature as the "climate" of the planet. If the Earth were a blackbody with respect to radiation in the infrared portion of the spectrum, we would expect that the radiation energy per unit time per unit surface area to space would be given by the Stefan–Boltzmann law, $\sigma_{SB} \cdot T_R^4$, where the subscript R indicates that the temperature is the "radiation temperature" and it is in kelvin; σ_{SB} is the Stefan–Boltzmann constant, 0.56687×10^{-7} Wm^{-2}K^{-4}. T_R is an area average and it has been assumed that $\overline{T_R^4} \approx \overline{T_R}^4$, where the overline indicates (global) area average. The global average temperature fluctuates on all timescales, but we want an average of this in the early parts of this chapter. We imagine a long record or time series of global averages and we average along the record to get a mean climate.

Alternatively, we could imagine a large number of identical Planet Earths simultaneously, but started with randomized initial conditions, and we could average *across* their records, a technique called *ensemble* averaging. Records of the individual identical planets are called the *ensemble members*. The ensemble average is implied by T_R throughout this section.

The global average temperature T_R is determined by the (long-term) balance of the rate of solar energy absorbed by the Earth and the rate at which energy is emitted to space by the Earth–atmosphere system. The amount absorbed per unit time is given by $\sigma_\odot a_p \pi R_e^2$, where $\sigma_\odot$ is the *total solar irradiance* (known as the *solar constant* in older literature) (≈ 1360 W m^{-2}), a_p is the planetary average *coalbedo*[4] (≈ 0.70) and R_e is the

---

[2] *Fudge factors* are a slang and sarcastic expression indicating the authors' distaste for the use of too many adjustable parameters in modeling. We shall draw attention to these from time to time in this book. See for example, Montgomery *et al.* (2001).

[3] Some readers only interested in the surface temperature models may wish, after this chapter, to skip directly to Chapter 5.

[4] The coalbedo is 1 minus the albedo.

Earth's radius. The amount of energy emitted to space per unit time is the Earth's surface area times the Stefan–Boltzmann constant. The balance occurs when the temperature satisfies

$$4\pi R_e^2 \sigma_{SB} \overline{T}_R^4 = \pi R_e^2 \sigma_\odot a_p \tag{2.1}$$

or

$$\overline{T}_R \approx \left(\frac{\sigma_\odot}{4} \frac{a_p}{\sigma_{SB}} \right)^{\frac{1}{4}}. \tag{2.2}$$

Using the values indicated, we compute a value of about 255 K for the global average temperature. As expected, this is well below the measured value of 287 K.

2.1.2 Budyko's Empirical IR Formula

On the basis of radiative transfer estimates, Budyko (1968) suggested the empirical terrestrial radiation formula

$$I \doteq A + B \cdot (T - 273), \tag{2.3}$$

where the "$\doteq$" sign means that it is a statistical or regressional relationship, I is the outgoing radiation to space from the top of the atmosphere in W m^{-2}, T is in kelvins (we include the -273 in the formula to appear the same as in earlier literature) and A and B are empirical coefficients that can be computed from real atmospheric conditions or estimated from satellite data. We shall take them from satellite data to be

$$A = 218 \text{ W m}^{-2}, \tag{2.4}$$
$$B = 1.90 \text{ W m}^{-2} \text{ K}^{-1} \tag{2.5}$$

(see Chapter 1). Figure 2.1 shows a plot of satellite-observed data of infrared radiation to space for various surface temperatures. These temperature data are collected contemporaneously and cospatially with the infrared data coming from different locations (latitudes) and different seasons.

The values in Figure 2.1a were derived from the data without regard or correction made for the presence or absence of clouds. In other words, cloud effects are included

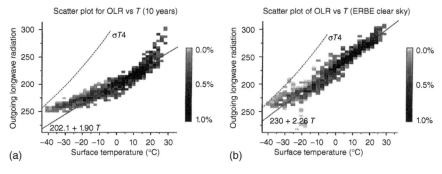

Figure 2.1 Density plots of month-average infrared radiation to space measured by satellite versus the same month's surface temperature at the same location. (a) Same as in Figure 1.11, repeated here for comparison: infrared data density versus surface temperature (°C) for the whole sky, including cloudy areas. (b) *IR* versus *T* (°C) pixels (picture elements) for which there are no clouds, hence the label "clear sky." (Figure modified from Graves *et al.* (1993) (© American Geophysical Union, with permission).)

in some sense. Figure 2.1b had the cloudy pixels (picture elements) removed from the data set prior to averaging and therefore that graph is labeled "clear sky." The relationship in the clear sky data is more linear and exhibits tighter fit to the straight line. The slope of the "all sky" case is noticeably lower (Graves *et al.*, 1993).

The way these data were used to obtain A and B is crude, and we should be aware of this as we proceed. Note that a Taylor expansion of the Stefan–Boltzmann formula about 273 K (0 °C) gives the first two terms,

$$A_{BB} = 314.9 \text{ W m}^{-2}, \tag{2.6}$$

$$B_{BB} = 4.61 \text{ W m}^{-2} \text{ °C}^{-1}. \tag{2.7}$$

Slightly different values are obtained if we choose to expand about $T = 255$K which is convenient in some problems. For now, let us take (2.3) literally and proceed, but this serves as a first reminder to us of the fact that EBMs are in some sense "schematic." The global average temperature computed with the empirical values of A and B formulas instead of the blackbody formula is 16.7 °C to be compared with the observed value of 14.4 °C. Why the dramatic improvement over the blackbody calculation? Instead of radiation to space upwelling directly from the surface, sensible and latent heat, leaves the surface, warming the air above at the expense of the surface temperature. The air above consists of slabs of radiatively active substances, these radiating upward toward space and downward toward the ground. The temperature of the air decreases on the average as we go vertically (as we will demonstrate in the next chapter), hence the radiation eventually emitted to space is from colder slabs of matter than the Earth's surface. Hence, the radiating temperature is about 255 K but the surface temperature is some 30 °C warmer.

$$A + B \cdot (T_{eq} - 273) = Qa_p; \tag{2.8}$$

$$T_{eq} = \frac{Qa_p - A}{B} + 273 \tag{2.9}$$

with $Q = \sigma_\odot/4$.

2.1.3 Climate Sensitivity

It is interesting to ask how the global average temperature $T_{eq}(Q)$ changes if the solar constant is changed by 1%. We define the sensitivity parameter $\beta_\odot$ by

$$\beta_\odot \equiv \frac{Q}{100} \frac{dT}{dQ} = \frac{\sigma_\odot}{100} \frac{dT}{d\sigma_\odot}. \tag{2.10}$$

For the model just presented,

$$\beta_\odot^{IR} = \frac{(A + B \cdot (T - 273))}{100 \cdot B} = \frac{Qa_p}{100 \cdot B}, \tag{2.11}$$

where the superscript, IR, indicates that the model includes only infrared radiative effects (e.g., a_p does not depend on T). The above gives a value of $\beta_\odot^{IR} = 1.25°$, which means that the global average temperature would change $\pm 1.25°$ for a $\pm 1\%$ change in the solar constant. This formula gives us an idea of how sensitive $\beta_\odot$ is to the value of B. For example, the blackbody Earth at 15 °C gives

$$\beta_\odot^{BB} = 0.63 \text{ K}. \tag{2.12}$$

Absorbing layers in the atmosphere, as indicated by the reduced value of B or the decreased slope of the IR(T) versus T in Figure 2.1, lead to an increased sensitivity to external forcing.

2.1.4 Climate Sensitivity and Carbon Dioxide

A more modern definition of *equilibrium climate sensitivity* (when the system is started in equilibrium or steady state and we wait for equilibrium to be established after the perturbation is applied) is for the doubling of CO_2 in the atmosphere. We can modify the parameter A in Budyko's radiation formula (2.3) to account for changes in CO_2 concentration. For this, we refer to a paper by Myhre *et al.* (1998) in which accurate radiation transfer computer codes are used to calculate the change in outgoing IR due to changes in CO_2 concentration. We may write the formula as

$$\Delta A = -5.35 \ln \frac{[CO_2]_t}{[CO_2]_0} \quad (\text{W m}^{-2}), \tag{2.13}$$

where the subscripts denote the value of CO_2 concentration at time t and at time 0.[5] For doubling CO_2 we find $\Delta A = -5.35 \ln 2 = -3.71$ W m^{-2}.

The perturbation to the energy balance leads to the response

$$(\Delta T)_{2 \times CO_2} = -\frac{\Delta A}{B} \sim 2.0 \text{ K}. \tag{2.14}$$

Note that if we had used B_{BB}, the value for the blackbody Earth, the sensitivity would be less than half this value. This is because there are positive feedbacks in the model that uses Budyko's formula for the IR. These feedback mechanisms cause the empirical value of B to be smaller for the Budyko global-average model. The most obvious of them is the water vapor. Water vapor is a strong greenhouse gas. As we increase the temperature, the absolute humidity of the air in the column will increase. This automatic operation of the climate system results in the positive feedback.[6] Note that the coefficients in Budyko's IR formula are computed using monthly averages. During the month-long averaging period, the atmospheric column has time to equilibrate the water vapor in the column. The subject of climate feedbacks will be introduced in Section 2.4 and discussed in more detail in Chapter 6.

The Myhre *et al.* (1998) paper gives simple forms for other greenhouse gases such as CH_4 and N_2O.

2.2 Time Dependence

When a time-dependent imbalance (a *forcing*) exists between the incoming and outgoing rates of energy, we can expect the surface temperature field to respond. The rate of

---

5 Myhre *et al.* (1998), use a very accurate "line-by-line" (or high spectral resolution) radiative transfer code to obtain this formula.
6 Current general circulation models (GCMs) used in the Intergovernmental Panel on Climate Change (IPCC) Report Number 5 (2013) suggest that the average value across about 20 GCMs from about that many countries currently operating around the world is about $(\Delta T)_{2 \times CO_2} = 3.0 \pm 1.5$ K. This means that our EBM is somewhat less sensitive than most GCMs. This discrepancy is likely to be related to a difference in feedback mechanisms, probably involving the treatment of clouds. Such an active mechanism is outside the purview of EBMs.

change of temperature is proportional to the difference in incoming and outgoing flux densities.

$$C\frac{dT}{dt} = -A - B \cdot (T - 273) + Qa_p, \qquad (2.15)$$

where C is the effective heat capacity per unit area on the sphere.[7] Let us take our toy planet to consist of a uniform thin shell whose radius is that of the Earth. The thickness and composition of the shell will determine the heat capacity per unit area over the sphere. A closer analog to the Earth is one where the heat capacity depends strongly on position. The effective heat capacity per square meter is pretty small over land for seasonal or monthly timescales because the heat energy does not penetrate much into the solid ground and large over ocean surfaces. This is primarily because the ocean's stirring in the first few tens of meters mixes the heat energy to those depths and leads to a relatively large heat capacity. In the later chapters, we will see how the ocean's heat capacity depends on the Fourier frequency composition of the forcing. At longer timescales of disturbance, the heat penetrates even further below and the adjustment time is elongated even more.

For the present class of idealized problems, we will take the value of C to be uniform over the Earth and to have a value equal to about half the heat capacity at constant pressure of the column of dry air over a square meter ($\sim 10^7 \text{J K}^{-1} \div 2$). Later we will see why this seems to be an appropriate value over land and we will also consider the case of a mixed-layer ocean that has a value of C about 60 times larger.

Now imagine that the temperature $T(t=0)$ is out of equilibrium. The differential equation (2.15) has a solution

$$T(t) = T_{eq} + (T(0) - T_{eq})e^{-t/\tau_0} \qquad (2.16)$$

that can be demonstrated by substitution into (2.15). The decay time constant is given by $\tau_0 = C/B$. The perturbed climate relaxes to the equilibrium solution with a decay time of τ_0, which, for the all-land planet, is about $2.5 \times 10^7 \text{s} \approx 30$ days as shown in Figure 2.2. It is easy to see why this happens. If $T(0) > T_{eq}$, there is more radiation energy flux density to space than the amount absorbed from the Sun. This means that $A + B \cdot (T - 273)$ will be greater than Qa_p and therefore the planet will cool until equilibrium is established. If the initial condition is that $T(0) < T_{eq}$, the temperature will increase until the inequality no longer holds. Note that the adjusting temperature does not overshoot the equilibrium mark. This property is characteristic of EBMs. It would not happen above the boundary layer of the atmosphere.

2.2.1 Frequency Response of Global Climate

A standard tool in analyzing a *linear system*[8] is to consider the response of the system to periodic forcing. In particular, if we force the system with a sinusoidal heating variation, we want to know the amplitude and phase of the sinusoidal response as a function of forcing frequency and the other parameters in the system. At low frequencies, we should see the response expected from a static perturbation, but at higher frequencies, we can expect a diminished amplitude of response.

---

7 The heat capacity is the proportionality constant between a heat influx $\Delta H(t) = (-A - B \cdot (T - 273) + Qa_p)\Delta t$ over the interval Δt and the response $C\Delta T(t)$.
8 A linear system is one for which the response amplitude is a linear function of the strength of the forcing.

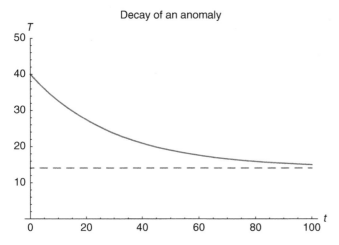

Figure 2.2 Exponential decay of an anomaly of the global average temperature as computed by a globally and annually averaged EBM. The initial temperature is $T(0) = 40\,°C$ and the equilibrium temperature is $T_{eq} = 14\,°C$ as indicated by the horizontal dashed line. The solid curve is the solution as a function of time for the case $\tau_0 = 30$ days.

Suppose the global system is forced away from its equilibrium by a sinusoidal heat source

$$H(t) = H_f \cos 2\pi f t, \tag{2.17}$$

where $H(t)$ is the amount of the heating perturbation per unit area per unit time and f is the frequency of the oscillating heat source. For intuitive purposes, we could think of the Sun as a variable star. We will follow a familiar engineering practice of using complex notation for the oscillating quantities. This is facilitated by the notation

$$H(t) = \mathrm{Re}\{H_f e^{i2\pi f t}\}, \tag{2.18}$$

where $\mathrm{Re}\{\cdot\}$ indicates the real part of the complex number in the curly brackets. The departure of the temperature from its static equilibrium value satisfies the forced linear equation

$$C\frac{dT_f(t)}{dt} + BT_f(t) = H_f e^{i2\pi f t}. \tag{2.19}$$

We proceed by insertion of the trial solution $T_f(t) = T_f e^{i2\pi f t}$, with T_f a complex number whose complex phase indicates the phase lag of the climate behind the forcing, and it does not depend on time. After canceling the exponentials, we have as a solution

$$T_f = \frac{H_f/B}{2\pi i f \tau_0 + 1}, \tag{2.20}$$

where $\tau_0 = C/B$ as before is the relaxation time of the system. The modulus or amplitude of the response at frequency f is

$$|T_f| = \frac{H_f/B}{\sqrt{1 + (2\pi f \tau_0)^2}}; \tag{2.21}$$

and the phase lag (in radians) is given by

$$\phi_f = \arctan 2\pi f \tau_0. \tag{2.22}$$

These relations are shown in Figure 2.3. Note that in the *high-frequency limit*,

$$f \gg 1/2\pi \tau_0; \tag{2.23}$$

the phase lag approaches $\pi/2$ (one quarter cycle).

Note that in the foregoing, we found a solution that worked, but what about the initial condition? That part (the so-called *particular* or *transient solution*) will decay away as

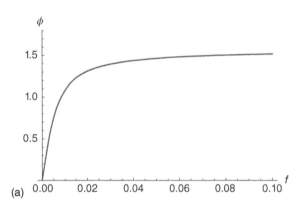
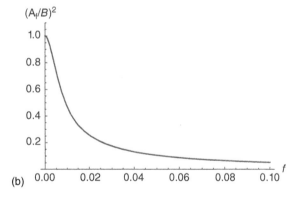
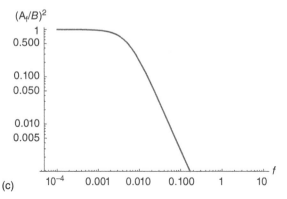

Figure 2.3 Response of global average temperature to sinusoidal forcing at different frequencies. (a) Phase lag ϕ_f as a function of frequency of the driver amplitude ($|T_f|/|T_f^{max}|$). The phase lag approaches $\pi/2$ (quarter cycle) as the frequency is increased to high levels compared to $1/\tau_0$. (b) The amplitude squared of the response, normalized to unity at $f = 0$. (c) Amplitude squared plotted (log–log) versus frequency, f in units of τ_0^{-1}. The log–log slope of -2 for high frequencies indicates an f^{-2} power law regime. For an all-land planet, τ_0 is about a month and for an all-ocean planet τ_0 is a few years.

$T(0)e^{-t/\tau_0}$ just as in the previous section. After several multiples of τ_0, the solution will converge on the periodic one studied above.

2.2.2 Forcing with Noise

Now consider the situation where the system is forced from its static steady state by a noise term.[9] This is quite plausible considering that fluctuations of such quantities as weather and/or cloudiness as well as other perturbations to the energy budget occur at timescales that are short compared to τ_0:

$$\tau_Z \ll \tau_0. \tag{2.24}$$

The equation governing the response is given by

$$C\frac{dT(t)}{dt} + BT(t) = Z(t), \tag{2.25}$$

where $Z(t)$ is a random or stochastic function with a short autocorrelation time (compared to τ_0). In this chapter, we will adopt a discrete time variable rather than a continuous one in order to avoid certain difficulties at short timescales (see, e.g., Gardiner, 1985). The time variable will then be $t_0, t_1, \ldots, t_{N-1}$, with intervals $t_{n+1} - t_n = \Delta t$. Reformulating the differential equation above in (approximate) finite difference form,[10] we have

$$T_{n+1} = \lambda T_n + z_n; \tag{2.26}$$

where $T_n = T(t_n)$, $\lambda = 1 - \Delta t/\tau_0$, and $z_n = \Delta t\, F(t_n)/C$. We may specify that $0 \leq \lambda < 1$. Note that z_n depends explicitly on Δt; a fact that leads to difficulties in the continuous time formulation. We specify that the forcing z_n have zero ensemble mean[11] and that it is *white noise* (no correlation from one time step to the next):

$$\langle z_n \rangle = 0; \tag{2.27}$$

$$\langle z_n z_m \rangle = \sigma_z^2 \delta_{mn}; \tag{2.28}$$

where $\delta_{mn} = 1$, if $n = m$, otherwise zero.[12] First consider the homogeneous problem with $z_n \equiv 0$. In this case,

$$T^h_{n+1} = \lambda T^h_n. \tag{2.29}$$

We can try the solution $T^h_n = a^n$, where the superscript h denotes the homogeneous solution, after which we find that $a = \lambda$. For a given initial condition T_0, we find

$$T^h_n = T_0 \lambda^n, \tag{2.30}$$

---

9 The idea of noise forcing in a climate model comes from Hasselmann (1976).
10 Begin with the difference equation: $\frac{\Delta T}{\Delta t} = -\frac{T_n}{\tau_0} + F_n$. Rearranging, we have

$$T_{n+1} = \left(1 - \frac{\Delta t}{\tau_0}\right) T_n + F_n \Delta t.$$

Then set $\lambda = 1 - \frac{\Delta t}{\tau_0}$.
11 The ensemble mean or expectation value of a random variable (called a *variate*, denoted by r) is denoted by the angle brackets $\langle r \rangle$ (if the range of the sum includes m).
12 The symbol δ_{nm} is the Kronecker delta symbol, $\delta_{mn} = 0$ if $m \neq n$ and $\delta_{mn} = 1$ if $m = n$. It has the useful property that $\sum_n f_n \delta_{nm} = f_m$.

with

$$\lambda = 1 - \frac{\Delta t}{\tau_0}. \tag{2.31}$$

We can use the formula $x^n = \exp(n \ln x)$ to form

$$\lambda^n = \exp\left\{\left[\ln\left(1 - \frac{\Delta t}{\tau_0}\right)\right]n\right\} \approx \exp\left(-\frac{n\Delta t}{\tau_0}\right) = \exp\left(-\frac{t}{\tau_0}\right); \tag{2.32}$$

where we used the formula $\ln(1 + \epsilon) \approx \epsilon$ for small ϵ (i.e., $\Delta t/\tau_0 \ll 1$) to obtain the result

$$\lim_{\Delta t \to 0} T_n^h = T_0 e^{-t/\tau_0}. \tag{2.33}$$

The last formula for the decay of an anomaly agrees with our earlier result derived from the continuous form of the differential equation. Next consider the nonhomogeneous case (particular solution). We introduce the integrating factor $1/\lambda^{n+1}$:

$$\frac{T_{n+1}}{\lambda^{n+1}} - \frac{T_n}{\lambda^n} = \frac{z_n}{\lambda^{n+1}}. \tag{2.34}$$

The left-hand side has the interesting form $x_{n+1} - x_n$. If we sum this difference over the index n from 0 to $N-1$, we obtain

$$(x_1 - x_0) + (x_2 - x_1) + (x_3 - x_2) + \cdots + (x_N - x_{N-1}) = x_N - x_0, \tag{2.35}$$

where $x_n = T_n/\lambda^n$, as all the intermediate terms cancel out. Now summing from 0 to $N-1$ on (2.34) and afterward multiplying through by λ^N yields

$$T_N = T_0 \lambda^N + \lambda^N \sum_{n=0}^{N-1} z_n \lambda^{-(n+1)}. \tag{2.36}$$

It is comforting to note that the ensemble average of a decaying anomaly $\langle T_N \rangle$ is precisely the same as the homogeneous solution decay $T_0 \lambda^N$, as $\langle z_n \rangle = 0$.

Next consider the properties of the temperature for $N \gg \tau_0/\Delta t$ (this means that all knowledge of initial conditions, T_0, has died away); we refer to this limit as *climatology*. First we note that $\langle T_N \rangle \to 0$, as we have subtracted the mean (T_{eq}) already. We say that T_N is the departure from the mean or the *anomaly*. The *lagged covariance* is then

$$\langle T_N T_{N+l} \rangle = \lambda^{2N+l} \sum_{n=0}^{N-1} \sum_{m=0}^{N+l-1} \langle z_n z_m \rangle \lambda^{-(m+1)-(n+1)}$$

$$= \lambda^{2N+l} \frac{\sigma_z^2}{\lambda^2} \sum_{n=0}^{N-1} (\lambda^{-2})^n$$

$$= \lambda^{2N+l} \frac{\sigma_z^2}{\lambda^2} \frac{1 - \lambda^{-2N}}{1 - \lambda^{-2}}; \tag{2.37}$$

where, in the last formula, we used the summation formula for a finite-length geometric series

$$\sum_{n=0}^{N-1} x^n = \frac{1 - x^N}{1 - x}. \tag{2.38}$$

The formula (2.37) may be simplified for large N as

$$\langle T_N T_{N+l} \rangle \Rightarrow \sigma_{ss}^2 \lambda^l \quad \text{(climatology limit)} \tag{2.39}$$

with

$$\sigma_{ss}^2 = \frac{\sigma_z^2}{1-\lambda^2} = \frac{\sigma_z^2}{1-(1-\frac{\Delta t}{\tau_0})^2} \approx \frac{\sigma_z^2}{1-1+2\frac{\Delta t}{\tau_0}}, \tag{2.40}$$

$$\approx \frac{\tau_0}{2\Delta t}\sigma_z^2 \tag{2.41}$$

which is the variance in the limit of climatology. Hence, we find that the temperature approaches a stationary time series for $N \gg \tau_0/\Delta t$ and the *lagged autocorrelation*

$$\rho(m) \equiv \frac{\langle T_N T_{N+m}\rangle}{\sigma_{ss}^2} \tag{2.42}$$

$$= \lambda^m, \tag{2.43}$$

where $m = \tau/\Delta t$ is the lag in discrete steps of interval Δt. The formula for $\rho(m)$ takes exactly the same form as the decay of an anomaly where τ is the lag in temporal units. For small Δt, we can write

$$\left(1 - \frac{\Delta t}{\tau_0}\right)^{\frac{\tau}{\Delta t}} \Rightarrow e^{-\frac{\tau}{\tau_0}}, \quad \text{as } \Delta t \to 0. \tag{2.44}$$

This last statement suggests that in the limit of small steps, the decay of the lagged autocorrelation is exponentially decaying. The lagged autocorrelation (as opposed to the lagged covariance) is particularly interesting as it does not depend on the strength (or variance) of the noise forcing, z_n, in this linear problem.

2.2.3 Predictability from Initial Conditions

Consider the evolution of a state $T_0 = T(0)$ toward climatology. To what extent can we say with any certainty what a single member of the ensemble of evolving trajectories will be? A reasonable way of expressing our knowledge of the confinement of an individual trajectory to a narrow band of possibilities is to look at the spreading of a bundle of trajectories as they leave a common initial state (North and Cahalan, 1981). Next we examine this issue for the simple linear model with noise forcing.

We seek the variance about the ensemble mean at step N. This can be written as

$$\langle (T_N - T_0\lambda^N)^2\rangle = \lambda^{2N} \sum_{n=0}^{N-1}\sum_{m=0}^{N-1} \langle z_n z_m\rangle \lambda^{-(n+1)}\lambda^{-(m+1)}. \tag{2.45}$$

Using the relation (2.28), we can collapse the double sum to a single one:

$$\langle (T_N - T_0\lambda^N)^2\rangle = \lambda^{2N}\sigma_z^2 \sum_{n=0}^{N-1} \lambda^{-2n-2} = \lambda^{2N-2}\sigma_z^2 \sum_{n=0}^{N-1}(\lambda^{-2})^n. \tag{2.46}$$

The last sum is of a finite geometrical series, see (2.38). This allows us to rearrange the terms to

$$\langle (T_N - T_0\lambda^N)^2\rangle = \frac{\sigma_z^2}{1-\lambda^2}(1-\lambda^{2N}); \tag{2.47}$$

leading to our final result:

$$\langle (T_N - T_0\lambda^N)^2\rangle = \sigma_{ss}^2(1-\lambda^{2N}). \tag{2.48}$$

38 | 2 Global Average Models

The last formula says that the spread of the bundle of subensemble members with initial condition $T = T_0$ is initially zero, but as time evolves from the initial condition, it fills out to a constant width (see Figures 2.4 and 2.5). The thickness of the bundle is a measure of *predictability* conditioned on the value of the initial anomaly. The negative of the logarithm of the ratio of the width of the bundle at time step N to the width in the climatology range ($N \to \infty$) is related to the *information* content of the bundle as it evolves (Leung and North, 1990; see also Kleeman, 2011; Roulston and Smith, 2002). The characteristic time for filling out the thickness toward its saturation value occurs in half the time as the

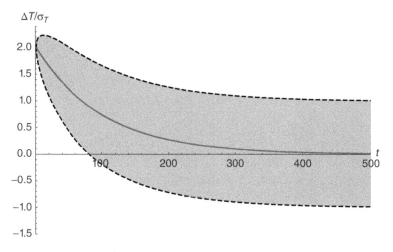

Figure 2.4 Heavy Line: The decay of mean climate (global average temperature) toward the steady-state climate from an anamoly of two times the standard deviation. Dashed curves enclose shading of the envelope of the standard deviation of individual ensemble members about the mean decay curve. Units in the vertical are in standard deviations of the reference climate. Units on the abscissa are autocorrelation times for the reference climate, in time steps N. In this case, $\tau_0 = 0.5$. The Δt for the numerical integration was $\tau_0/500 = 0.5/500 = 1/1000$.

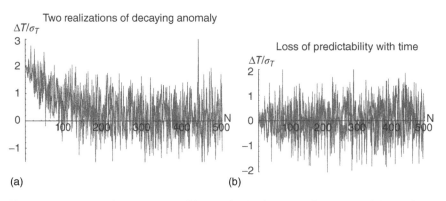

Figure 2.5 (a) Two realizations (ensemble members) of numerically computed decays from an initial anomaly of two times the standard deviation of climatology. The model parameters are the same as in the previous figure. (b) A single realization with $T_0 = 0$. The model parameters are the same as in the previous figure and (a). Note the widening of the envelope of extremes as time progresses. This is equivalent to the loss of predictability with evolution to climatology.

characteristic time of decay toward equilibrium[13] (climatological mean) $\approx \tau_0$. This has implications about the decay of information in a forecast. Figure 2.4 shows the decay of the subensemble mean solution with an envelope that denotes the spread (σ_{T_n}) of the pdf (probability density function) about the mean. Figure 2.5a shows a bundle (or ensemble) of two realizations computed numerically evolving from the same initial condition (in this specific example, two times the climate standard deviations from the mean climate). Figure 2.5b shows a single realization where the initial condition is $T_0 = 0$. Here we see the width of the envelope of variability grow from zero to saturation just as in the previous figures. We can see from the figures that we have information in two ways:

1. The departure of the ensemble mean from zero; if this separation is large, we can see a large *signal-to-noise ratio.*
2. The thickness of the bundle is also a measure of how well the spread is constrained.

When the spread is narrow we know a lot, when it is wider, we know less. We see from the formula and the figures that the spread quickly fills out (in half a decay time) to the climatological width. Note that in Figure 2.5b there is predictability even though the signal-to-noise ratio is zero. A measure of predictability or a "skill score" needs to take both features into account. For example, once the decay to the climatological mean and its width is that of climatology, we say the predictability is zero.

In forecasting, there are three simple methods:

1. Persistence: tomorrow will be the same as today.
2. Climatology: tomorrow will be that of the climatological long-term average.
3. Finally, there is damped persistence where one exponentially relaxes the initial condition toward the climatological mean with the relevant time constant. This latter is the case for the autoregressive model of order one that we have developed for the global average climate model (van den Dool, 2007).

The *skill* of modern forecasting tools is measured relative to a reference no-brainer forecast, which can be any of the three simple methods listed. Perhaps the best measure of predictability is the *transinformation* (see Leung and North, 1990 and Kleeman, 2011 for derivations). A formula for the transinformation for the simple AR1 process of this section is given by

$$I(T, T_0) = -\frac{1}{2} \log(1 - e^{-2t/\tau_0}). \qquad (2.49)$$

This function starts at $+\infty$ at $t = 0$, and decays toward zero with a characteristic time of $\tau_0/2$.

2.2.4 Probability Density of the Temperature

We ask about the pdf for the temperature generated by the stochastic model (2.26) whose solution is given by (2.36). Recall that z_n in (2.36) is to be a normally distributed variate with individual z_n independent of one another. Since T_N is a linear combination of these normally distributed variables, it can be shown[14] that it is itself a normally

---

[13] This make use of $\lambda^{2n} \to e^{-2t/\tau_0}$ as t and τ_0 become large compared to Δt.
[14] The proof follows quickly from use of the *moment generating function*: $M(t) = \langle e^{tX} \rangle$, for details see any book on probability theory, for example, Bulmer (1979).

distributed variate. In fact, if the series in (2.36) has many terms, then T_N will be normally distributed even if z_n were not normally distributed by the central limit theorem (assuming the variance is finite; an example that does not qualify is the Cauchy distribution $f(T) \propto 1/(1 + T^2)$).

2.3 Spectral Analysis

We have now examined the response of the simple linear system to sinusoidal forcing in time, and we have also seen how the system behaves under noise forcing. There is a connection to these two subjects. First consider the noise, z_n; $n = \ldots, -2, -1, 0, 1, 2, \ldots$. We can transform the noise time series into its Fourier representation:

$$\tilde{z}(f) = \sum_{n=-\infty}^{\infty} z_n\, e^{2\pi i f n}. \tag{2.50}$$

The inverse of the Fourier transformation is

$$z_n = \int_{-f_N}^{f_N} \tilde{z}(f) e^{-2\pi i f n}\, df, \tag{2.51}$$

where $f_N = 1/2N$ is the highest frequency resolved by the discrete time steps. This is like an ordinary Fourier series that is discrete in frequency and continuous in time, except that here the continuous function is in the frequency domain and the discrete is in the time domain. The function $\tilde{z}(f)$ is continuous on the finite interval $(-f_N, f_N)$, while the time sequence is discrete and runs in integer steps from minus infinity to plus infinity. The *basis functions* for the decomposition are

$$\psi_n(f) = e^{2\pi i f n}. \tag{2.52}$$

These basis functions are *orthonormal*[15]:

$$\int_{-\frac{1}{2}}^{\frac{1}{2}} \psi_n^*(f) \psi_m(f)\, df = \delta_{nm}. \tag{2.53}$$

The *completeness relation*[16] is

$$\sum_{n=-\infty}^{\infty} \psi_n^*(f) \psi_n(f') = \delta(f - f'), \tag{2.54}$$

where $\delta(f - f')$ is the Dirac delta function.[17]

---

[15] *Orthonormal* means that the functions are orthogonal as in the next equation and they are normalized such that the coefficient of δ_{nm} in the next equation is unity.

[16] *Completeness* implies that there are enough orthogonal functions in the basis to represent the functions of interest as a series of components. An example of completeness is when one wishes to expand a vector in three dimensions: it takes all three basis vectors **i, j**, and **k** to represent a vector in three dimensions. It can be shown that the completeness relation guarantees that there are enough basis vectors. However, rigorous proof of completeness for an arbitrary basis set is usually rather difficult. Fortunately, this is not a problem for the basis sets we will encounter in this book.

[17] The Dirac delta function is a sharp spike at the location where its argument is zero. It has the remarkable property $\int_a^b f(x)\delta(x - x')\, dx = f(x'); x' \in (a, b)$. It can be represented in many different ways, but a simple one is $\delta(x) = \lim_{n \to \infty} \frac{1}{n\pi} \left(\frac{\sin nx}{x} \right)^2$.

2.3.1 White Noise Spectral Density

Now take the z_n to be an infinitely long discrete sequence of independent normally distributed random variables (so-called *white noise*) with mean zero ($\langle z_n \rangle = 0$) and standard deviation unity ($\sigma_z = 1$). Consider the transform (2.50) as a continuous random function on the interval $(-\frac{1}{2}, \frac{1}{2})$. Clearly, $\langle \tilde{z}(f) \rangle = 0$ as $\langle z_n \rangle$ is also zero. The covariance of $\tilde{z}(f)$ can be computed by taking $\langle \tilde{z}^*(f)\tilde{z}(f') \rangle$ using (2.51). The steps are a little tricky here because the math needs to take into consideration singularity problems, but we can quote the result

$$\langle \tilde{z}^*(f)\tilde{z}(f') \rangle = \sigma_z^2 \delta(f - f'). \tag{2.55}$$

The delta function whose argument is $(f - f')$ tells us that the covariance of $\tilde{z}(f)$ with itself evaluated at a different frequency vanishes; as $f \to f'$, the covariance becomes infinite. This turns out to be a useful and powerful property of all *stationary time series* (those for which the mean and variance are constant with n and the lagged covariance depends only on lag). The coefficient of the delta function, σ_z^2 is called the *spectral density* for white noise. Note that it is a constant function of frequency, a key property of white noise.

2.3.2 Spectral Density and Lagged Correlation

The spectral density, $S_T(f)$, of a stationary process $T(t)$ is defined as

$$\langle \tilde{T}^*(f)\tilde{T}(f') \rangle \equiv S_T(f)\delta(f - f'). \tag{2.56}$$

We can now find an interesting relationship between the spectral density and the lagged covariance function, R_ℓ, where ℓ is the lag:

$$R_\ell = \langle T_n T_{n-\ell} \rangle. \tag{2.57}$$

Note that as the time series is stationary, R_ℓ does not depend on n.

We start by taking the Fourier transform (abbreviated FT) of R_ℓ:

$$\sum_{\ell=-\infty}^{\infty} \langle T_n T_{n-\ell} \rangle e^{2\pi i \ell f}. \tag{2.58}$$

Next insert the FT for each quantity inside the brackets:

$$\sum_\ell \int \int \langle \tilde{T}^*(f')\tilde{T}(f'') \rangle \exp\left(2\pi i(nf' - nf'' - \ell f' + \ell f'')\right) df' \, df'', \tag{2.59}$$

and each integral runs from $-\frac{1}{2}$ to $\frac{1}{2}$. Next carry out the sum over ℓ:

$$\int \int \langle \tilde{T}^*(f')\tilde{T}(f'') \rangle \exp\left(2\pi i n(f' + f'')\right) \delta(f - f'') df' \, df''. \tag{2.60}$$

Now do the integral over f'':

$$\int \langle \tilde{T}^*(f')\tilde{T}(f) \rangle e^{2\pi i n(f - f')} df'. \tag{2.61}$$

Now substitute the definition of the spectral density (2.56):

$$\int S_T(f)\delta(f - f') e^{2\pi i n(f - f')} df' = S_T(f). \tag{2.62}$$

This last is our desired result, that is, the FT of the lagged correlation coefficient is the spectral density for a stationary process. A similar result holds for the continuous time case. This important result for the spectral density can be stated as follows:

$$S_T(f) = \sum_{\ell=-\infty}^{\infty} \langle T_n T_{n-\ell} \rangle e^{2\pi i \ell f}. \tag{2.63}$$

2.3.3 AR1 Climate Model Spectral Density

We can now examine the spectral density of the noise-forced linear global climate model. Here we are interested in the large N limit after the initial conditions have decayed away and we are left with the climatological regime. As a starting point, we recall that the AR1 climate model has as its lagged correlation coefficient $R_\ell = \lambda^{|\ell|}$ (see (2.43)). The magnitude sign in the exponent indicates that the lagged correlation function is an even function of ℓ (plus or minus lags lead to the same correlation). To obtain the spectral density, we must take the FT (following Papoulis, 1984, p. 290):

$$\sum_{\ell=-\infty}^{\infty} \lambda^{|\ell|} e^{2\pi i f \ell} = \sum_{\ell=-\infty}^{-1} \lambda^{-\ell} e^{2\pi i f \ell} + \sum_{\ell=0}^{\infty} \lambda^{\ell} e^{2\pi i f \ell} \tag{2.64}$$

$$= \sum_{\ell=-\infty}^{-1} (\lambda^{-1} e^{2\pi i f})^{\ell} + \sum_{\ell=0}^{\infty} (\lambda e^{2\pi i f})^{\ell}. \tag{2.65}$$

Now use the formulas for geometric series:

$$\sum_{n=-\infty}^{-1} x^n = \frac{1}{x-1} = \frac{1}{\lambda^{-1} e^{2\pi i f} - 1}, \tag{2.66}$$

$$\sum_{n=0}^{\infty} w^n = \frac{1}{1-w} = \frac{1}{1 - \lambda e^{2\pi i f}}. \tag{2.67}$$

Next add them up and rearrange:

$$S_T(f) = \frac{1 - \lambda^2}{1 + \lambda^2 - 2\lambda \cos 2\pi f}. \tag{2.68}$$

The last formula is graphed in Figure 2.6. Compare this with the continuous time case (Figure 2.3c). The minus 2 power law holds over a wide range, but fails at high frequencies near $f = \frac{1}{2}$, the highest frequency available in a discrete time ($\Delta t = 1$) model. This highest available frequency is actually $\frac{1}{2\Delta t}$ more generally. It is called the *Nyquist frequency*. See Papoulis (1984), or most any book on time series analysis, especially one including spectral analysis.

2.3.4 Continuous Time Case

In the continuous time case, we have the governing equation (2.25). To solve this equation, we must use the continuous Fourier transform on the infinite domain in space as well as time. Some properties of continuous white noise follow:

$$\langle Z(t) \rangle = 0; \quad \langle Z(t) Z(t') \rangle = \sigma_Z^2 \delta(t - t'); \tag{2.69}$$

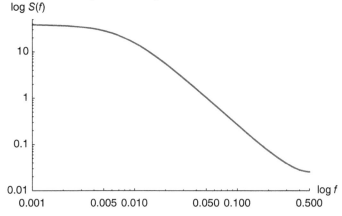

Figure 2.6 Log–log graph of the spectral density for a discrete time AR1 climate model. In this model, the EBM is modeled as an AR1 process. In this case, the global climate is driven by a white noise forcing, imitating weather. Note that the EBM *filters out* the high frequencies of the driving force (white noise) which has a *flat* spectrum.

the Fourier transformation (FT) of the time series is

$$\tilde{T}(f) = \int_{-\infty}^{\infty} T(t) e^{2\pi i f t} \, dt; \tag{2.70}$$

and the inverse FT:

$$T(t) = \int_{-\infty}^{\infty} \tilde{T}(f) e^{-2\pi i f t} \, df. \tag{2.71}$$

Applying the FT to (2.25), we find

$$(B - 2\pi i f C)\tilde{T}(f) = \tilde{Z}(f). \tag{2.72}$$

Rearranging and taking expectations:

$$\langle \tilde{T}^*(f)\tilde{T}(f') \rangle = \frac{\sigma_Z^2 \delta(f - f')}{B^2 + 4\pi^2 C^2 f f'}. \tag{2.73}$$

Finally,

$$S_T(f) = \frac{\sigma_Z^2 / B^2}{1 + 4\pi^2 \tau_0^2 f^2}. \tag{2.74}$$

Finally, you can now see the reason for plotting in Figure 2.3c. The spectral density for the global average temperature (in this model) is the response to white noise. White noise delivers sinusoidal forcing at all frequencies with the same amplitude, but random phases (remember the white noise has a flat spectral density or *variance spectrum*). The spectral density tells us how the variance of the fluctuations is distributed across frequencies. It is intuitively appealing to think of the response spectral density to be a "filtered" product of the white noise spectrum by the filter $(1 + 4\pi^2 \tau_0^2 f^2)^{-1}$. This filter removes the high-frequency components of the noise forcing being fed into the response.

2.4 Nonlinear Global Model

2.4.1 Ice-Albedo Feedback

Water vapor feedback is probably the most important feedback mechanism for the global climate.[18] There are several others including cloud feedback and lapse rate feedback. Another important one is the ice-albedo feedback mechanism. We envision a zonally symmetric ice cap centered at the North Pole. When the polar ice sheets expand, they cause the planet to be more reflective to sunlight, thereby reducing a_p. In fact, to "close" the system we make the coalbedo just a function of global temperature, $a_p(T)$. This leads us to the nonlinear energy balance equation with only one dependent variable:

$$C\frac{dT(t)}{dt} = -A - B \cdot (T(t) - 273) + Qa_p(T(t)). \tag{2.75}$$

Equilibrium solutions occur when the left-hand side is set to zero.

$$A + B \cdot (T_{eq} - 273) = Qa_p(T_{eq}). \tag{2.76}$$

What are reasonable forms for $a_p(T)$? For large T, the coalbedo will be that of an ice-free Earth and therefore it will have a value of about 0.70, whereas, when the planetary average temperature is below freezing, the planet will be iced over and its coalbedo will be close to 0.35. A convenient way to parameterize such a function is with the hyperbolic tangent, which is unity for large argument and minus unity for large negative argument.

$$a_p(T) = a_i + \frac{1}{2}(a_f - a_i)(1 + \tanh \gamma(T - T_{ref})), \tag{2.77}$$

where T_{ref} is the transition temperature from ice albedo to ice-free albedo, a_i is the coalbedo for ice-covered surface, a_f for an ice-free surface, and γ is a parameter that controls how steep the transition is in going from the ice-covered range to the ice-free range. Figure 2.7 shows a graph of $a_p(T)$ for several values of γ.

The energy balance (2.76) is a nonlinear transcendental equation whose roots give the solutions T_{eq}. A convenient way to proceed is to solve the equation for Q and find the values of T_{eq} that correspond to these values of solar constant.

$$Q = (A + B \cdot (T_{eq} - 273))/a_p(T_{eq}). \tag{2.78}$$

One can find the roots of this equation by plotting

$$A + B \cdot (T - 273) \text{ and } Qa_p(T - 273)$$

as functions of T on the same graph (Figure 2.8). The intersections of these curves are the steady-state solutions satisfying the energy balance equation. Note that as A is reduced (analogous to increases in greenhouse gases), the intercept of the straight line is reduced, making the temperature corresponding to the uppermost root increase. Similarly, increasing the total solar irradiance Q causes the dashed curve to be amplified and again the temperature of the uppermost root to increase. As the solar constant is decreased, roots I and II merge together at a critical point and there remains only one root corresponding to the ice-covered Earth. Similarly if A is increased (equivalent

---

18 The definition of equilibrium climate sensitivity is slightly ambiguous because the different feedback mechanisms can have very different timescales. For example, water vapor feedback is usually associated with weather timescales (<1 month), while glacial ice feedback may take centuries to equilibrate.

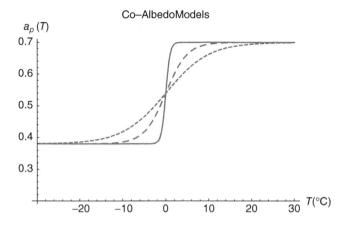

Figure 2.7 Three coalbedo models $a_p(T)$, each with a different "abruptness" of the sensitivity of ice extent to T shown as a function of global average temperature. The solid line represents a very abrupt transition to total ice cover, while the others are less abrupt. The abruptness is controlled by the parameter γ. The steepest curve is for $\gamma = 1$, the intermediate case is $\gamma = 0.2$, and the weakest case is for $\gamma = 0.1$.

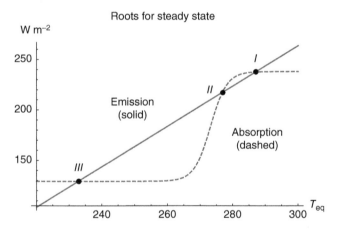

Figure 2.8 The steady-state solutions can be found by plotting $A + B(T - 273)$ (the straight line) versus $Qa_p(T - 273)$ (the curved, dashed line). The three roots, *I*, *II*, and *III* are indicated by the intersections of the two curves where steady state occurs. Root *I* corresponds (roughly) to the present climate, root *II* is an intermediate (shown in the text to be unstable), and root *III* is a snowball Earth, completely ice covered.

to lowering the concentration of a greenhouse gas), the same thing happens with a resulting snowball Earth.

A solution curve is plotted in Figure 2.9 as a function of the *control parameter Q*. This graph is called the *operating curve*. Note that with the nonlinear model, the situation is much more complicated (and interesting) than in the simple linear case. There are three solutions for a fixed value of Q over a small range, labeled *I–III*. Solution *I* corresponds roughly to the present climate of the Earth, while *III* corresponds to an ice-covered planet (the so-called *deep-freeze* solution). The intermediate solution, *II*, is strange, as it has the peculiar property that the global temperature *decreases* as the solar constant

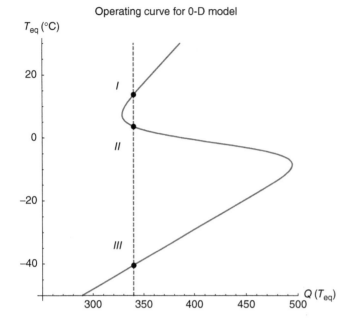

Figure 2.9 A plot of $Q = \sigma_\odot/4$ versus T for the nonlinear model (curved line), called the *operating curve*. The vertical dashed line is the present value of Q. Roots *I*, *II*, and *III* represent steady-state climates. The uppermost root *I* corresponds roughly to the present climate, and the lowest root *III* is an ice-covered planet solution. Values of the constants are $A = 218$ W m^{-2}, $B = 1.90$ W m^{-2} K^{-1}, $a_i = 0.70$, $a_j = 0.38$, and $\gamma = 1$. The intermediate partially ice-covered planet solution has the peculiar property that as the Sun is brightened, the planet becomes cooler. Note that if we start on branch III and increase the TSI ($Q(T_{eq})$), we have to go all the way to nearly 500 W m^{-2} before we get to a jump up to the ice-free planet (branch I). Then we have to slip back down to our present climate. This is a problem called the *faint Sun paradox*. Another problem is that if we start at the present climate I and lower the TSI or equivalently encounter a mega-volcano that keeps on erupting and screening sunlight, we could drop to the dreaded *snowball Earth*!

is *increased* (increasing the solar constant is equivalent to shifting the dashed line to the right). This oddity is resolved in the next section.

Can you imagine how Mikhail Budyko and/or William Sellers might have felt when they (independently) discovered this bizarre behavior that if the solar constant is lowered just a few percentage points, the planet might ice over? Their simultaneous discoveries came in the late 1960s, based on one-dimensional climate models that we will treat in Chapters 5 and 7. In the present chapter, we see that such exotic effects can even be found in the simplest of EBCMs.

2.4.2 Linear Stability Analysis: A Slope/Stability Theorem

Consider a climate whose equilibrium temperature T_{eq} is a solution of (2.76), corresponding to a certain value of the solar parameter Q_0. Let the initial condition be away from equilibrium, $T(0) \neq T_{eq}$. The time dependence of the temperature will be an evolving solution of (2.75). We denote the temperature of the perturbed state as

$$T(t) = T_{eq} + \delta T(t). \tag{2.79}$$

For the rest of this section, we consider $\delta T(t)$ to be small so that second orders (in powers of δT) and higher can be neglected compared to the first order. Then

$$C\frac{d}{dt}\delta T(t) \approx [-B + Qa'_p(T_{eq})]\delta T(t). \tag{2.80}$$

An alternative expression to that in the brackets can be found by differentiating the expression defining the equilibrium condition (2.76) with respect to Q (considering Q to be a function of T_{eq}; i.e., the functional dependence is the operating curve):

$$-a_p\frac{dQ}{dT_{eq}} = [-B + Qa'_p(T_{eq})], \tag{2.81}$$

which allows us to write

$$\frac{d}{dt}\delta T(t) = -\lambda_{eq}\delta T(t); \tag{2.82}$$

where

$$\lambda_{eq} = \frac{a_p(T_{eq})}{C}\frac{dQ}{dT_{eq}}. \tag{2.83}$$

The solution to this homogeneous linear problem is

$$\delta T(t) = \delta T(0)e^{-\lambda_{eq}t}. \tag{2.84}$$

Positive λ_{eq} leads to solutions that decay to the local equilibrium solution. Similarly, negative λ_{eq} leads to *unstable* solutions that *run away* from the local equilibrium solution exponentially.[19] The only factor in the definition of λ_{eq} that can change sign is the slope dQ/dT_{eq} of the operating curve. If the local slope is positive, the solution at that point will be stable, otherwise it will be unstable. This result is known as the *slope-stability theorem*.

2.4.3 Relaxation Time and Sensitivity

Another interesting point to be made here is that the size of λ_{eq} determines the relaxation time for the linearized model including ice-albedo feedback. When dT_{eq}/dQ is large (e.g., near the doubling-back point or *bifurcation* point), the relaxation time ($=1/\lambda_{eq}$) will be long. Similarly, a linearized model forced by noise near a bifurcation will have large climate variance and long autocorrelation times (also $=1/\lambda_{eq}$). In fact, we have

$$\mathcal{T}_T = 1/\lambda_{eq}. \tag{2.85}$$

The latter is an important property as it relates the autocorrelation time $\mathcal{T}_T$ of the present fluctuating climate to its sensitivity

$$\mathcal{T}_T = \frac{100C}{a_p(T_{eq})Q_0}\beta_\odot, \tag{2.86}$$

where $\beta_\odot$ is the sensitivity parameter (2.4), a result that is independent of the variance or "strength" of the forcing noise, σ_F^2. This tells us that the more sensitive the climate, the longer will be its response time and the longer will be its autocorrelation time. Turned

---

[19] This elegant proof was transmitted to author GRN by Robert Dickinson in a letter *ca.* 1976.

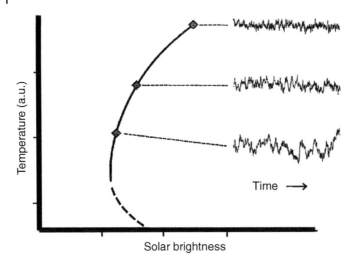

Figure 2.10 Schematic of an operating curve with a bifurcation where the slope → ∞. The amplitude of the fluctuations increases and the autocorrelation time (same as the relaxation time for simple models) becomes larger as one approaches the bifurcation. (Crowley and North (1988). Reproduced with permission of AAAS.)

around, we find that the longer the characteristic time, the more sensitive the climate is to *external* perturbations. Moreover, if the forcing is held fixed (weather-induced noise), the amplitude will increase for an increasing sensitivity. The aforementioned relation is a special case of the so-called *fluctuation–dissipation theorem* (Leith, 1975; North et al., 1993). Figure 2.10 shows a schematic of such a march toward the bifurcation as the total solar irradiance is decreased. A final note is that there is nothing special here about the linear IR formula. It could have had a form $I(T)$ and our results would still hold with the substitution $B = dI/dT$.

2.4.4 Finite Amplitude Stability Analysis

In some cases, such as in the zero and one-dimensional models, one can carry the stability analysis even further by constructing a *potential function* or *Lyapunov function* for the problem; that is, we can find a function $F(T)$ called the *potential function* such that the time derivative of the temperature is proportional to its gradient in the temperature coordinate:

$$\frac{dT}{dt} = -\frac{1}{C}\frac{dF}{dT}. \tag{2.87}$$

For the zero-dimensional model, we can find the potential function by direct integration (see North et al., 1979):

$$F(T) = AT + \frac{1}{2}BT^2 - Q\int_0^T a_p(T')\,dT'. \tag{2.88}$$

This potential function for global-averaged models is sketched in Figure 2.11. A useful property of the potential function is that it is always decreasing along solution trajectories as the climate relaxes after a perturbation. This is easily demonstrated

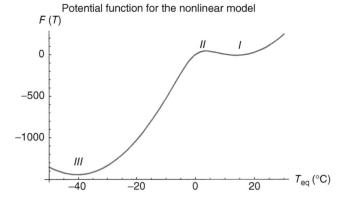

Figure 2.11 The potential function for the global model. The extrema represent equilibrium climates. The value of γ in this figure is 0.9 in order to deepen the minimum (as an aid in visualization) at *I*; otherwise, the parameter values are the same. Steady-state solutions at minima are stable (or metastable, meaning they may eventually be unstable for large enough flexions as in *I*). The extremum at *II* is unstable as is shown in the text.

(using the chain rule and the potential function):

$$\frac{dF}{dt} = \frac{dF}{dT}\frac{dT}{dt} = -C\left(\frac{dT}{dt}\right)^2. \tag{2.89}$$

The significance of the last step is that one can conduct a qualitative analysis of the stability problem even for finite-amplitude perturbations. Relative minima correspond to stable solutions and relative maxima correspond to unstable solutions, as is easily verified by considering a small perturbation away from such extrema. One learns how large a finite perturbation needs to be to push the system over the hill *II* in Figure 2.11 out of the shallow minimum *I* representing the present climate and into the deep minimum *III* corresponding to the deep freeze. We will leave to later chapters the question of noise forcing the nonlinear systems. One can be tantalized, however, by the possibility that noise forcing can eventually kick a solution fluctuating about *I* in the figure over the hill at *II* and into the deep freeze. How long does this take for reasonable noise forcing?

2.4.5 Potential Function and Noise Forcing

Equation (2.25) can be written as

$$\frac{dT}{dt} + \frac{1}{C}\frac{dF}{dT} = \sigma_Z Z(t), \tag{2.90}$$

a form known as a *Langevin equation* (e.g., Gardiner, 1985). The variance of the noise is normalized as follows:

$$\langle Z(t)Z(t')\rangle = \sigma_Z^2 \delta(t-t'). \tag{2.91}$$

Physically, this system is the same as the (one-dimensional) equation of motion of a particle being disturbed by noise with T replaced by the particle's velocity. This is the analogy with Brownian motion when the potential function $F(T)$ is a quadratic function of T (i.e., it is the drag force on the particle and it is proportional to T). When the potential function has a cup-shaped minimum as in Figure 2.11, and near a particular minimum, the particle (climate state) is trapped if the noise is not too strong. If the potential function can be locally approximated by a parabola, the state will be a random variable that is normally distributed as was argued for the Langevin equation earlier in this chapter. To the extent that the neighborhood of the minimum is skewed, we can expect non-normal statistics for the temperature random variable.

To go further with this class of problems, one must turn to some fancy methods. One approach makes use of the Fokker–Planck equation (FPE) (e.g., Hasselmann,1976; Gardiner, 1985).[20] We will state some results with at best hueristic proofs (see Gardiner, 1985, for more complete discussions). The FPE is a governing equation for the pdf as it evolves with time. A few examples for the linear system are shown in Figures 2.4 and 2.5 for the case in which $F(T)$ is parabolic, which occurs when the Langevin equation is linear. In solving those equations, we found the trajectories for individual ensemble members and studied their statistical properties. We were able to deduce that if the noise forcing is normally distributed, the solutions are also normally distributed about the decaying relaxation curve. We were also able to deduce that the bundle of subensemble members saturated at the width of climatology.

The FPE provides a means to examine the time evolution of a pdf directly instead of looking at individual ensemble members. Suppose the PDF is given by $f(T, t)$ and at all times

$$\int_{-\infty}^{\infty} f(T, t) \, dT = 1, \quad \text{for all time.} \tag{2.92}$$

The theory applies when the ensemble members satisfy Langevin equation:

$$\frac{dT}{dt} = g(T) + h(T)Z(t), \tag{2.93}$$

where, in our particular linear model case, $g(T) = -\frac{T}{\tau_0}$ and $h(T) = \frac{\sigma_Z}{C}$.

But what about the case of the nonlinear model with noise forcing represented by (2.90)? The FPE reads as follows:

$$\frac{\partial f}{\partial t} = -\frac{\partial}{\partial T}(g(T)f(T, t)) + \frac{1}{2}\frac{\partial^2}{\partial T^2}(h^2(T)f(T, t)). \tag{2.94}$$

The factor $g(T)$ is called the *drift coefficient* and $h(T)$ is known as the *diffusion coefficient*. The first governs the motion of the center of the distribution as it shifts with time, while the second causes the ensemble members to diffuse apart from one another. Consider the linear-noise-forced climate model first. Then $g(T) = \frac{T}{\tau_0}$ and $h(T) = \frac{\sigma_Z}{C}$. In the linear case, the FPE reads as

$$\frac{\partial f}{\partial t} = \frac{1}{\tau_0}\frac{\partial}{\partial T}Tf(T, t) + \frac{\sigma_Z^2}{2C^2}\frac{\partial^2}{\partial T^2}f(T, t). \tag{2.95}$$

First, consider the steady state, that is, $\frac{\partial f}{\partial t} = 0$. The steady-state solution is labeled $f_{ss}(T)$. Then we can integrate once to obtain

$$BTf_{ss}(T) + \frac{\sigma_Z^2}{2C}\frac{df_{ss}}{dT} = \text{constant}. \tag{2.96}$$

---

20 Hasselmann (1976) is probably the first to mention the FPE in connection with the climate problem.

Take the constant to be zero. As we will see presently, this is equivalent to taking the origin to be at $T_{ss} = 0$ and centered at the minimum of $f_{ss}(T)$. We have $\frac{df_0}{dT}\big|_{T=T_{ss}} = 0$. Now divide through by f_0 and multiply through by dT:

$$\frac{df_{ss}}{f_{ss}} = -\frac{2C^2}{\sigma_Z^2 \tau_0} T\, dT. \qquad (2.97)$$

Finally, integrating and rearranging, we have

$$f_{ss}(T) = K\, e^{-\frac{C^2}{\sigma_Z^2 \tau_0} T^2}. \qquad (2.98)$$

This tells us that the steady-state solution for the FPE is just the normal distribution.

Earlier in this chapter, we studied the case of the noise-forced global model in which an ensemble of trajectories evolve from a sharp initial condition (refer again to Figures 2.4 and 2.5). The probability density distribution function for this initial state is

$$f(T, t = 0) = \delta(T - T_{init}). \qquad (2.99)$$

As the distribution evolves, the mean (and mode) value of $f(T, t)$ drifts toward $T = T_{ss}$, the steady-state mean, and the variance inflates slowly to σ_{ss}, the steady-state or climatological variance. The steady-state distribution has variance $= \tau_0 \sigma_Z^2/(2C^2)$. This is very reminiscent of the thermalization of a collection of molecules as they relax toward the Maxwell–Boltzmann distribution of molecular velocities in the kinetic theory of gases. In the same vein, one thinks of the entropy of such an ensemble in going from a state of very high information content to one approaching its maximum value. In the case of climate prediction, it is *information* that is lost as the ensemble slowly relaxes toward climatology where there is no predictive power (or information content) remaining. Climatology here refers to no information in terms of relative entropies.

The problem becomes more interesting when the potential function is not simply quadratic. For example, consider the upper minimum in the potential function $F(T)$ in Figure 2.11. A Langevin equation of this potential function would be nonlinear and it opens questions about how long such a state can remain trapped in the shallow bowl identified with the present climate. How large must the variance of the noise be and how long would it take to cause the climate state to jump out of the upper minimum and fall into the dreaded "deep freeze"? This problem is similar to the problem of tunneling that is famous in quantum theory for explaining α-particle radioactivity.

There is a large literature on the FPE in physics (this book may be among the few to mention it in the climate science literature). The equation is generally hard to solve for nontrivial cases, so the investigator must resort to numerical methods. It is comforting that most analyses of large-scale temperatures are normally distributed, suggesting that we are not near an ugly threshold, but there may be cases in dynamical meteorology that one is near the threshold for a so-called catastrophe (e.g., the phenomenon of blocking).

The mathematics of stochastic processes is difficult because of the mathematical pathologies associated with white noise at small intervals. We have skirted around these by using continuous time descriptions only when such singularities are absent or unimportant. These peculiar problems are of no interest in climate because the white noise in question has a finite autocorrelation time (few days). We also avoided the problems by using discrete steps in the time domain such as monthly or annual averages.

2.4.6 Relation to Critical Opalescence

Critical Opalescence refers to a phenomenon observed in phase transitions. It was first explained by the Polish physicist in Smolochowski (1906). As a system is near a critical point such as droplets about to freeze spontaneously at −40 °C, fluctuations that are large in amplitude and in spatial scale occur as the system is gradually taken over by the bifurcation in Gibbs energy at the point of freezing. The Gibbs energy function looks somewhat like in Figure 2.11 (see Figure 9.3 in Callen, 1985). When this happens, the two minima in Figure 2.11 merge together forming a very broad minimum before merging into a single minimum. The system in the intermediate state just at the bifurcation will exhibit large fluctuations. There are many laboratory illustrations in video form of critical opalescence that can be viewed on the Internet. The term opalescence comes from the reflected light that suddenly flashes just at the phase transition. Another historical note is that in 1910, Einstein published a paper on the subject related to the effect of the blue sky being the result of fluctuations high in the atmosphere that are necessary for the blue color to be observed below. The point here is that as a control parameter leads to the solution approaching a bifurcation, fluctuations become very large. The phase transition is very similar to the bifurcation encountered in the ice cap model. See the schematic in Figure 2.10.

2.5 Summary

This chapter has been about an idealized planet whose climate is described by a single number, its temperature T. The equilibrium temperature is determined by the balance of radiation absorbed from the Sun and that emitted by the Earth to space. Radiation to space is taken to be a linear function of the surface temperature that turns out to be a pretty good approximation over the range of interest. Climate forcing in this case means an action that disturbs the balance of radiation at the top of the atmosphere. If the climate is disturbed and then left free, it will return exponentially to equilibrium (or more precisely, steady state) with a time constant τ_0 that is about 30 days if the Earth were all-land and perhaps a few years if the Earth were covered by a mixed-layer only ocean. The equilibrium sensitivity of climate can be defined as the change of temperature for a change of 1% in the total solar irradiance or for more commonly a doubling of CO_2. The doubling experiment must be carried out from a steady state to the new steady state. For an all-land or all mixed-layer ocean planet this adjustment to the new steady state is comparable to the autocorrelation or relaxation time. For the real planet, the deep ocean coupling and its timescale(s) have to be reckoned with.

Since weather fluctuations are of timescale of a few days, they can be considered white noise in the energy balance wherein the shortest timescale operating for an all-land planet is about 1 month. This order-of-magnitude gap in timescales allows us to ignore many details of the weather's influence on the statistics of large-scale climate fluctuations. When we introduce white noise in the energy balance, we can find solutions to the linear problem that turn out to be the familiar first-order autoregressive (AR1) models. All the properties such as the autocorrelation time, spectral density, and predictability characteristics can be derived in simple forms. The relaxation time and the autocorrelation times are the same, τ_0. Equilibrium climate sensitivity and variance are proportional to the relaxation time, τ_0.

Some nonlinear global models with ice cap feedback can be solved analytically. In this case, three distinct solution branches are found as a function of the control parameter, the total solar irradiance. One corresponds to the present climate, another is an ice-covered or "snowball" Earth, and a third is intermediate. The latter proves to be unstable as the slope of that branch is negative, violating the stability criterion of the slope stability theorem, which will be proved in Chapter 7. The nonlinear model can also be studied analytically for finite amplitude anomalies by introducing a potential function. The FPE is introduced and discussed briefly as a means of solving for the time dependence of the probability distribution of an anomaly in nonlinear models.

Suggestions for Further Reading

Chapter 3 of Pierrehumbert (2011) emphasizes physics in a comprehensive discussion of global average models for both the Earth and the other planets. Percival and Walden (1993) present a complete picture of Fourier spectral analysis. Mathematical methods are covered in Arfken and Weber (2005). Papoulis (1984) covers random variables as needed by electrical engineers (and climate scientists).

Fourier analysis is a beautiful subject only touched upon in the most operational way in this book. The books by Körner (1989) and Gasquet and Witomski (1991) do not cut corners as we have done so many times in this text. These books cover the convergence properties of Fourier series and their cousins (Legendre, Hermite, etc.) and in so doing take us through some of the darkest and deepest corners of mathematical analysis and the role of Fourier analysis in the great history of analysis. Körner's index contains no entry for *delta function*, G & W only refer to Dirac's impulse function. Neither book needs delta functions for their proofs. Some say "the operational methods work well until they don't."

Information theory is covered in the books by Kullback (1968) and Cover and Thomas (1991).

Exercises

2.1 Linearize the Stefan–Boltzmann law in the form

$$\sigma_{SB} T_R^4 \doteq A + B(T_R - 288),$$

where σ_{SB} is the Stefan–Boltzmann constant and T_R is the radiation temperature in kelvins. Compare the resulting A and B with those in (2.4) and (2.5).

2.2 (a) Assuming that the Earth is a black body, what would be the equilibrium temperature of the Earth for the current value of solar irradiance?
 (b) Show that $\beta^{BB} = 0.70$ at the equilibrium temperature in part (a).

2.3 Show that the solution of

$$C \frac{dT}{dt} + A + B(T - 273) = Q a_p$$

is given by

$$T(t) = T_{eq} + (T(0) - T_{eq})e^{-t/\tau}, \quad \tau = C/B,$$

where T_{eq} is the equilibrium temperature, that is, $dT_{eq}/dt = 0$. Determine the equilibrium temperature T_{eq}.

2.4 Let us consider a simple one-dimensional EBM in the form

$$C\frac{dT}{dt} + BT = Z(t),$$

where $Z(t)$ is a white noise time series with variance σ_Z^2. Show via Fourier transform (FT) of the governing equation that $\tilde{T}(f)$, the Fourier transform of $T(t)$, is given in the form

$$(B - i2\pi fC)\tilde{T}(f) = \tilde{Z}(f),$$

and the spectral density function of $T(t)$ is given by

$$S_T(f) = \frac{\sigma_Z^2/B^2}{1 + 4\pi^2\tau^2 f^2}, \quad \tau = C/B.$$

2.5 Let us consider a one-dimensional nonlinear EBM:

$$C\frac{dT}{dt} = -A - B(T - 273) + Qa_p,$$

where the coalbedo $a_p(T)$ is a function of T.

(a) Show that a small perturbation from equilibrium temperature $\delta T = T - T_{eq}$ satisfies the linearized equation

$$C\frac{d}{dt}\delta T \approx [-B + Qa_p'(T_{eq})]\delta T,$$

where the equilibrium temperature satisfies

$$A + B(T_{eq} - 273) = Qa_p(T_{eq}).$$

(b) Show from the result in part (a) that the following equations should be satisfied:

$$-a_p(T_{eq})\frac{dQ}{dT_{eq}} = -B + Qa_p'(T_{eq}).$$

Further, show that

$$\frac{d}{dt}\delta T = -\lambda_{eq}\delta T, \quad \lambda_{eq} = \frac{a_p(T_{eq})}{C}\frac{dQ}{dT_{eq}}.$$

2.6 Let us consider a one-dimensional linear energy balance equation

$$C\frac{dT}{dt} + A + BT = Qa_p(T).$$

(a) Show that

$$\frac{dT}{dt} = -\frac{1}{C}\frac{dF}{dT}$$

yields an implicit solution (potential function)

$$F(T) = AT + \frac{1}{2}BT^2 - Q\int_0^T a_p(T')dT'.$$

(b) Show that

$$\frac{dF}{dt} = -C\left(\frac{dT}{dt}\right)^2.$$

What does this relationship mean in terms of the temporal change in the potential function?

3

Radiation and Vertical Structure

The average vertical temperature profile of the atmosphere in steady state depends upon a balance of the divergence of the energy flux densities at each level of the atmosphere. In this chapter,[1] we deal with global averages, but we retain the vertical dimension, z. In vector notation, the divergence of a vector flux density in one (vertical) dimension is simply

$$\nabla \cdot \vec{\mathbf{F}} = \frac{\mathrm{d} F_z}{\mathrm{d} z}. \tag{3.1}$$

Flux density $\vec{\mathbf{F}}(\mathbf{r})$ is a vector indicating the direction and magnitude of the flow of energy per unit area perpendicular to that same vector in units W m^{-2}. The energy flux densities include those of radiative transfer and of flux densities of sensible and latent heat by vertical atmospheric motions. Because of the global averaging, there is no contribution from horizontal transports. Basic to the derivation of this profile is an understanding of the interaction of radiation with matter. This chapter is no substitute for a full course on radiative transfer, and readers are referred to the many excellent books on the subject for detailed treatments (see the suggested reading notes at the end of the chapter). Here we restrict ourselves to introducing and developing the minimal tools needed for the solution of a few key problems, that, although very idealized, provide a heuristic basis for the establishment of the vertical temperature profile. In particular, we focus our attention on a cloudless air column in which no scattering or absorption of visible and near infrared radiation energy occurs in the atmosphere. All interaction of the Earth–atmosphere system with sunlight occurs at the ground where some radiation energy is absorbed and some is reflected out to space. We will further specify that in the infrared portion of the electromagnetic spectrum (wavelengths in the range 0.750–200 μm), the atmosphere is *gray*; that is, its absorptivity (and equivalently[2] its emissivity) are constant as a function of radiation frequency. We will find that this is sufficient detail to recover many qualitative aspects of the vertical structure of the Earth's atmosphere in radiative balance. Our simple model in this chapter ignores clouds and their interaction with radiation altogether.

---

1 Reminder: This chapter is not essential for understanding the EBMs of Chapters 5 onwards. The same is true of Chapter 4.
2 The emissivity of a substance and its absorptivity for an infinitesimal segment of a given wavelength band are equal (Kirchoff's law). If the equality were not true, there would be a violation of the laws of thermodynamics.

Energy Balance Climate Models, First Edition. Gerald R. North and Kwang-Yul Kim.
© 2017 Wiley-VCH Verlag GmbH & Co. KGaA. Published 2017 by Wiley-VCH Verlag GmbH & Co. KGaA.

These idealizations will lead us to an analytical solution for the vertical structure of an atmosphere in *radiative equilibrium*[3] with no vertical or horizontal transfer of heat by conduction, air motions, or release of latent heat. This solution has some peculiar features that compel us to delve further into the problem. In particular, the solution has a much warmer ground temperature (the *skin temperature*) than the air just above it. This discontinuity means the layer near the ground is unstable. Moreover, even if there were no discontinuity at the ground, the air temperature profile is unstable to convection, as its lapse rate exceeds the criterion for instability. We could partially mend the problem by adding vertical heat conduction that can be used to remove the discontinuity at the surface, but it does not cure the instability to convection. Finally, we pass to the radiative–convective models for a gray atmosphere. In our treatment, the effects of convection are imitated by an instantaneous adjustment of the temperature profile to a standard lapse rate such that no unstable layers are allowed to develop. The most delicate aspect in solving these models is application of the boundary conditions at the top and bottom of the atmosphere.

As with the many models treated in this book, the gray, nonscattering atmosphere should be thought of as a *heuristic* model of the vertical temperature profile because its differences from the real world are significant. Nevertheless, it constitutes a first approximation to a real atmosphere and it has the advantage of a step-by-step approach revealing, one at a time, the complications that must be dealt with in a more realistic simulation. Many more realistic radiative–convective models have been developed and studied numerically over the last four decades (e.g., the classic papers by Ramanathan and Coakley, 1978; and Manabe and Strickler, 1964). These radiative–convective models attempt to account for atmospheric scattering and absorption of solar radiation, cloud interactions with the radiation field, and the wavelength dependencies of the interactions. Such models have been the source of many important preliminary estimates of the effects of various greenhouse gases and aerosol particles. In addition, radiative–convective models serve as a laboratory for the development and testing of fast numerical schemes that eventually wind up in general circulation models.

3.1   Radiance and Radiation Flux Density

The study of radiative transfer in a continuous medium requires careful definitions of the fields involved. The concepts of *radiance* and *radiant energy flux density* will be introduced first along with some explicit representations using an approximation due to Eddington. Our treatment is not complete; only the concepts needed in solving for the radiative transfer parameters and the temperature profile in a plane-parallel stratified medium transparent to visible radiation and gray[4] in the infrared will be introduced.

---

3 Radiative equilibrium refers to an atmospheric state in which the temperature profile is determined by the flow of radiant energy and, in turn, the radiative energy flux densities are determined by the atmosphere's freely adaptable thermal and (usually fixed) composition profile. The profiles of radiation energy flux densities and temperature are in steady state. In this model, there are no vertical flux densities of atmospheric sensible or latent heat content.
4 *Gray atmosphere* in the infrared means that there is no wavelength dependence on the absorption (=emissivity) in that portion of the electromagnetic wave spectrum.

3.1 Radiance and Radiation Flux Density

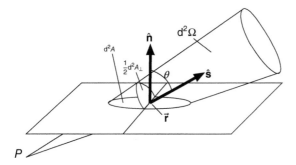

Figure 3.1 Illustration of the spectral radiance of radiation. Radiation is emitted from the point P in a cone in the direction $\hat{s}$ that passes through the plane surface that is normal to the unit vector $\hat{n}$. The center of the cone as it intersects the plane surface is the tip of the position vector $\vec{r}$. At the point $\vec{r}$ where beam's center passes through the plane, its cross section is a circle whose area is $d^2A_\perp$. The beam intersects the plane surface with an ellipse of area d^2A. The solid angle subtended by the cone is $d^2\Omega$. Both points P and $\vec{r}$ are referred to a fixed origin not shown.

Following Wendisch and Yang (2012), consider a pencil of radiation passing through a surface depicted in Figure 3.1 as the flat plane (but it need not be flat, merely flat in an infinitesimal neighborhood of the intersecting beam). We may think of the plane as being at the disposal of the observer, that is, it may be oriented at the will of the observer. The diverging beam intersects the plane with infinitesimal elliptical area d^2A centered at the point $\vec{r}$, where $\hat{s}$ is a unit vector pointing along the direction of the pencil of radiation. The infinitesimal area $d^2A_\perp$ is perpendicular to the beam. This beam is emerging from a point P (whose tail is at the origin and whose head is located at the point in question) far away (P is below and at the tail of the vector $\vec{r}$ in Figure 3.1) from the head of the vector $\vec{r}$. The intersection of the beam and the surface is d^2A. Note that $d^2A \geq d^2A_\perp$. In fact,

$$(\hat{s} \cdot \hat{n})d^2A = \cos\theta \, d^2A = d^2A_\perp. \tag{3.2}$$

The radiation is emitted from the originating point P into a cone of solid angle $d^2\Omega$. When it reaches the surface denoted by its perpendicular unit vector $\hat{n}$, it fills the infinitesimal area d^2A on the surface.

The energy passing through the infinitesimal area of the surface d^2A along the unit vector $\hat{s}$ into the cone corresponding to $d^2\Omega$ in time dt in the frequency interval $(\nu, \nu + d\nu)$ is

$$d^6E_\nu = I_\nu(P, \hat{s}) \cos\theta \, d^2A \, d^2\Omega \, d\nu \, dt. \tag{3.3}$$

The coefficient $I_\nu(P, \hat{s})$ is called the *spectral radiance*.[5] It represents the radiant energy per unit area of the reference surface per unit of solid angle into which it is sent. The qualifier *spectral* refers to the fact that it is per unit frequency ν of the source. We can think of it as

$$I_\nu(P, \hat{s}) = \frac{d^6E_\nu}{\cos\theta \, d^2A \, d^2\Omega \, d\nu \, dt}. \tag{3.4}$$

Note the importance of the factor $\cos\theta$. It indicates the radiant energy *per unit area of the surface*, not of the beam's cross-sectional area perpendicular to the beam.

---

5 In some books, the term *intensity* is used instead of radiance.

For definiteness, consider an example form for $I_\nu(P,\hat{s})$ that is especially useful in vertically stratified atmospheres (use Cartesian coordinates, (x,y,z), z upward, with corresponding unit vectors $(\hat{i},\hat{j},\hat{k})$):

$$I_\nu^{\text{Edd}}(P,\hat{s}) = I_{0\nu}(z) + I_{1\nu}(z)\hat{k}\cdot\hat{s}. \tag{3.5}$$

The reference plane we have chosen is the horizontal x–y plane elevated a distance z above the ground. Here the spatial dependence is denoted by the point $\vec{r}$ and is in the coefficients $I_{0\nu}(z)$ and $I_{1\nu}(z)$ at the level z. By symmetry, there is no x or y dependence of I_ν^{Edd}. The angular dependence is explicitly given in the coefficient $\hat{k}\cdot\hat{s} = \cos\theta$, where θ is the zenith angle. We have used the superscript Edd, since this is only a two-term truncation of an infinite series in powers of $\hat{k}\cdot\hat{s}$, or better yet, degrees of Legendre polynomials[6]: $I_{n\nu}(z)P_n(\cos\theta)$. The one-term truncation is known as *Eddington's approximation*, which we will employ here. Note that $\theta = \cos^{-1}(\hat{k}\cdot\hat{s})$ is the zenith angle of the ray directed along $\hat{s}$. The term in the expansion proportional to $\hat{k}\cdot\hat{s}$ takes into account the anisotropy of the spectral radiance field to the lowest order. Such an up–down anisotropy is crucial in describing the thermal-radiative heating and cooling of a stratified atmosphere.

The *flux density*, $F_\nu(P)$, is the total radiation energy flowing across unit area perpendicular to the unit vector $\hat{k}$ per unit time. Note that $F_\nu(P)$ depends explicitly on the point in space (in this case the level, z) in question as well as the orientation of a reference direction, $\hat{k}$. Let the infinitesimal surface element $d^2 A_\perp$ be oriented at an angle $\theta = \cos^{-1}(\hat{n}\cdot\hat{s})$ to the area element $d^2 A$ with perpendicular unit vector $\hat{n}$. The projection factor for $d^2 A$ onto $d^2 A_\perp$ is $\cos\theta$. We are led to

$$F_\nu(P) = \int_\Omega I_\nu(P,\hat{s})\,(\hat{k}\cdot\hat{s})\,d^2\Omega_s. \tag{3.6}$$

To illustrate the flux concept in a concrete example, consider the Eddington form given above. First, we list the values of some useful solid angle integrals:

$$\begin{aligned}
&\int_{4\pi} d^2\Omega_s = 4\pi, & &\int_\cap d^2\Omega_s = 2\pi, \\
&\int_\cup d^2\Omega_s = 2\pi, & &\int_{4\pi}(\hat{s}\cdot\hat{k})d^2\Omega_s = 0, \\
&\int_\cap(\hat{s}\cdot\hat{k})d^2\Omega_s = \pi, & &\int_\cup(\hat{s}\cdot\hat{k})d^2\Omega_s = -\pi, \\
&\int_{4\pi}(\hat{n}\cdot\hat{s})^{\text{odd}}\,d^2\Omega_s = 0, & &\int_{4\pi}(\hat{n}\cdot\hat{s})(\hat{s}\cdot\hat{m})d^2\Omega_s = \tfrac{4\pi}{3}\hat{n}\cdot\hat{m},
\end{aligned} \tag{3.7}$$

where $\hat{k}$ is a unit vector in the vertical (z) direction, $\cap$ stands for the upper hemisphere ($0\leq\theta\leq\pi/2$); $\cup$ stands for the lower hemisphere ($\pi/2\leq\theta\leq\pi$); and $\hat{n}$ and $\hat{m}$ are arbitrary constant unit vectors. The flux density for the Eddington spectral radiance (3.5) with respect to an infinitesimal area parallel to the x–y plane at level z is

$$F_\nu^{\text{Edd}}(P) = \frac{4\pi}{3}I_{0\nu}(z). \tag{3.8}$$

It is often convenient in stratified-atmosphere applications to find the upward contribution to the flux separately from the downward part. This is achieved by integrating

---

[6] More about Legendre polynomials in Chapter 5.

the spectral radiance over the upper hemisphere in the solid angle integration and the lower hemisphere separately. In the Eddington approximation (defined in (3.5)), we may write as follows:

$$F_\nu^\uparrow(z) = \pi I_{0\nu}(z) + \frac{2\pi}{3} I_{1\nu}(z), \tag{3.9}$$

$$F_\nu^\downarrow(z) = \pi I_{0\nu}(z) - \frac{2\pi}{3} I_{1\nu}(z). \tag{3.10}$$

The equations for $F_\nu^\uparrow(z)$ and $F_\nu^\downarrow(z)$ (or equivalently $I_{0\nu}(z)$ and $I_{1\nu}(z)$) are the two dependent variables in the problem. When the series is truncated at the first order in powers of $\cos\theta$, it is called the *two-stream approximation* for the radiation field.[7] Higher-order approximations take more powers into account but necessarily involve more equations that we would have to deal with.

3.2 Equation of Transfer

In this section, we introduce the equation of radiation energy transfer describing how the spectral radiance is modified as radiation passes through the atmosphere. We proceed by finding how the radiance is diminished by absorption and scattering out of the beam and how it can be augmented by emission and scattering into the direction of the beam.

3.2.1 Extinction and Emission

Consider the beam of radiation along a direction $\hat{s}$ and for an increment of distance δs. In the process of passage over this distance, an amount of radiation will be removed in proportion to the amount entering an infinitesimal cylindrical volume whose axis is the segment δs, the length of the segment δs itself, and an extinction coefficient κ_ν^{ext} that depends on the medium:

$$\delta I_\nu^{\text{ext}} = -\kappa_\nu^{\text{ext}} I_\nu \, \delta s. \tag{3.11}$$

In general, κ_ν^{ext} will consist of two additive parts, one due to scattering and one due to absorption. Moreover, κ_ν^{ext} will be proportional to a linear combination of the densities of the scatterers and absorbers along the segment δs weighted by their individual extinction coefficients.

In addition to losses of spectral radiance along the path, there can be sources of spectral radiance. In the same cylindrical volume element whose axis is the segment δs, radiation energy can be added to the beam through emission or scattering of radiation energy into the beam that would otherwise have been moving in some other direction. The augmentation is also proportional to the length of the segment δs,

$$\delta I_\nu^{\text{aug}} = +\kappa_\nu^{\text{aug}} J_\nu \, \delta s, \tag{3.12}$$

where J_ν is called the *source function* and κ_ν^{aug} is a coefficient proportional to the sum of densities of individual source particles and molecules and their individual source

---

7 A derivation of a slightly different form of the two stream approximation can be found in Houghton (1986).

strengths. We are now in position to write an equation governing the evolution of the spectral radiance of the radiation along the path.

$$\frac{dI_\nu}{ds} = -\kappa_\nu^{ext} I_\nu + \kappa_\nu^{aug} J_\nu. \tag{3.13}$$

Not all gases absorb or emit radiation in the infrared spectral range. In fact, in the real atmosphere O_2, N_2, and Ar do not absorb in the infrared portion of the spectrum. The *radiatively active* molecular species are H_2O, CO_2, CH_4, O_3, and some other trace gases found in the natural atmosphere. Some of these molecules (especially H_2O) scatter incoming solar radiation as well. Cloud particles also absorb strongly in the infrared and scatter in the visible. Aerosol particles can absorb but their more important roles are in scattering solar radiation and in their roles in cloud droplet nucleation.

3.2.2 Terrestrial Radiation

The important special case is that for which scattering can be neglected. This might hold approximately for the infrared portion of the spectrum in the lower atmosphere in the absence of cloud particles. In such a case, we assume the molecules in a volume element are in thermal equilibrium with respect to molecular collisions and that they are continuously emitting radiation according to Planck's law. In this case,

$$(\kappa_\nu^{aug} J_\nu)_{thermal} = \kappa_\nu^{ext} B_\nu(T), \tag{3.14}$$

where the Planck radiation function is given by

$$B_\nu(T) = \frac{2h\nu^3}{c^2(e^{h\nu/(kT)} - 1)}, \tag{3.15}$$

and

T = Kelvin temperature; (3.16)
h = Planck's constant = 6.626×10^{-34} J s; (3.17)
k = Boltzmann's constant = 1.381×10^{-23} J K^{-1}; (3.18)
c = Speed of light in vacuum = 3.00×10^8 m s^{-1}. (3.19)

Figure 3.2 shows the distribution of radiation over different frequencies in terahertz (10^{12} cycles s^{-1}) for three different source temperatures. An important formula[8] to keep in mind is

$$\int_0^\infty B_\nu(T) d\nu = \frac{\sigma_{SB} T^4}{\pi} \equiv B(T), \tag{3.20}$$

where the Stefan–Boltzmann constant is $\sigma_{SB} = 5.670 \times 10^{-8}$ W m^{-2}K^{-4}. Note that in the emission case, the coefficients for absorption and emission are the same, a consequence of one of Kirchoff's laws for a substance in thermodynamic equilibrium – which holds in the troposphere, but might not in the upper atmosphere where atmospheric molecular

---

[8] After a change of variable, the integral can be solved by reducing it to $\int_0^\infty \frac{x^3 \, dx}{e^x - 1} = \frac{\pi^4}{15}$.

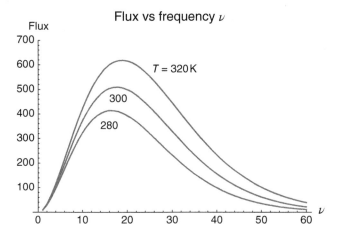

Figure 3.2 Planck's radiation function $B_\nu(T)$ ($\times 10^{14}$) in J/m² versus frequency (in units of 10^{12} cps) for $T = 280, 300, 320$ K. The abscissa is in terahertz (10^{12} *cycles* s^{-1}).

number densities are small.[9] In the plane-parallel case, the function $B(T)$ is a function of height z, as the temperature is; it will occasionally be expressed as $B(z)$. Thus, we have the emission rate is tightly controlled by the local temperature of the emitting substance.

In a steady-state, stratified atmosphere, κ_ν is a function of z. This allows us to make a convenient change in the coordinate measuring distance along the path. The differential *optical path length* is defined by

$$d\tilde{\tau}_\nu = -\kappa_\nu \, ds. \tag{3.21}$$

The sign convention is chosen such that the larger ds the more attenuation of the beam $d\tilde{\tau}$ along the path.

$$\frac{dI_\nu}{d\tilde{\tau}_\nu} = I_\nu - B_\nu. \tag{3.22}$$

This equation governs the evolution of I_ν along a specific path, perhaps at an angle θ to the vertical.

3.3 Gray Atmosphere

Recalling that our goal is to derive the properties of a gray atmosphere in thermal–radiative equilibrium, we specialize the problem to the case of a gray atmosphere. A gray atmosphere is one in which the *absorptivity* (=*emissivity* by Kirchoff's laws on systems in radiative–thermal equilibrium[10]) κ_ν is independent of frequency ν. The gray property is far from true in the real atmosphere and the

---

9 Local thermodynamic equilibrium (LTE) depends on the rate of collisions of the molecules compared to the rate of absorption and emission of photons. If collisions are not frequent enough, the upper quantum levels may not be populated to an equilibrium level. It need not concern us in the models considered in this book.
10 A system without this property can be shown to violate the second law of thermodynamics.

approximation is made mainly for mathematical convenience. We will consider a more realistic dependence[11] on frequency of absorption/emission in Chapter 4. Once the gray approximation is adopted, we have $d\tilde{\tau}_\nu = d\tilde{\tau}$ and in (3.22) we can drop the ν dependence in $\tilde{\tau}_\nu$. Then we can integrate over all ν to obtain

$$\frac{dI}{d\tilde{\tau}} = I - B(\tilde{\tau}), \tag{3.23}$$

where $B(\tilde{\tau}) = \sigma T^4(\tilde{\tau})/\pi$ is given by (3.20). We remind the reader that our problem as we have posed it does not consider atmospheric scattering or absorption of the solar radiation except for the portion that is absorbed by the ground. The portion that is not absorbed by the ground, the *surface albedo*, is reflected directly back to space. The albedo as seen from above the atmosphere in our model is equal to the surface albedo.

3.4 Plane-Parallel Atmosphere

Consider next specialization to the case of a plane-parallel atmosphere. All properties are strictly a function of z, the distance above the ground. We will restrict ourselves to the thermal emission case treated earlier. We neglect scattering of radiation out of or into the beam.

We define the *optical depth* at level z in the gray atmosphere as

$$\tau = \int_z^\infty \kappa(z') dz'. \tag{3.24}$$

By tradition, the optical depth is defined to increase as we descend into the atmosphere from above along the local zenith. Then a small increment of optical path $d\tilde{\tau}$ along a direction $\hat{s}$ is related to the corresponding optical depth increment by

$$d\tau = (\mathbf{k} \cdot \hat{\mathbf{s}}) d\tilde{\tau} = d\tilde{\tau} \cos\theta, \tag{3.25}$$

where θ is the zenith angle. For the plane-parallel thermal emission atmosphere, we may now write

$$\cos\theta \frac{dI}{d\tau} = I(\tau) - B(\tau). \tag{3.26}$$

This is a tremendous simplification of the problem, but we still have to worry about the angular integrations involving the zenith angle θ. Here is where the Eddington approximation comes in. The only vertical dependence in what follows occurs through the τ-dependence. Hence, we will use τ instead of z as the vertical independent variable. At the top of the atmosphere (TOA), ($\tau = 0$) and at the bottom of the atmosphere, ($\tau = \tau_*$), the *optical thickness* of the whole atmosphere. We may think of the optical thickness as being caused primarily by water vapor whose concentration has a typical scale height of 1–2 km (the atmospheric boundary layer). Similar to pressure, optical depth is a monotonically decreasing function of z (Figure 3.3).

---

11 Pierrehumbert derives and discusses approximation methods for rough estimations of infrared radiation transfer that are convenient in examination of planetary atmospheres. Pierrehumbert (2011).

Figure 3.3 Schematic diagram of the plane-parallel atmosphere. The vertical distance from the surface is indicated by z and the direction of a ray is indicated by the unit vector ŝ. The zenith angle of the ray is θ. The shading of layers is used to indicate that the medium is optically thinner as z increases.

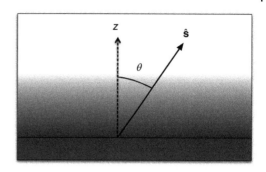

3.5 Radiative Equilibrium

The equation for radiative transfer along a ray in a plane-parallel atmosphere is given by (3.26). In order to obtain upward and downward fluxes, we must consider upper- and lower-hemisphere solid-angle integrations over the solutions of (3.26). The ultimate question we wish to ask is what is the temperature profile of the atmosphere governed by the above system – the solution is contained inside the function $B[T(z)]$ $(= \sigma T(z)^4)$. If we find $B[T(z)]$ as a function of z (or equivalently τ, given the vertical distribution of absorptivity), we invert it to obtain T as a function of z. In addition to finding the solution to (3.26), we must impose boundary conditions at the top and bottom of the atmospheric column. Also, we must force the vertical temperature structure to be in radiative equilibrium at each level. This last means the net local heating rate must vanish by a cancelation of the net rates of radiation flux absorbed and emitted in a thin slab. This condition can be expressed in terms of the flux passing into and out of an infinitesimally thin horizontally oriented slab.

In our model (actually, the model dates to Schwarzschild (1906), also described in Houghton (1986) and the history is summarized in Goody and Yung, 1989), the ground acts like a blackbody in the infrared, radiating upward a flux density

$$F^{\uparrow}_{\text{surf}} = F^{\uparrow}(\tau_*) = \pi B(T_*), \tag{3.27}$$

where T_* is the ground temperature and τ_* represents a point just above the ground (perhaps a few millimeter). The surface temperature is determined by a balance of radiation flux densities. No conduction of heat to the air or evaporative cooling is allowed in this model. The absorbed solar radiation is Qa_p, where Q is the total solar irradiance (TSI)[12] divided by 4 and a_p is the (planetary) coalbedo which has the nominal value of 0.68 as in the last chapter.

$$\pi B(T_*) = Qa_p + F^{\downarrow}(\tau_*). \tag{3.28}$$

Consider next the equation of transfer (3.26) at an arbitrary level z and referred to an arbitrary direction, ŝ. Using $\hat{\mathbf{k}} \cdot \hat{\mathbf{s}} = \cos\theta$:

$$(\cos\theta)\frac{d}{d\tau}[I_0(\tau) + I_1(\tau)\cos\theta] = I_0(\tau) + I_1(\tau)\cos\theta - B(\tau), \tag{3.29}$$

---

[12] *The Solar Irradiance* or $\sigma_\odot$ is the amount of radiant energy from a parallel beam sunlight reaching a perpendicular 1 m² surface at the TOA and averaged through the annual cycle. Its current value is 1340 W m⁻². In earlier literature, the TSI was referred to as the *solar constant*.

where we have inserted the Eddington approximation for I and we have dropped the subscript ν because we are treating a gray atmosphere. It is convenient to consider solid angle integrals with respect to $d^2\Omega$ over the last equation. In fact, one method of proceeding is to multiply by powers of $\cos\theta$ and then performing the integrals. In this way, we obtain separate differential equations for $I_0(z)$ and $I_1(z)$. Applying a few of the integration formulas in (3.7), we obtain the two basic formulas for the Eddington approximation: the first is a straightforward integral over all solid angles (4π) of (3.29); in the second, we first multiply (3.29) through by $\cos\theta$, then integrate over 4π steradians, again using (3.7):

$$\frac{1}{3}\frac{d}{d\tau}I_1(\tau) = I_0(\tau) - B(\tau); \qquad (3.30)$$

$$\frac{d}{d\tau}I_0(\tau) = I_1(\tau). \qquad (3.31)$$

This pair of equations would determine $I_0(\tau)$ and $I_1(\tau)$ if $T(\tau)$ (or equivalently $B(\tau)$) were specified. But we want to solve the problem wherein the temperature adjusts itself to be in steady-state equilibrium with the radiation field. In order to close the system, we need to impose the condition of thermal equilibrium. This is the statement that no net heat per unit time is deposited in any infinitesimally thin slab. In other words, the upward and downward flux density of energy leaving an infinitesimally thin slab must be zero at all times (note that upward and downward flux densities of photons crossing a horizontal plane do not cancel, as each is a stream of photons that interact with matter, and that the flux density is defined with respect to a unit upward vector $\hat{\mathbf{n}} = \hat{\mathbf{k}}$). This condition is given by

$$\frac{d}{d\tau}(F^\uparrow - F^\downarrow) \propto \frac{d}{dz}(F^\uparrow - F^\downarrow) = 0, \qquad (3.32)$$

as otherwise there would be removal or augmentation of either the upward or downward flux density and such a perturbation would induce a change in the temperature of the local medium. According to (3.9) and (3.10), this means

$$\frac{dI_1(\tau)}{d\tau} = 0. \qquad (3.33)$$

The set (3.30), (3.31), and (3.33) represent three first-order differential equations for the three unknown functions of z: $I_0(z), I_1(z)$, and $B(T(z))$. It is useful at this point to summarize the assumptions and approximations that led us to this state.

- The atmosphere is clear to solar radiation; all absorption and/or reflection takes place at the ground.
- In gray atmosphere with no clouds in the infrared: $\tilde{\tau}_\nu \to \tilde{\tau}, I_\nu \to I$.
- Layers of atmosphere are gray thermal radiators: $(\kappa_\nu^{\text{aug}} J_\nu) \to \kappa B(T)$.
- Plane-parallel stratified atmosphere: $I(P, \hat{\mathbf{s}}) \to I(z, \cos\theta)$.
- The Eddington approximation for the IR: $I(z, \cos\theta) = I_0(z) + I_1(z)\cos\theta$.
- The divergence of net (upward plus downward) radiation flux densities has to vanish to assure steady state.

These statements allow us to find the functional form of the solution but with integration constants that will await evaluation:

$$I_1(\tau) = c_1 = \text{constant}. \qquad (3.34)$$

3.5 Radiative Equilibrium

According to (3.30), we have

$$I_0(\tau) = B(\tau). \tag{3.35}$$

Then according to (3.31), we have

$$\frac{d}{d\tau} I_0(\tau) = c_1, \tag{3.36}$$

or

$$I_0(\tau) = c_1 \tau + c_0, \tag{3.37}$$

where the integration constants c_1 and c_0 must be chosen to satisfy the boundary conditions at the TOA. First, no infrared is entering the atmosphere from above:

$$\text{at TOA:} \quad \tau = 0, F^{\downarrow}(\tau = 0) = 0. \tag{3.38}$$

Second, flux density of radiation energy leaving the TOA must equal the rate of solar radiation absorbed per unit perpendicular area by the system:

$$\text{at TOA:} \quad \tau = 0, F^{\uparrow}(\tau = 0) = Qa_p. \tag{3.39}$$

Applying (3.38) and (3.10) yields the integration constant

$$c_0 = \frac{2}{3} c_1. \tag{3.40}$$

Using the second TOA condition leads to

$$c_0 = \frac{Qa_p}{2\pi}, \tag{3.41}$$

and

$$c_1 = \frac{3}{4} \frac{Qa_p}{\pi}. \tag{3.42}$$

Now that we have the two constants, we write the solution in the interior ($0 \leq \tau \leq \tau_*$).

$$I_0(\tau) = B(\tau) = \frac{Qa_p}{2\pi} \cdot \left(\frac{3}{2}\tau + 1 \right); \tag{3.43}$$

$$I_1(\tau) = \frac{3}{4} \frac{Qa_p}{\pi}; \tag{3.44}$$

and the solution for the flux densities:

$$F^{\uparrow}(\tau) = \frac{Qa_p}{2} \cdot \left(\frac{3}{2}\tau + 2 \right); \tag{3.45}$$

$$F^{\downarrow}(\tau) = \frac{3}{2} \frac{Qa_p}{2} \tau. \tag{3.46}$$

One can check that these satisfy the upper boundary conditions.

Next we turn to the surface properties (lower boundary conditions). The surface radiates upward and absorbs energy from the Sun and from the downwelling thermal radiation from the layers of atmosphere above.

$$\pi B_* = Qa_p + F^{\downarrow}(\tau = \tau_*). \tag{3.47}$$

We then find that

$$B_* = \frac{Qa_p}{\pi} \cdot \left(1 + \frac{3}{4}\tau_*\right) = \frac{\sigma T_*^4}{\pi}, \qquad (3.48)$$

where B_* is the blackbody radiation upward coming directly from the hot surface and τ_* is the optical depth taken just above the surface. It is also the total optical depth of the column of air above. This last is an expression for the surface temperature, T_*.

An interesting peculiarity is that the temperature at the surface is not the same as that at the bottom of (but still inside) the atmosphere.

$$\pi B_* - \pi B(\tau_*) = \sigma T_*^4 - \sigma T_{\text{air}}^4 = \frac{Qa_p}{2}. \qquad (3.49)$$

The formula (3.43) gives the radiative equilibrium temperature T as a function of altitude τ or equivalently z. The positive difference (3.49) shows that there is a discontinuity between the temperature just above ground and that of the actual ground. The ground is always warmer than the adjacent air in a system whose atmosphere is governed solely by radiative equilibrium without any other agent for vertical heat transport. The expressions (3.43)–(3.46), combined with the ground results (3.48) and (3.49), constitute the solution of the radiative equilibrium problem as we have posed it so far. It is important to note that the surface temperature as revealed by B_* is an increasing function of τ_*, the total optical thickness of the atmosphere. This is our first encounter with the so-called greenhouse effect.

3.6 Simplified Model for Water Vapor Absorber

Next consider the case of a gray atmosphere with $\kappa(z)$ an exponential function. We pretend that the absorbing substance is gray water vapor

$$\kappa(z) = \kappa_0 e^{-z/H_w}, \qquad (3.50)$$

where the scale height H_w can be taken to be about 1.6 km (this varies with latitude on the real Earth).

$$\tau(z) = \int_z^\infty \kappa(z)\,dz = \kappa_0 H_w e^{-z/H_w}. \qquad (3.51)$$

The total optical depth is $\tau_* = \kappa_0 H_w$. Note further that

$$\frac{d\tau}{dz} = -\frac{\tau}{H_w} \; ; \; \text{or } \tau = \tau_* \, e^{-z/H_w}. \qquad (3.52)$$

We differentiate $B(\tau)$ in the Stefan–Boltzmann law and in (3.43) to obtain

$$\frac{dB(\tau)}{d\tau} = \frac{4B(\tau)}{T}\frac{dT}{dz}\frac{dz}{d\tau} = \frac{3Qa_p}{4\pi}. \qquad (3.53)$$

After rearrangements,

$$\frac{dT}{dz} = -\frac{3T}{4H_w}\frac{\tau}{(3\tau + 2)}, \qquad (3.54)$$

which is a compact form for the lapse rate in the radiative equilibrium atmosphere as a function of τ. As $\tau \to 0$, $\frac{dT}{dz} \to 0$. Figure 3.4 shows the dependence of κ_0 as a function of

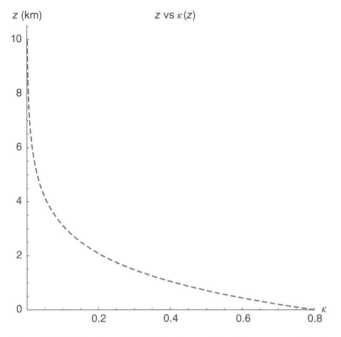

Figure 3.4 A simple model for the absorptivity of water vapor (horizontal axis) as a function of altitude (vertical axis). Altitude z in km versus absorptivity $\kappa(z) = \kappa_0\, e^{-z/H_w}$ for optical depth $\tau_* = 1.20$, scale height of water vapor $H_w = 1.6$ km, $\kappa(0) = \tau_* H_w$.

z for our simplified water vapor model with $\tau_* = 1.20$. The dependence is exponential with a height scale of 1.6 km. Other parameters are indicated in the figure caption.

Consider the vertical dependence of temperature for this model. Figure 3.5 shows a plot (dashed line) of the temperature (K) versus height z for the same model parameters. We adjusted τ_* to get a reasonable surface temperature. Several interesting features are apparent in Figure 3.5. There is a discontinuity in the temperature at the ground. Other aspects of the solution differ from the real atmosphere. For example, the slope of the curve is very negative and is much larger than the dry adiabatic lapse rate,[13] $\Gamma_d = 10$ K km^{-1}; the dry adiabat is shown as a thin straight line in the figure. This means the lower atmosphere is unstable to adiabatic convective overturning up to about 2.2 km where the slope equals that of the adiabat. In the real atmosphere, heat will be transferred from the surface to the lowest layers of air by convection. In addition, there will be evaporation contributing to the surface energy budget. Hence, to make our model more realistic, we must add some vertical convection process that adjusts the surface temperature downward and fills out the temperature profile until it becomes stable. Combining the radiative and convective/conductive vertical heat transfer in an equilibrium calculation of the temperature profile is the subject of the rest of this chapter.

The vertical dashed line in Figure 3.5 is at the *brightness temperature* of $T_b = 255$ K. Note that $T_b = (Qa_p/\sigma)^{\frac{1}{4}}$ in order to maintain a balance of incoming and outgoing flux densities. If one were looking down at the atmosphere from above, one would infer that the level of intersection with the temperature profile to be a characteristic level of the

---

13 By convention the *lapse rate* is the negative of dT/dz.

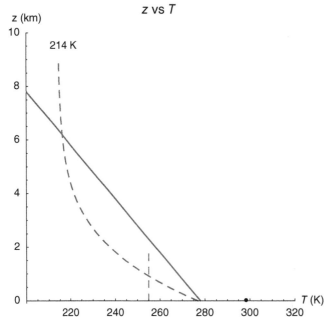

Figure 3.5 Depiction of altitude z versus radiative equilibrium temperature (K) (dashed curve) for $\tau_* = 1.20$, $H_w = 1.5$ km, $Qa_p = 238$ W m^{-2}. The ground temperature is indicated by the big dot at 298 K. Note the discontinuity between the temperature just above the surface and that at the surface. The vertical thin-dashed line is the *brightness temperature* (255 K) that intersects the thermal profile at 0.91 km. The thin straight line is a dry adiabat, indicating that the profile is unstable up to about 2.2 km.

emission. In this case it is 0.91 km, which is well below the 1.6 km characteristic height of the water vapor absorptivity profile, which is the e-folding length of the function $I_0(z)$. But the upward flux density has a contribution from $I_1(z) = +\frac{3Qa_p}{4\pi}$ as well and this lowers the characteristic level of emission. A more important factor is the strong negative vertical gradient of temperature in the boundary layer. Upward flux density is not just the concentration of emitters but is proportional to $T^4(z)$ as well.

Given the radiative equilibrium solution, we examine the effective values of A and B we obtain. This gives some insight into the IR formula discussed in the first two chapters. A useful approximation in analytical calculations comes from linearizing $B(T)$ to form

$$\pi B(T) = \sigma T_K^4 \approx A_0 + B_0 T, \qquad (3.55)$$

where

$$A_0 = 314.9 \text{ W m}^{-2}; \qquad (3.56)$$
$$B_0 = 4.61 \text{ W m}^{-2}(°C)^{-1}; \qquad (3.57)$$
$$T = T_K - 273.2 = \text{Celsius Temperature}; \qquad (3.58)$$

as it is a reasonable approximation over the range of interest in climate problems. Figure 3.6 shows the approximation over the range $\pm 30°C$.

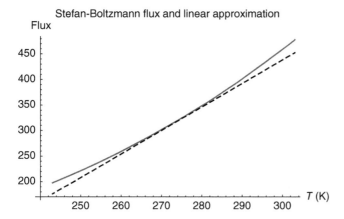

Figure 3.6 Comparison of the Stefan–Boltzmann form σT^4 (solid curve) with the linear form $A_0 + B_0(T_K - 273.2)$ (dashed line).

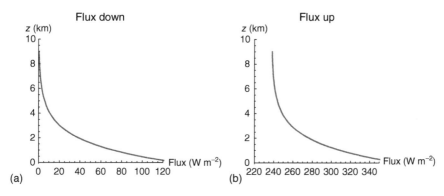

Figure 3.7 (a) Altitude z versus flux density of downward radiation $F^\downarrow(z)$; (b) altitude z versus flux density of upward radiation $F^\uparrow(z)$. Both figures are for the gray water vapor absorber in radiative equilibrium. Note the change of horizontal scale in the two panels.

Figure 3.7 shows the upward and downward flux densities. Note that the slopes are the same for each (see 3.45 and 3.46), keeping the net flux density divergence zero (proportional to net heating rate). The upward flux density is $Qa_p = 238$ W m^{-2} greater than the downward flux density (Figure 3.7). The outgoing flux density at the TOA is Qa_p, which is the rate of absorbed incoming radiation at the surface. We see that the ground not only receives the $Qa_p = 238$ W m^{-2}, but in addition the 214 W m^{-2}, as seen in Figure 3.7b. This is the reason for the higher temperature of the surface.

The empirical values (as measured from satellites) of A and B are around 202 W m^{-2} and 1.90 W m^{-2} (°C)$^{-1}$. If we use the formula connecting B_* to $Qa_p = F^\uparrow(\tau \to 0)$, we find that

$$A_{RE} = \frac{A_0}{1 + \frac{3}{4}\tau_*}, \tag{3.59}$$

$$B_{RE} = \frac{B_0}{1 + \frac{3}{4}\tau_*}. \tag{3.60}$$

Taking $\tau_* = 1.20$, we obtain values of $A_{RE} = 165.7$ W m^{-2} and $B_{RE} = 2.43$ Wm^{-2}(°C)$^{-1}$, both improvements over the blackbody values, but far from the empirical values. There is a further reduction in B_{RE} if we take into account the upward convection of heat by the atmosphere as will be seen later.

The linear results for A_{RE} and B_{RE} are based upon the ground temperature derived from the radiative equilibrium model. In this model, we adjusted the optical thickness ($\tau_* = 0.25$) so that the ground temperature agreed with observations (287 K). There are many things wrong with this model. Aside from the assumption that the atmosphere is gray and does not absorb or scatter sunlight, we have neglected heat transport to the air by contact with the surface. First, we have neglected the cooling of the surface by evaporation and heat conduction, both of which are enhanced by turbulence. Both processes deposit the heat in the boundary layer and above through convective overturning. Radiative equilibrium is impossible in the real lower atmosphere because the temperature profile above the ground is too steep (falls off too rapidly) for convective stability. Thus the hot surface in contact with adjacent air just above will also lead to intense convection until the discontinuity is removed (Table 3.1).

3.7 Cooling Rates

In our simple model, the heating rate is given by

$$q^{\uparrow}(z) = \frac{dF^{\uparrow}}{dz} = \frac{3}{4}Qa_p\frac{d\tau}{dz} = \frac{3}{4}Qa_p\frac{\tau_*}{H_w}e^{-z/H_w}, \tag{3.61}$$

and it is the same as the cooling rate $q^{\downarrow}(z)$ in equilibrium. The heating rate $q^{\uparrow}(z)$ in this case is just proportional to the total optical thickness of water vapor τ_* and it falls off exponentially with height. In more complicated models, one often expresses the heating rate in terms of degrees per day. For this, we need the density of the air, $\rho(z) = \rho_0 e^{-z/H}$ with $\rho_0 = 1.2$ kg m^{-3}, $H_w = 1.6$ km, $H \sim 10$ km, and the specific heat of air at constant pressure, $c_p = 1004$ J kg^{-1} K^{-1}.

$$\frac{dT}{dt} = \frac{3}{4}Qa_p\frac{\tau_*}{\rho_0 c_p H_w}e^{-\left(\frac{1}{H_w}-\frac{1}{H}\right)z}. \tag{3.62}$$

Table 3.1 Coefficients of the infrared linear radiation rule under different conditions.

Blackbody	$A_0 = 314.9$ W m^{-2}
	$B_0 = 4.61$ W m^{-2} °C^{-1}
All sky (satellite)	$A_{AS} = 202$ W m^{-2}
	$B_{AS} = 1.90$ W m^{-2} °C^{-1}
Clear sky (satellite)	$A_{CS} = 230$ W m^{-2}
	$B_{CS} = 2.26$ W m^{-2} °C^{-1}
Radiative equilibrium	$A_{RE} = 165.7$ W m^{-2}
	$B_{RE} = 2.43$ W m^{-2} °C^{-1}
Nominal values used in this book	$A = 190$ W m^{-2}
	$B = 2.00$ W m^{-2} °C^{-1}

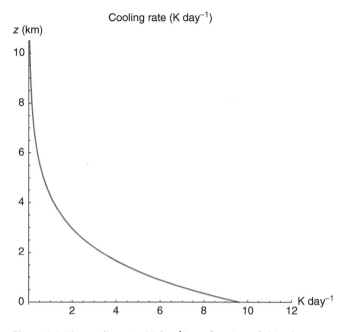

Figure 3.8 The cooling rate (K day^{-1}) as a function of altitude as computed for the radiative equilibrium model for gray water vapor. In this figure, the scale height of water vapor is $H_w = 1.6$ km.

Figure 3.8 shows the cooling rate in kelvins per day as given in (3.62). Note that the scale height for the cooling rate H_q is

$$\frac{1}{H_q} = \frac{1}{H_w} - \frac{1}{H}, \tag{3.63}$$

which is dominated by the smaller H_w, the scale height of the absorber/emitter. The cooling rate is offset exactly by the heating rate in radiative equilibrium and is found by taking the vertical derivative of $F^\downarrow(z)$.

3.8 Solutions for Uniform-Slab Absorbers

In this section, we treat the particular case of a uniform slab of absorptivity imitating the boundary layer of thickness $H_w = 1.6$ km and $\kappa_0 = 0.75$. The optical depth as a function of z in this case is given by

$$d\tau = -\kappa_0 \, dz \, ; \quad z < H_w \tag{3.64}$$

or

$$\tau = \kappa_0 (H_w - z) \, ; \quad z < H_w, \tag{3.65}$$

and $\tau = 0$ for $z \geq H_w$. The optical thickness of the atmosphere is

$$\tau_* = 1.20, \tag{3.66}$$

the same as in the previous section.

The solutions for the radiance components are given by

$$I_1^{RE}(z) = \frac{3}{4}\frac{Qa_p}{\pi}; \quad \text{for all } z, \tag{3.67}$$

$$I_0^{RE}(z) = \begin{cases} \left(\frac{3}{2}\kappa_0(H_w - z) + 1\right)\frac{Qa_p}{2\pi}; & 0 < z < H_w, \\ \frac{Qa_p}{2\pi}; & z > H_w. \end{cases} \tag{3.68}$$

By use of the formulas of the last section, we obtain a solution for the radiative-equilibrium temperature profile, and it is shown in Figure 3.9a.

The temperature profile is given by

$$T^{RE}(z) = \begin{cases} \left(\frac{Qa_p}{2} - A_0\right)\frac{1}{B_0}, & z \geq H_w, \\ \left(\left(\frac{3}{2}\kappa_0(H_w - z) + 1\right)\frac{Qa_p}{2} - A_0\right)\frac{1}{B_0}, & 0 < z < H_w. \end{cases} \tag{3.69}$$

Note that just above surface,

$$T_C(z \to 0+) = \left(Qa_p\left(\frac{3}{4}\tau_* + \frac{1}{2}\right) - A_0\right)\frac{1}{B_0}. \tag{3.70}$$

But at the material surface,

$$T_C(z = 0) = T_* = \left(Qa_p\left(\frac{3}{4}\tau_* + 1\right) - A_0\right)\frac{1}{B_0}, \tag{3.71}$$

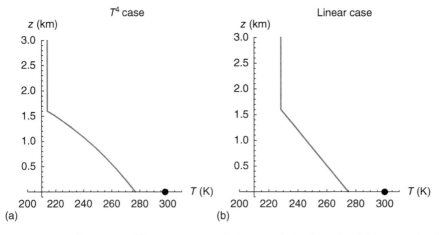

Figure 3.9 An illustration of the error in using the linear radiation law of Budyko versus the blackbody law $B(T(z)) = \sigma T^4/\pi$ in uniform slab solutions. (a) Using the conversion from $B(T(z)) = \sigma T^4/\pi$. (b) The case where we use instead the linear rule $\sigma T^4 = A_0 + B_0(T - 273)$. In both cases, $\kappa_0 = 0.75$, $\tau_* = 1.20$.

with $\tau_* = \kappa_0 H_w$ and using the subscript C to denote Celsius temperature. These forms show that the discontinuity at the surface is

$$\Delta T_C = \frac{Qa_p}{2B_0}. \tag{3.72}$$

The temperature T above and at the surface do not depend on H_w. The only dependence is on the optical thickness, τ_*. Also the gap, ΔT, is independent of τ_*. The lapse rate of temperature in the slab for the linear case (3.69) is 29 K km^{-1} (Figure 3.9b), which means this atmospheric profile is extremely unstable. For the nonlinear case, the lapse rate at the surface is 28 K km^{-1} and at $z = 1$km it is 19 km^{-1}. The largest difference is for the temperature above the boundary layer. In the nonlinear case (see Figure 3.6), it is 214 K the same as in the exponential κ_0 case of the previous section, whereas for the linear case (231 K), the departure from the tangent point (273 K) is too large for the correct upper air value (see Figure 3.6).

3.9 Vertical Heat Conduction

Next we consider some models with vertical heat conduction. This is still a simplification over the case of convection which occurs in the real atmosphere, but this class of models has some intermediate properties that serve to illustrate the bridge between radiative equilibrium atmospheres and those with realistic vertical heat transfer that involves turbulent convection. We introduce a vertical heat conduction presumably caused by eddy processes. This is, of course, simplistic but it can be set up in such a way as to keep the problem linear and thus it will help us to understand a few additional points without resort to heavy use of the computer. Such a formulation does not take into account the stability of the atmosphere, conduction merely acts here to carry heat vertically from the hot ground and lower layers up into the layers aloft.

Consider a vertical heat flux density due to conduction (eddy diffusion).[14]

$$q_{\text{cond}} = -K\frac{dT}{dz}, \tag{3.73}$$

where K is a down-gradient heat conduction coefficient, sometimes called the *thermal conductivity*[15] which will be taken here as a constant independent of z. For equilibrium, the vertical derivative of the heat flux should be balanced by the vertical derivative of the radiation energy flux.

$$-\frac{d}{dz}\left(K\frac{dT}{dz}\right) + \frac{d}{dz}(F^\uparrow - F^\downarrow) = 0. \tag{3.74}$$

---

14 A better choice would be to use the gradient of *potential temperature* rather than the gradient of conventional temperature. We ignore this here, as we will only be interested in qualitative effects near the ground in this section.
15 The thermal conductivity K is not the *molecular thermal conductivity*, but rather a macroscopic coefficient whose mode of transport is eddy motions of the fluid atmosphere, in contrast to the molecular case where the transport is effected by individual molecular collisions. So-called eddy diffusion or conductivity in an unstable stratified layer is orders of magnitude larger than in the molecular case.

This last is readily integrated from z to z_{top}, where z_{top} refers to the top of the optically active part of the atmosphere, to yield

$$-K\frac{d}{dz}T(z) + \frac{4\pi}{3}I_1(z) = Qa_p, \qquad (3.75)$$

where the right-hand side comes from evaluation of the left at $z \to z_{top}$. Here we have imposed the conditions that at $z \to \infty$, $F^\downarrow \to 0$ and the upward flux must exactly balance the total solar flux absorbed per unit area per unit time, Qa_p.

The system of equations governing the flow of heat upwards via radiation and thermal conduction are as follows:

$$\frac{d}{dz}T(z) = \frac{4\pi}{3K}I_1(z) - \frac{Qa_p}{K}; \qquad (3.76)$$

$$\frac{d}{dz}I_0(z) = -\kappa(z)I_1(z); \qquad (3.77)$$

$$\frac{d}{dz}I_1(z) = -3\kappa(z)\left(I_0(z) - \frac{A_0 + B_0 T(z)}{\pi}\right); \qquad (3.78)$$

where $\pi B(T) = (A_0 + B_0 T)$ is used in (3.78). The error in the approximation over the range of interest is shown in Figure 3.6. The balance of flux divergences (material heating and radiance) is expressed by (3.76). The second and third equations in the aforementioned hold for the radiance field. These last three equations are to be solved simultaneously on the interval from $z = 0$ to $z = +\infty$ subject to the boundary conditions that

$$F^\uparrow(z \geq z_{top}) = Qa_p, \qquad (3.79)$$

$$F^\downarrow(z \geq z_{top}) = 0. \qquad (3.80)$$

In terms of $I_0(z = z_{top}) \equiv I_0^{toa}$ and $I_1(z = z_{top}) \equiv I_1^{toa}$, these become

$$I_0^{toa} + \frac{2}{3}I_1^{toa} = \frac{Qa_p}{\pi}, \qquad (3.81)$$

$$I_0^{toa} = \frac{2}{3}I_1^{toa}, \qquad (3.82)$$

or

$$I_0^{toa} = \frac{1}{2}\frac{Qa_p}{\pi}, \qquad (3.83)$$

$$I_1^{toa} = \frac{3}{4}\frac{Qa_p}{\pi}. \qquad (3.84)$$

One more boundary condition must be imposed, namely, the surface energy balance. This takes the form

$$\pi B_* = Qa_p + K\frac{dT}{dz}\bigg|_{z=0} + F^\downarrow(z = 0+), \qquad (3.85)$$

with

$$F^\downarrow(z = 0+) = \pi I_0(0) - \frac{2\pi}{3}I_1(0). \qquad (3.86)$$

This surface boundary condition informs the system that the Earth's surface is a black radiator and that all the outgoing radiation originates there. Note that in contrast with the radiative equilibrium situation, we now have a vertical heat-transfer term proportional to the vertical temperature gradient at the surface. The system is posed as three linear differential equations for three unknowns (I_0, I_1, T) with two boundary conditions at the top (z_{top}) and one at the bottom $(z = 0)$. In general, the coefficients in the system of differential equations are functions of z. In the following, we illustrate this by solving for a special case.

3.9.1 $K > 0$

Solutions can be obtained for the mixed radiative conduction problem. In this section, we consider the special case of a uniform slab of optically gray material analogous to water vapor which is $z_1 = H_w$ thick and lying just infinitesimally above the surface. For graphical display purposes, we choose $H_w = 1.6$ km. In this case,

$$\kappa(z) = \begin{cases} \kappa_0, & z \leq z_1, \\ 0, & z > z_1. \end{cases} \tag{3.87}$$

When $K > 0$ and in the same range of z, $\kappa_0 > 0$, analytical solutions are possible but they are very complicated requiring pages of formulas from Mathematica, such that it is not instructive to pursue them here. The book by Liou (1992) has a section in which this case is treated numerically.

Some approximate solutions can be obtained, as most of the water vapor absorbed is in the boundary layer. Let us make some drastic assumptions that will allow us to extract analytical solutions. First assume that the boundary layer containing all the absorber is very thin and that it is well stirred within. This means the temperature is essentially constant in the boundary layer, which we will call the *slab*. Above the slab, the atmosphere is transparent to infrared radiation, $\kappa_0 = 0$, and heat can be transported vertically only by thermal diffusion (same as thermal conductivity mathematically). The requirement of steady state requires that

$$\frac{d}{dz} K \left(\frac{dT}{dz} + \Gamma \right) = 0, \tag{3.88}$$

where we have introduced an average adiabatic lapse rate Γ. Integrating twice leads to

$$T(z) = \left(\frac{C_1}{K} - \Gamma \right) z + C_2, \tag{3.89}$$

where C_1 and C_2 are integration constants. Note that if the thermal conductivity K is very large, the temperature has the slope of the adiabatic lapse rate.

3.10 Convective Adjustment Models

Since the lapse rates of radiative equilibrium solutions are always unstable near the surface, we expect convective overturning to occur carrying heat energy into the layers well above the boundary layer. The last section gives us an idea how this works. Consider the case where we approximate the boundary layer as a thin slab wherein all the infrared absorption occurs. We approximate the convection as conduction with K very large. As seen in the previous section, this means the temperature profile will relax to the adiabatic profile (which we take to be 6.0 K km^{-1}, a value chosen to reflect the contribution from latent heat release). After steady state is established and the adiabatic profile is filled out, there is no further heat transfer by the convection because if there were, there would be no way to radiate the heat convected to space from these layers. The absorbed solar radiation quickly heats the boundary layer and the radiation energy emerges only from the top of the boundary layer. Above the boundary layer, the radiance components $I_0(z) = \frac{Qa_p}{\pi}$ and $I_1(z) = \frac{3}{4}\frac{Qa_p}{\pi}$ are constant and satisfy the boundary conditions at the TOA: $F^\uparrow = Qa_p$ and $F^\downarrow = 0$. At the surface, we assume the ground temperature is equal to the slab temperature because of the large mixing in the slab. The absorption of solar radiation is Qa_p in the slab–ground system. Figure 3.10 shows a schematic of the solution.

Figure 3.11, taken from the classic paper by Manabe and Strickler (1964), shows how in an isotherm hot and cold profiles relax to equilibrium. On the left, the profile relaxes to a pure radiative equilibrium, while on the right, the relaxation is to a radiative convective equilibrium. It is interesting that the characteristic relaxation time in each figure is about 1 month. The Manabe–Strickler model includes the detailed spectral properties of the infrared spectrum and also includes solar scattering and absorption of the solar incoming radiation. Additionally, the heating due to O_3 in the stratosphere is included. It is a remarkable fact that both the radiative equilibrium and the convective-adjustment models (by eye) have relaxation times of about 30 days. It is comforting that our global average models with atmosphere only use this characteristic

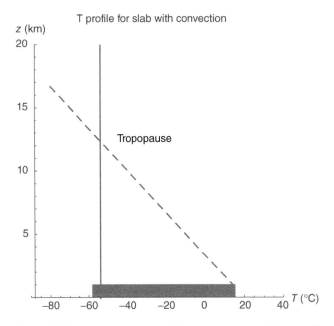

Figure 3.10 The temperature profile for a moist slab of air in a thin planetary boundary layer (nominally 1.6 km, but here thinner). There is no infrared absorber above the slab, only thermal convection or very large thermal conduction that leads to a linear adiabatic profile (dashed line). The radiative equilibrium temperature profile (−54 °C) would hold aloft above the intersection with the adiabat. The intersection determines the tropopause height in this simple model. In this figure, $\kappa_0 = 0.90$.

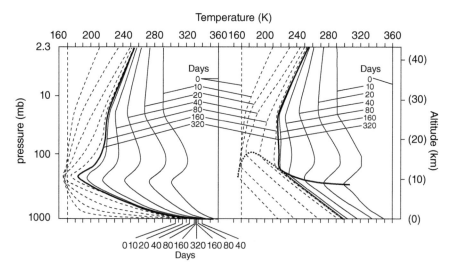

Figure 3.11 Relaxation to an equilibrium profile in a very detailed radiative convective model by Manabe and Strickler (1964). The profiles indicate the time evolution from a cold and a hot isothermal atmosphere. On the left is a relaxation to a pure radiative equilibrium model, while on the right the model includes convective adjustment. (©Amer. Meteorol. Soc., with permission.)

time of 1 month (independently). Using our results from Chapter 2, we might infer that the sensitivity of radiative equilibrium and convective adjustment models are identical.

3.11 Lessons from Simple Radiation Models

We have seen that, for some simple cases, we can find analytical solutions for gray atmospheres. In the case of pure radiative equilibrium (no thermal fluxes other than radiation transfer), these solutions provide the equilibrium temperature profile. In these cases, the profile is always unstable to convection. We can amend the solution to allow convection by applying what is known as a *convective adjustment*. In this chapter, we found the steady-state solutions for radiative–convective equilibrium when the radiation absorption is confined to a thin planetary boundary layer where the water vapor absorber is assumed to be confined. We did not examine the case of other greenhouse cases nearly all of which are well mixed throughout the global atmosphere. In the case of carbon dioxide, for example, the altitude at which radiation to space effectively occurs over a scale height is in the neighborhood of the tropopause. But water vapor is the dominant greenhouse gas and our model gives us some insights into its influence on the atmospheric profile.

The models solved in this chapter are essentially linear, especially when we make the linear approximation, $IR = A_0 + B_0 T$. It is very interesting that, in these cases, the equilibrium ground temperature is a linear function of the TSI $4Q$ and also the optical depth of the absorber τ_*. And as the total outgoing radiation flux to space in equilibrium must equal Qa_p, we find that the outgoing flux must be proportional to the ground temperature.

$$Qa_p = IR = \frac{A_0 + B_0(T_C)}{1 + \frac{3}{4}\tau_*}, \tag{3.90}$$

or
$$T_C = \frac{Qa_p \cdot (1 + \frac{3}{4}\tau_*) - A_0}{B_0}, \tag{3.91}$$

and for $\tau_* = 1.20$, we obtain

$$B_{RE} = 2.43 \text{ W m}^{-2} \text{ K}^{-1}. \tag{3.92}$$

The value of B is very important as the sensitivity of global average temperature to external perturbations is inversely proportional to B. This last value is to be compared with the value between 1.90 W m^{-2} K^{-1} (all sky) and 2.20 W m^{-2} K^{-1} (clear sky only) based upon a regression of the outgoing IR measured from satellites and the seasonal cycle and different geographical locations. The lower empirical value is derived from the whole sky, including clouds and the larger value comes from data wherein the cloudy pixels are removed. Of course, in most of this book, we choose to use the "nominal" value of 2.00 W m^{-2} K^{-1}. Reduction of the blackbody value of B by a factor of ~ 0.53 suggests that even the presence of some optical thickness increases the sensitivity of climate to external perturbations. The additional enhancement of the radiative equilibrium sensitivity to that of the empirical value range is probably due to a combination of water vapor and lapse rate feedback. Climate feedbacks to sensitivity will be discussed in Chapter 4.

Let us consider one final problem in this chapter. Suppose the temperature is increased by 1.0 °C by increasing Qa_p appropriately. Then according to the Clausius–Clapeyron equation[16] for the vapor pressure of water, such an increase in temperature would increase the amount of water vapor in the boundary layer by 7%. This would lead to an increase of τ_* by the same percentage to a value of 1.284. The additional water vapor in this simple radiative equilibrium model (see Figure 3.5) would lead to an increase in temperature of 3.1 K, with an increase of the effective level of emission to 1.01 km for the same brightness temperature of 255 K. As we will see in Chapter 4, the increase due to doubling CO_2 is about 1.0 K. If the current model were correct, we might expect an additional increase of 3.1 K due to the water vapor feedback in the boundary layer, a rather large value compared to some more accurate estimates we will find in Chapter 4. Conventional wisdom suggests a value closer to 1.0 K for water vapor feedback to be added to the 1.0 K from direct forcing. Strictly speaking, the two feedbacks do not add so simply as we will also see in Chapter 4. The strength of this and other feedbacks is one of the several important foci of climate research today.

3.12 Criticism of the Gray Spectrum

While very educational, the gray spectrum for greenhouse gases such as water vapor gives some incorrect impressions. Figure 3.12 comes from a radiative transfer model in which there are no clouds and no greenhouse gases except water vapor. The abscissa is the wave number ($\tilde{\nu} = \frac{1}{\lambda} = \frac{\nu}{c}$ with the speed of light, $c = 3 \times 10^8$ m s^{-1}), in units of cm^{-1}, a traditional way of displaying infrared spectra. The atmosphere in this model is not in

---

16 The Clausius–Clapeyron equation is $\frac{de_s}{e_s} = \frac{L}{R}\frac{dT}{T^2} \approx 0.0692$, where e_s is the saturation vapor pressure, T is temperature (here, 280 K); dT is the change in temperature (here, 1 K); L is latent heat of water (2.50×10^6 J kg^{-1}); and R is the gas constant for water vapor (461.5 J K^{-1} kg^{-1}).

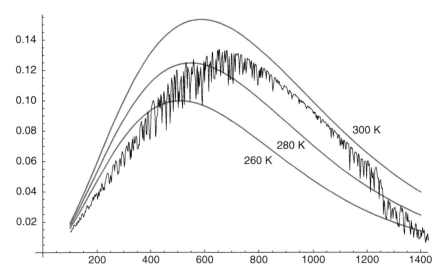

Figure 3.12 The outgoing spectral radiance of water vapor (black zig-zag line) as a function of wave number ($\tilde{\nu} = \frac{1}{\lambda} = \frac{\nu}{c}$ with the speed of light, $c = 3 \times 10^8 \text{m s}^{-1}$) as seen from space looking down from 70 km as modeled by the MODTRAN radiative transfer code (details in Chapter 4). There are no other greenhouse gases in the atmosphere in this simulation. In this case, the ground temperature is set at 300 K, with temperature decreasing vertically at a prescribed lapse rate for the tropics. Thick, smooth, gray lines are blackbody curves at 300, 280, and 260 K. (This figure was constructed from the calculator found on the website: http://geoflop.uchicago.edu/forecast/docs/Projects/modtran.orig.html.)

thermal equilibrium with the solar input. It simply shows the outgoing radiation in a case where the ground is at 300 K and the temperature decreases linearly in the vertical. We will discuss the MODTRAN code in the next chapter. Returning to Figure 3.12, we see three smooth curves along with the very irregular outgoing radiation. The radiation from about 800 cm^{-1} to about 1250 cm^{-1} originates almost at the ground with a brightness temperature of about 295 K. This interval of wave numbers is called the *water vapor window*, as there is practically no absorption or emission from water molecules in this range. Above 1200 cm^{-1}, the brightness temperature is much colder indicating that the emission is occurring at a higher level in the atmosphere. The same occurs at wave numbers below about 700 cm^{-1} and below 400 cm^{-1}. It appears that the emission is at about 255 K, a temperature near the tropopause, and well above the boundary layer.

This is hardly the picture we painted earlier in the chapter where, with the gray atmosphere, the emission level was at about 0.91 km above the ground, and in the radiative equilibrium case, it corresponded to an emission temperature of 255 K. In the real atmosphere, of course, this is the temperature in the neighborhood of the tropopause which is at about 10 km in mid-latitudes and as high as 18 km in the tropics. This warming of the atmosphere between the boundary layer top and the tropopause in the real world is due to convection of heat (latent and sensible) by air parcels that are lifted from near the surface to height where the air is stable to dry as well as moist adiabatic perturbations. But even this convection and warming cannot fill the upper troposphere with water vapor because most of it will condense and fall out as precipitation. Only very tiny amounts of water vapor can survive to the tropopause. Something more must be at work to make an outgoing spectrum as shown in Figure 3.12.

The answer to this problem lies in the very uneven spectrum of water vapor. It turns out that the spectrum is a series of millions of discrete lines all across the infrared portion of the spectrum. When averaged over a small interval (the version of MODTRAN used here averages over intervals of 2 cm^{-1}), the absorptivity or emissivity appears to be finite, but the tinier the interval, the more one is able see the gaps between lines. It turns out that even small concentrations of water vapor (a few molecules per million of air molecules) well above the boundary layer can absorb significant amounts of infrared radiation. This can lead to brightness temperatures corresponding to very cold parts of the atmosphere. Hence, the concentration changes of carbon dioxide and water vapor high in the atmosphere can play a significant role in driving climate change and in providing feedback due to the response of water vapor to the initial driver.

3.13 Aerosol Particles

Another gross omission in our treatment of the vertical properties of the atmosphere is our neglect of aerosol[17] particles. These particles pose one of the most difficult unresolved problems in climate science. Their main influence is in their interaction with the incoming solar radiation. They include such important species as cloud particles (both liquid water and ice), cloud condensation nuclei, dust, organic debris from decaying leaves and other biogenic sources, minerals airborne from deserts and ancient lake beds (playas), soot, air pollution from factory chimneys or transportation devices, electrolyte solutions with solutes such as sulfur, and nitrogen oxides. Most of the particles tend to be spherical or nearly so, exceptions being dust, ice, and minerals. Many are hydrophilic, with adsorbed water encasing nuclear material. The size of these hydrophilic particles depends on the relative humidity of the local environment. All of the particles interact with sunlight either scattering or absorbing. Spherical particles scatter mainly, but some absorb especially if they have a carbonaceous nucleus inside. Clear water spheres (including many nonspherical species) have a characteristic size parameter, the ratio of the circumference of the particle to the wavelength of the light being considered $x = \left(\frac{2\pi r}{\lambda}\right)$. Large particles ($x \gg 1$) scatter light primarily in the forward direction (e.g., when the Sun shines on a dark cloud, it indicates that the water droplets are large); another example is that you can see through a macroscopic water object (the light passes directly to your eye). Small cloud particles ($x \ll 1$) scatter in all directions (Rayleigh scattering) scatter isotropically (e.g., white clouds). In addition to the many texts on radiation transfer in the atmosphere, the book by Pierrehumbert (2011) is strong in physical reasoning in its examination in both the visible and infrared radiation interacting with the atmosphere and its constituents and especially with the climate problem in mind.

A further even more inscrutable problem with aerosol particles is their interaction with clouds. We know that aerosol particles (mostly salt) over the oceans are larger than those over land. The number density of aerosol particles is larger over land. This leads to clouds that are rather different over ocean than over land. The cloud particles over ocean are larger and the number density is smaller. Scattering of sunlight by oceanic clouds will

---

17 Strictly speaking the atmosphere is an aerosol analogous to a colloidal suspension such as milk or coffee. The *aerosol particles* are the particles that are suspended in the air.

be different from those over land. How will climate change factor in this matter? It may be yet another subtle feedback process that we do not yet fully understand.

In this book, we will not consider aerosol particles directly in our further modeling efforts. Aerosol particles will only be included as given information in the coalbedo of the atmospheric column.

Notes for Further Reading

Some recently published books on radiative transfer in the atmosphere include Coakley and Yang (2014), Wendisch and Yang (2012), Petty (2006), and Liou (2002). The classics include Goody and Yung (1989) and Chandrasekhar (1960). Pierrehumbert (2011) covers approximation methods for simplified atmospheric modeling of the Earth and the other planets.

Exercises

3.1 The average distance of the Sun from the Earth is $R_S = 1.496 \times 10^8$ km and the radius of the Earth is $R_E = 6.370 \times 10^3$ km. Assuming that the Earth's orbit around the Sun is a circle, calculate the solid angle representation of the Earth's surface exposed to the Sun's radiation.

3.2 Consider a beam of radiation in an arbitrary direction $\hat{\mathbf{s}}$ from the center of a unit sphere as depicted in Figure 3.13. The magnitude of the radiation is unity (=1) in all directions. Calculate the total amount of radiation in the vertical direction $\hat{\mathbf{k}}$ using spherical geometry, and confirm that

$$\int_{4\pi} (\hat{\mathbf{s}} \cdot \hat{\mathbf{k}}) \, d^2\Omega = \pi.$$

3.3 Show by integration in spherical coordinates that

$$\int_{\Omega} (\hat{\mathbf{s}} \cdot \hat{\mathbf{k}})^{\text{odd}} \, d^2\Omega = 0 \quad \text{and} \quad \int_{4\pi} (\hat{\mathbf{s}} \cdot \hat{\mathbf{k}})(\hat{\mathbf{s}} \cdot \hat{\mathbf{n}}) \, d^2\Omega = \frac{4\pi}{3} \hat{\mathbf{k}} \cdot \hat{\mathbf{n}}.$$

Figure 3.13 Figure for Exercise 2.

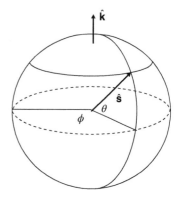

3 Radiation and Vertical Structure

3.4 Show by using solid angle integration formulae that

$$F_v^\uparrow(z) = \pi I_{0v}(z) + \frac{2\pi}{3} I_{1v}(z) \quad \text{and} \quad F_v^\downarrow(z) = \pi I_{0v}(z) - \frac{2\pi}{3} I_{1v}(z).$$

3.5 Given the Planck radiation function defined by

$$B_v(T) = \frac{2hv^3}{c^2(e^{hv/(kT)} - 1)},$$

show that

$$\int_0^\infty B_v(T) dv = \frac{\sigma T^4}{\pi}.$$

How is the Stefan–Boltzmann constant defined?

3.6 Derive (3.30) and (3.31) from (3.29).

3.7 Show (3.45) and (3.46) by proving (3.43) and (3.44).

3.8 Show for the radiative equilibrium model that

$$T(z) = \left(\frac{(3\tau + 2)Qa_p/4}{\sigma}\right)^{\frac{1}{4}}.$$

3.9 Based on the radiative equilibrium model, what would be the temperature just above the surface? What would be the difference between the surface temperature and air temperature just above the surface, that is, $\Delta T = T_*(\tau \to t_*)$? Use $Q = 365 \text{W m}^{-2}$, $a_p = 0.7$, and $\tau_* = 1.2$.

3.10 In the radiative equilibrium model, determine the altitude in kilometers of the brightness temperature. What would be the height where the atmospheric column becomes stable?

4

Greenhouse Effect and Climate Feedbacks

4.1 Greenhouse Effect without Feedbacks

In Chapter 3, we examined the vertical distribution of temperatures in the atmosphere for some very simplified atmospheric models.[1] In particular, we made the assumption that the spectral dependence of the absorption of infrared radiation by gases in the atmosphere were "gray," that is, they had no dependence on wavelength. In this chapter, we consider a more realistic absorption spectrum of the gases in the atmosphere. It will be seen that the details of the spectral dependence of the infrared absorption are important in the perturbation of the radiative balance by changes in the concentration of greenhouse gases (GHGs) such as water vapor, carbon dioxide, ozone, and methane.

4.2 Infrared Spectra of Outgoing Radiation

The infrared absorption spectrum of polyatomic molecules is complicated because such molecules have many internal natural (resonant) frequencies in the infrared frequency range. Typically, for a *GHG*, there are thousands of such lines (classically speaking, the differences between quantum energy levels represent resonant frequencies) across the infrared band of frequencies. It is conventional in infrared spectroscopy to use *wavenumber* instead of frequency. Wavenumber, denoted $\tilde{\nu}$, is proportional to frequency:

$$\tilde{\nu} = \frac{\nu}{c} = \frac{1}{\lambda}, \tag{4.1}$$

where λ is wavelength, ν is frequency (Hz), and c is the speed of light. The wavenumber is usually expressed in units of cm^{-1}. More than a century of laboratory and theoretical research on molecular absorption spectra have resulted in a very detailed understanding of the relevant parameters such as line intensities and the pressure and temperature dependencies of line widths. The main constituents of the atmosphere N_2, O_2, and Ar do not absorb appreciably in the infrared because the two diatomic molecules (being composed of an identical pair of atoms) do not have a permanent dipole moment and the Ar atom has no modes of rotation or vibration in the infrared. Electronic transitions as opposed to vibrational and rotational transitions are at much higher

---

1 As with the previous chapter, this one can be skipped without loss of continuity to the one-dimensional EBMs of the next chapter. However, some readers might want to read Section 4.6 on climate feedbacks.

Energy Balance Climate Models, First Edition. Gerald R. North and Kwang-Yul Kim.
© 2017 Wiley-VCH Verlag GmbH & Co. KGaA. Published 2017 by Wiley-VCH Verlag GmbH & Co. KGaA.

frequencies (e.g., in the visible or ultraviolet). The presence of a dipole moment allows the electromagnetic field to couple directly to the molecule's mechanical motions and transfer energy to the internal mechanical modes of the molecule corresponding to the associated natural frequencies. The H_2O molecule has a strong permanent electric dipole moment allowing it to respond very efficiently to a passing electromagnetic wave. Of all the greenhouse molecules, only H_2O has pure rotational bands, whereas the other triatomics combine vibration and rotation. This strong absorption by the H_2O molecule makes it dominant in the blockage of infrared radiation in the atmosphere. Even a few parts per million by volume (ppm) are enough to render the upper troposphere essentially opaque in some portions of the spectrum.

Carbon dioxide (CO_2), methane (CH_4), nitrous oxide (N_2O), and ozone (O_3), do not have permanent dipole moments but they can have induced moments and this can lead to absorption of infrared radiation in the region where vibrational frequencies predominate.[2] When a molecule is excited into a higher vibrational mode, it has many rotational frequencies available. This can lead to a large variety of resonant frequencies where efficient transfer of energy occurs. Other important trace gases including chloroflourocarbons (CFCs) can have many resonant frequencies connected with rotation and vibration. These are the most familiar GHGs. Strictly speaking, the frequencies of these modes are the differences of the initial and final energy levels associated with the quantum states ($\Delta E = h\nu = hc\tilde{\nu}$, where h is Planck's constant). The infrared frequency band includes the mixed natural modes associated with both vibrational and rotational quantum states. Many of the emission/absorption frequencies of interest to us are from mixed vibrational and rotational transitions. In many cases, the discrete lines are spaced from one another at equal intervals of $\Delta\tilde{\nu}$, so simplicity of the spectral patterns and their direct association with molecular structure led historically to the use of frequency as opposed to wavelength. The use of wavenumber is because a century ago the wavelength λ could be measured very accurately with fine gratings and interferometry, but the speed of light was not known as accurately as today; hence, the spectra could be expressed in terms of a dependent variable that was very well measured, and the patterns of the spectra could be related to the underlying physics of molecules.

To appreciate the complexity of molecular spectra, Figure 4.1 shows the absorption lines from water vapor from 100 to 1500 cm^{-1}, an interval that spans the infrared range of interest here. The upper panel shows all the lines, while the lower panel is a kind of microscopic view of the lines between just 400 and 500 cm^{-1}. As shown, the lines are of very narrow width compared to 1 cm^{-1} (pure rotational lines without external broadening influences may be as narrow as 5×10^{-4} cm^{-1}, corresponding to natural lifetimes of the order of 10^3 s. This latter is of interest in astrophysics wherein very low density gas clouds are studied.). In the Earth's atmosphere where pressures are high, molecules striking the absorbing gas molecules shorten the lifetime of a state over its natural lifetime and cause the lines to broaden[3] (so-called *pressure broadening*). Hence,

---

[2] Discussions of molecular spectroscopy can be found in the books by Coakley and Yang (2014), Andrews et al. (1987), and Goody and Yung (1989).

[3] The uncertainty principle of quantum mechanics yields an estimate. We cannot know the exact energy of transition because of the relation $\Delta E \Delta t \geq \frac{h}{2\pi}$, where $\Delta E = h\Delta \nu$. Here Δt is the "lifetime" of the initial quantum state, but it could be shortened by a collision. If Δt is less, then ΔE will be larger. In terms of wavenumber $\tilde{\nu}$, we have $\Delta \tilde{\nu} = 1/c\Delta t$. The broadening of a spectral line has been studied for roughly a century. Good treatments can be found in Goody and Yung (1989) and Pierrehumbert (2011).

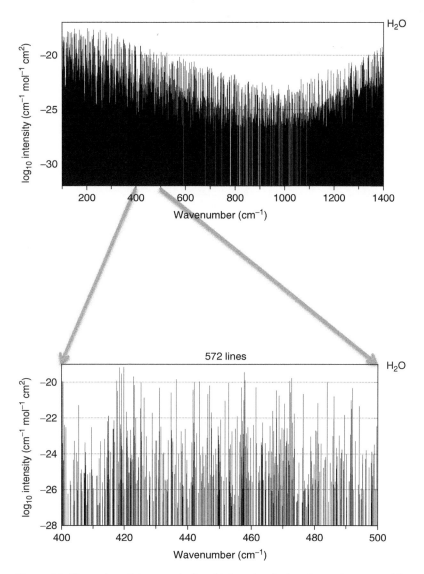

Figure 4.1 Illustration of the complexity of infrared radiation absorption spectral features. The abscissa is the wavenumber ((wavelength)$^{-1}$) in the traditional units of inverse centimeters (cm^{-1}). The ordinate indicates the intensity of the absorption in units (shown in the figure) but in logarithmic form. The upper panel shows a line absorption spectrum for water vapor across most of the infrared band. The lower panel shows a sub-band of spectral lines for water vapor between 400 and 500 cm^{-1}. This sub-band includes 572 lines. (These figures have been modified from the website spectralcalc.com.)

radiative transfer computer algorithms have to take atmospheric pressure (equivalent to altitude of the molecular absorber/emitter above sea level) into account. Molecular motions toward and away from the observer also cause a Doppler shift in frequency, which also gives a temperature dependence to the line width.

It is interesting to see how the line densities (number of lines per unit cm^{-1}) and their intensities of both H_2O and CO_2 vary in different wavenumber bands of the infrared.

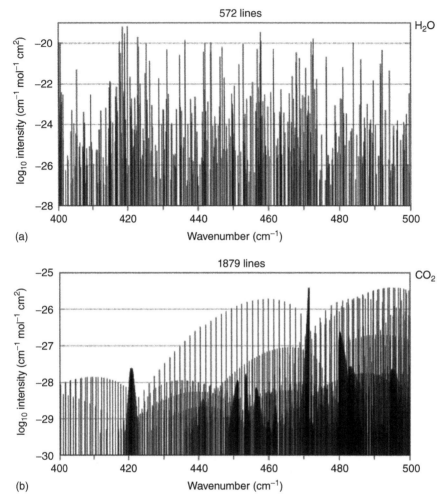

Figure 4.2 The difference between the water vapor and carbon dioxide absorption spectra in the important band between 400 and 500 cm^{-1}. Line spectra of H_2O (a) and CO_2 (b) in the spectral range 400–500 cm^{-1}. This is the "far infrared" region of the infrared spectrum. Although there are more CO_2 lines in this interval, notice that the intensity or emissivity of the H_2O lines is stronger by roughly four orders of magnitude. Note the large irregularity of the H_2O spectrum due to its dominance by pure rotational lines in contrast to the CO_2 spectrum which is composed of many equally spaced lines due to the mix of combined vibration–rotation transitions. (These figures have been downloaded and from the website spectralcalc.com.)

First consider the "far infrared" interval of 400–500 cm^{-1} as shown in Figure 4.2 taken from the website *spectralcalc.com*. Figure 4.2a shows spectral lines of H_2O and Figure 4.2b shows those of CO_2. The vertical extent of the lines show the absorption strength ($\propto$ absorption cross section) of the line on a $\log_{10}$ scale. The same holds for Figure 4.2b. The distribution of line spacing and line strengths of H_2O are very irregular owing to the many complicated rotational modes of the water vapor molecule. By contrast, those of CO_2 are very orderly and many of the lines are equally spaced from one another owing to the dominance of pure vibrational transitions in which no change occurs in the rotational state. The intensities have smooth rounded envelopes, giving the appearance of families of processes. There are about three times as many lines in this band for CO_2 than for H_2O but the line strength is approximately five orders of magnitude larger for the H_2O lines, attributable to the permanent electric dipole moment of that molecule. In this far-infrared region of the spectrum, water vapor is dominant in the absorption and emission of infrared radiation.

Figure 4.3 shows the lines for the range 600–700 cm^{-1}. This is the region we have designated as the "CO_2 ditch" where the molecules of CO_2 dominate the emission to space. The "ditch" is clearly identifiable in the satellite observations shown in Figure 4.4

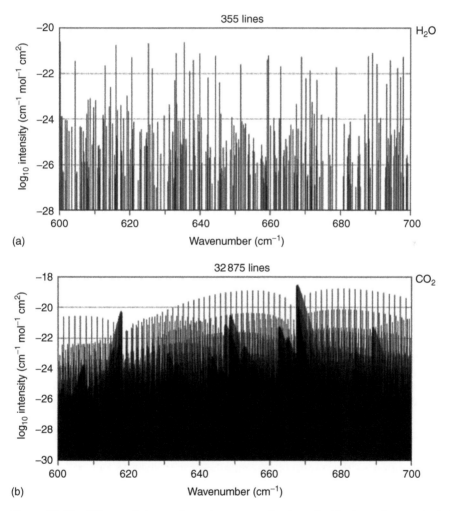

Figure 4.3 The difference between the water vapor and carbon dioxide absorption spectra in the important band between 600 and 700 cm^{-1}. Line spectra of H_2O (a) and CO_2 (b) in the spectral range 600–700 cm^{-1}. This interval spans the "CO_2 ditch" region where CO_2 has its major impact on outgoing infrared radiation flux density. Note the more comparable line strengths in this region, but also that there are spectral lines for CO_2 that are two orders of magnitude higher. (These figures have been downloaded and adapted from the website spectralcalc.com.)

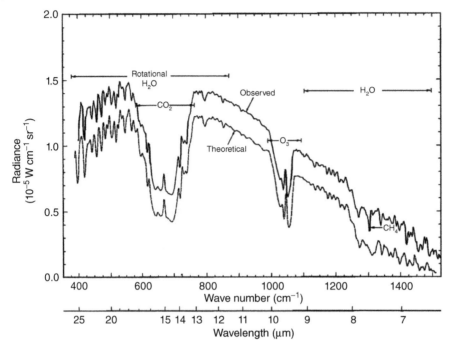

Figure 4.4 Data that emphasize the ability to compute radiative transfer with high spectral resolution to space for an atmosphere that is cloudless and for which the water vapor profile has been measured simultaneously from balloon measurements. Even before 1970, scientists had a very good means of computing this spectrum and measuring it from a satellite-based instrument. Spectrum of infrared radiance at the satellite (the upper solid line of data) taken in 1970 by the IRIS (infrared interferometer spectrograph) on board the Nimbus 3 satellite. The lower spectrum is the theoretically calculated expected spectrum. The flight is over the Gulf of Mexico on a cloudless overpass. The lower curve is a theoretical calculation of the expected spectrum, after knowing the temperature profile from *in situ* data. The data curve has been displaced upward by 0.2×10^{-5} W cm^{-1} sr^{-1}. (Conrath *et al.* (1970). Reproduced with permission of Wiley.)

(more about this figure in the following). We see in Figure 4.3 that the water vapor line strengths are comparable for both gases, but the line density of the CO_2 is two orders of magnitude over that of the H_2O. Understanding this band is crucial for understanding the greenhouse effect for the Earth's atmosphere.

For completeness, we show one more pair of panels, this time in the band 800–900 cm^{-1} shown in Figure 4.5. This is in the atmospheric window portion of the infrared spectrum. Note that there are only 192 H_2O lines in this panel, while there are 5454 lines in the CO_2 case. The line strengths are comparable. Together, there is very little absorption in this range of wavenumbers. It is right to call it a "window region." Actually, there is a continuum of absorption in the water vapor spectrum (not shown here), in which two molecules in close proximity interact with a photon. The effects of such a "three-body" collision are usually weak, but in the case of the window regions, it can be important (see details in Chapter 4 of Pierrehumbert, 2011). So the window is actually a bit hazy.

With the advent of the high speed computer, radiation transfer programs have been improved to the point that under ideal conditions such as no clouds or aerosols, we can

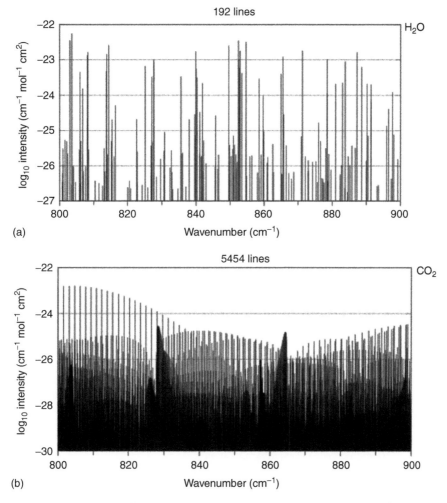

Figure 4.5 Line spectra of H_2O (a) and CO_2 (b) in the spectral range 800–900 cm^{-1}. This figure is in the spectral range of the "water vapor window" of the infrared. Both the CO_2 and the H_2O line strengths are weaker, and it is especially as the line density of H_2O is also so weak that the "window" description holds. (These figures have been downloaded and adapted from the website spectralcalc.com).

compute the spectrum of outgoing radiation to space with high accuracy if we know the profiles of humidity and temperature. Satellite observations with high-resolution infrared spectrometers have also verified the calculations. Figure 4.4 shows a remarkable indication of our ability to compute spectra based on radiative transfer calculations in 1970 (Conrath et al., 1970). The observed spectrum is displaced upward in the figure to make the comparison with the calculated spectrum easier. The giant "ditch" feature[4] between about 500 and 800 cm^{-1} is due to the absorption of upwelling infrared thermal

---

4 One of us (GRN) wondered from where he had come up with the term "ditch." In reviewing the literature in the final stages of this book, he came across the term in Pierrehumbert (2011). Obviously, he had forgotten its origin.

radiation by CO_2 molecules followed by radiation from these same molecules upward but at lower emission temperature. The emission temperature in the deepest part of the ditch is at about 220 K, roughly the temperature of the tropopause in the tropics. This very cold part of the tropical troposphere is key to understanding the greenhouse effect. This figure should convince anyone that we really are able to measure spectra from space and we can also calculate the upwelling radiance to very high accuracy when the temperature and humidity profiles are specified (in this case, by radiosonde measurements from a site in the Gulf of Mexico at a time when no clouds were present). If scientists could calculate spectra this well in 1970, it is no stretch of the imagination to say we can do even better today. Clouds and aerosols present other problems.

4.2.1 Greenhouse Gases and the Record

It might be well at this point to show the GHG increases over the last few hundred years. It is an accepted fact that the trace GHGs are increasing at an approximately exponential rate, and that the nearly certain source of the increase is anthropogenic. Figure 4.6 shows the increase of carbon dioxide in the atmosphere over the last few hundred years as inferred from examination of air bubbles embedded in datable layers found in Greenland ice cores. Instrumental records beginning in the late 1950s confirm the trend that started at the beginning of the industrial revolution, nominally *ca.* 1769 AD at the time James Watt patented his steam engine.

4.2.2 Greenhouse Gas Computer Experiments

In this section, we use an early version of the radiative transfer program MODTRAN that is available to the public. A simplified version of it is available in the form of a calculator at the website: http://climatemodels.uchicago.edu/modtran/, developed by David

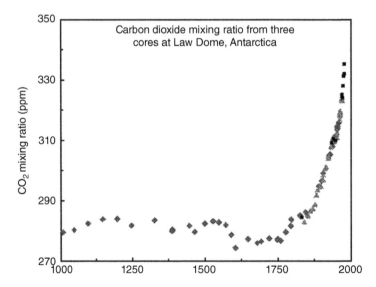

Figure 4.6 Carbon dioxide gas concentrations over the last 1000 years as measured from three ice cores from Law Dome in Antarctica. (Adapted from Oak Ridge National Laboratory cdiac.ornl.gov/trends/co2/graphics/lawdome.gif.)

and Jeremy Archer. The program allows the user to select a specific latitude belt (in our case, the tropics) as well as sky conditions (in our examples to follow, clear with no precipitation or clouds). The user can specify the (uniform) tropospheric mixing ratios of methane and carbon dioxide. Ozone's climatological profile can be scaled up or down by a constant factor. The surface temperature can be given but once entered into the calculator, the vertical dependence of temperature is constrained to the climatological profile at the latitude belt chosen (in this section, tropical). The program then assigns a vertical distribution of the GHGs, temperature, and air density appropriate to the latitude and mixing ratios of the GHGs as specified by the user. The vertical distribution of water vapor is adjusted to be in equilibrium with the climatological temperature profile as given in the calculator model (either fixed relative humidity or fixed vapor pressure). We chose the clear-sky tropics for our example because the tropical tropopause is very high (~ 18 km), causing very cold temperatures at its highest points (≈ 195 K) with an average lapse rate of nearly 10 K km^{-1}. Figure 4.7 shows the vertical dependence used by the calculator model for the clear-sky tropics. These conditions give a maximum greenhouse effect. A number of other minor GHGs are included in the standard program (such as N_2O, CH_4, and the CFCs) and are fixed at climatological profiles.

Figure 4.8 shows the spectrum of outgoing radiation (W m^{-2} cm) as a function of wavenumber (cm^{-1}). This first case includes all GHGs at or near their present concentrations. The infrared flux is for the clear sky in the tropics. Incidentally, all computations are conducted in this version of MODTRAN with spectral resolution[5] of 2 cm^{-1}. If there were no carbon dioxide (or other GHGs) in the atmosphere, the spectrum would be the dark gray curve denoting blackbody radiation at 300 K. Also shown in light gray is the blackbody curve for an emission temperature of 220 K. The large negative departure (the "ditch") between wavenumbers 600 and 800 cm^{-1} is due to the presence of CO_2. Notice that the emission around 680 cm^{-1} is at an emission temperature near 220 K, a temperature about 25 K warmer than that of the tropopause in the tropics (see Figure 4.8). The "spike" in the center (located at 667.4 cm^{-1}) of the band is due to a very strong narrow feature in the CO_2 absorption/emission spectrum. The reason for the spike is discussed in the following.

There are a number of interesting features besides the main CO_2 ditch in Figure 4.8. For example, there is another, but smaller, ditch at about 1100 cm^{-1} due to the GHG ozone (O_3). Note that the emission surrounding the ozone ditch from around 800 to about 1200 cm^{-1} hugs the dark gray 300 K emission spectrum for a blackbody whose surface is at 300 K, the surface temperature specified in this simulation. This broad band of strong emission to space is called the "atmospheric window," a range of wavenumbers where water vapor and other GHGs have almost no absorptivity so that the radiation in this band comes almost directly from the Earth's 300 K surface (this gets blocked by intervening clouds, if any are present, and this can be important). Left of the main CO_2 ditch is a wide band of emission that comes from cooler emission temperatures. This mainly comes from layers of water vapor in the boundary layer (the lowest 1–2 km of air where turbulent eddies dominate and most water vapor resides) and well above it. The

---

5 The most accurate calculations are the so-called line-by-line codes (LBL). These take into account each individual line, and its width and concentration along the path (variation with altitude). Such calculations are important for benchmarking approximate radiative transfer codes such as MODTRAN and testing against measurements, but LBL codes are too consumptive of computer time to be used directly in climate model calculations.

94 | *4 Greenhouse Effect and Climate Feedbacks*

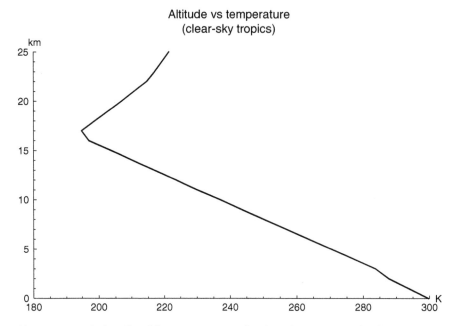

Figure 4.7 Vertical profile of the temperature in the clear-sky tropics used in this exercise and specified in the MODTRAN calculator model based on the website: http://geoflop.uchicago.edu/forecast/docs/Projects/modtran.orig.html. (Data from http://geoflop.uchicago.edu/forecast/docs/Projects/modtran.orig.html.)

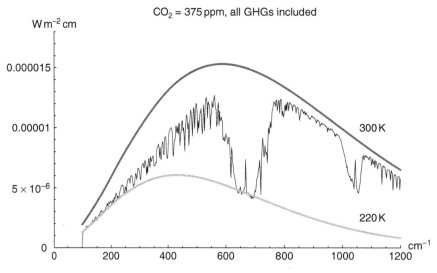

Figure 4.8 Outgoing infrared spectrum for the Earth's tropical clear-sky atmosphere with all greenhouse gases present. The abscissa is wavenumber in cm^{-1}, while the ordinate is $W\,m^{-2}\,cm$. The dark gray curve is the emission spectrum expected for a 300 K blackbody radiator and the light gray line is for a 220 K emitter. (Adapted from calculations based on the website: http://geoflop.uchicago.edu/forecast/docs/Projects/modtran.orig.html.)

same holds for wavenumbers beyond 1200 cm^{-1}, except that methane also plays a role there. The main point of Figure 4.8 is that the total contribution of all GHGs (including water vapor) reduces the outgoing radiation from that of a blackbody at 300 K by about 60 W m^{-2}.

As a second example, consider the planet with only one GHG, CO_2, at a nominal concentration of 375 ppm. Figure 4.9 shows this case wherein the same conditions as Figure 4.8 apply (temperature at the surface, 300 K, tropical clear sky), except that there are no other GHGs (including H_2O). We see the CO_2 ditch clearly but no other prominent features. Consider the bottom of the ditch. The bottom of the ditch follows the 220 K blackbody curve, except for the spike in its center. Why 220 K? This is because the tropopause temperature is at this level in this particular simulation. As shown in Figure 4.7, the temperature in the troposphere falls off nearly linearly from the surface (300 K) to the tropopause (195 K), then begins to increase steeply in the lower stratosphere. As we increase the concentration in the air column, the level at which the emission to space occurs rises. As this level rises, the emission temperature is lowered until the tropopause is reached. This explains the tilted, flat bottom of the ditch. The spike is due to a very strong CO_2 emission line (actually a convergence of many strong lines caused by a folding back and covering of a family of lines) at the center of the ditch. The emission level of this narrow band is well above the level corresponding to neighboring wavelengths in the ditch and it hits the tropopause when the CO_2 concentration reaches a mere 25 ppm. As the concentration of CO_2 in the air column increases, the spike's emission level increases; but being in the stratosphere, the corresponding emission temperature actually goes down. In other words, the growing spike tends to cool the atmospheric column as the concentration is increased. The decrease in outgoing radiation is not at the tilted flat bottom of the ditch but rather in the "wings" of the ditch.

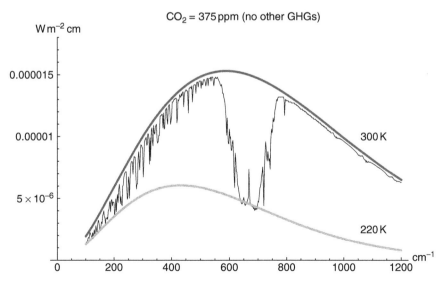

Figure 4.9 Graph of the spectrum of an atmosphere with CO_2 as the only greenhouse gas. (Adapted from calculations based on the website: http://geoflop.uchicago.edu/forecast/docs/Projects/modtran.orig.html.)

Figure 4.10 shows the result of doubling the concentration of CO_2 from 375 to 750 ppm in the case where no other GHGs are present. The range of wavenumbers focuses on the band of wavelengths covering the ditch in order to see more clearly the changes in the outgoing spectrum. As the CO_2 concentration is doubled, the flat portion of the ditch is relatively unaffected, but the wings of the ditch are clearly deepened to a small extent. Basically, the wings radiate from a lower layer in the atmosphere, and as the CO_2 concentration is increased, the level at which emission occurs is at a higher altitude. The changes in the wings seem tiny, but the integral over the ditch (and small contributions elsewhere) reveal (the Chicago website simulator tells us) that the infrared radiation is reduced by $4.4\,W\,m^{-2}$.

We show one more experiment where low concentrations of CO_2 are doubled twice as shown in Figure 4.11. Again the range of wavenumbers in the figure spans the CO_2 ditch. In this case, the emission from the ditch decreases with CO_2 doublings. This decrease limits the outgoing radiation and in order to restore radiative balance for the planet, the surface temperature must increase. As the doublings occur, the emission reduction spreads all across the ditch including both its floor and the wings. This is to be contrasted with the higher initial concentrations experiment of Figure 4.10. We can now explain the reversal of change in the floor of the ditch compared to the wings. In the lower concentration cases, the brightness temperature is well below the tropopause (except for the tiny spike in the very center of the ditch where the very strong absorption feature lies). As the concentration is increased, the strong absorption spike at the center becomes shallow (the spike grows), but the floor of the ditch deepens because the radiation to space is originating from higher and cooler layers of the troposphere. When

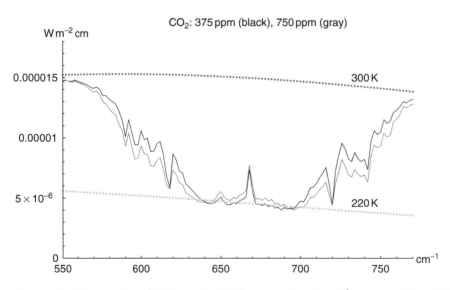

Figure 4.10 The outgoing infrared spectrum in the range 550–770 cm^{-1} at mixing ratios of CO_2 at 375 ppm (black) and doubled to 750 ppm (gray). The dark dotted curve is the emission spectrum expected for a 300 K blackbody radiator and the light dotted line is for a 220 K emitter. (Adapted from calculations based on the website: http://geoflop.uchicago.edu/forecast/docs/Projects/modtran.orig.html.)

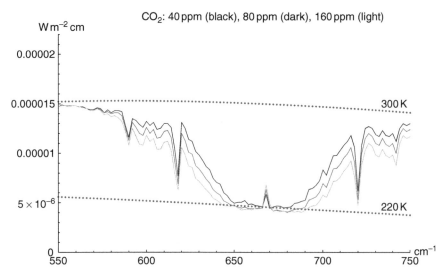

Figure 4.11 Another CO_2 doubling experiment but with the base at a concentration of 80 ppm. The outgoing infrared spectrum in the range 550–770 cm^{-1} at mixing ratios of CO_2 at 40 ppm (black), 80 ppm (dark), and 160 ppm (light). The upper dotted line is the emission spectrum expected for a 300 K blackbody radiator and the lower dotted line is for a 220 K emitter. (Adapted from calculations based on the website: http://geoflop.uchicago.edu/forecast/docs/Projects/modtran.orig.html.)

the concentration becomes large enough, the brightness temperature actually reaches a minimum and even turns around as the level of last radiation enters the inverted temperature profile of the lower stratosphere. It is interesting that the brightness temperature never falls to 195 K, which is the minimum temperature as seen in Figure 4.7. This is because the distribution of radiation origins in the continuous vertical profile is from a "smear" or integrated aggregate of infinitesimal levels and its effective value is never at the actual minimum of the distribution. We can illustrate this by considering a narrow band around 680 cm^{-1}, which is just to the right of the spike in the ditch. We can use the output from the Chicago website to compute the upwelling radiation at this wavenumber (actually a 2 cm^{-1} band) for a CO_2 concentration of 160 ppm at each altitude, starting from the ground. The result is shown in Figure 4.12. The curve is dominated by the temperature profile (cf. Figure 4.7), but notice that the curve approaches the emission level of temperatures warmer than the minimum reached at the tropopause. Figure 4.12 shows the absorptivity times the density. This quantity multiplied by the Planck distribution function is evaluated for the local temperature. The curve in Figure 4.12 was computed from the equation for the upwelling radiation flux at the Chicago website, based on the two-stream approximation (see Chapter 3). This profile is the amount of absorption that would take place from a beam of spectral width 2 cm^{-1} and centered at 680 cm^{-1} pointed downward at zenith. Note that the absorption and emission rates are related by the proportionality factor of the Planck distribution at the local temperature. The absorption rate of Figure 4.12 shows a peak at an altitude of about 16 km and it has a vertical width of about 5 km. This absorption layer is called the *Chapman layer* for this process. The thickness of the Chapman layer gives us an idea of

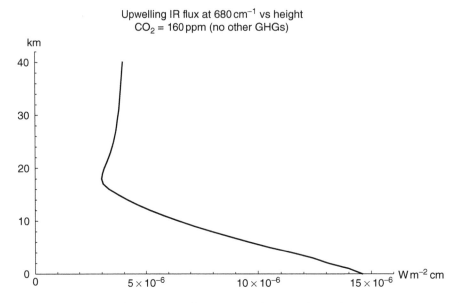

Figure 4.12 The upwelling radiation flux at 680 cm^{-1} versus altitude for a CO_2 concentration of 160 ppm. Note that the radiation at the top of the atmosphere (40 km here) is larger than that corresponding to the minimum in upwelling radiation at about 18 km. (Adapted from calculations based on the website: http://geoflop.uchicago.edu/forecast/docs/Projects/modtran.orig.html.)

the vertical averaging that takes place in absorption and emission. Note that this is in contrast with the Chapman layer of a cloud top where the layer is only of the order of a meter.[6]

So far, we have kept the surface temperature fixed; that is, we have not forced the incoming absorbed rate to balance the outgoing. Keeping the vertical profile rigid while simultaneously increasing the temperature at every altitude of the entire column renders an increase in the total outgoing radiation rate to space. This procedure brings the tropics into balance. In this way, we can restore the radiation balance at the top of the atmosphere (TOA), at the same time increasing the surface temperature. Strictly speaking, we must heat the tropics more than this, because some of the heat in the tropical column must be passed to mid-latitudes through the usual transports conducted by the general circulation of the atmosphere/ocean system.

To do the climate change experiment properly, we must cover the entire globe, simultaneously including the horizontal transport of heat. When we adjust the ground temperature, we must hold the water vapor mixing ratio fixed[7] to avoid feedbacks, because if only relative humidity were held fixed, we would increase the concentration

---

6 The mean free path for absorption of a photon in a cloud of droplets of radius $r_0 \sim 5.0$ μm is $(n\pi r_0^2)^{-1}$, where the number density of droplets is $n \sim 200$ cm^{-3}. This estimate leads to about 1 m for the optical depth in a cloud.

7 Actually, the MODTRAN code used here has some water vapor above about 12 km but it is held fixed during any changes.

of water vapor and therefore induce an additional greenhouse warming – called a *positive feedback*. Considering the tropics in isolation, we can iteratively change the ground temperature incrementally until the outgoing radiation is restored to the value it had when CO_2 was at 375 ppm. After this adjustment, we find that the required temperature increase is about 1.10 K, a value within a small percentage of the global average change found in LBL calculations (Myhre et al., 1998). Section 4.6 considers feedbacks in the climate system.

We can consider a few more cases to improve our insight into the changing greenhouse effect due to doubling CO_2. First consider the MODTRAN simulation when clouds are present (but still no other GHGs, which is slightly absurd as clouds are made of water droplets). The result is that for cloudy atmospheres in the tropics, we simply replace the broadband ground emission in Figure 4.8 with the temperature at cloud top – the ditch remains unaltered. In other words, the cloud tops in most cases are much lower than the emission level of the CO_2. The change in surface temperature to restore the outgoing radiation to its lower CO_2 concentration is left unchanged from the cloud-free case if we assume that the changes in the temperature aloft are carried all the way through the clouds to the ground. We can use MODTRAN in another series of experiments with summer and winter middle-latitude conditions. Perusing these cases (with no GHGs other than CO_2) reveals very little difference from the tropical case. We conclude that given the approximations inherent in MODTRAN, the change in temperatures will be between 0.9 and 1.1 K for a doubling of CO_2 from 375 to 750 ppm, and as the dependence on CO_2 mixing ratio is logarithmic, we can expect the doubling effect to be about the same for doubling from any base level.[8]

4.3 Summary of Assumptions and Simplifications

We have conducted some thought experiments leading us to believe that the increase in global average temperature due to a doubling of CO_2 is about 1.0 (± 0.1)°C. We have cautioned that the physical model and the approximations to it are pretty rough but have pedagogical value. We have to be careful not to apply the quantitative result to the real world without reviewing the assumptions that went into our calculation.

1. We used clear-sky-only in the calculations. While the presence of fixed (in their fractional coverage and altitudes), clouds (cloud tops are below the emission levels of both CO_2 and water vapor) would not have an appreciable effect on the response to doubling CO_2. Unfortunately, the assumption of "fixed" is not likely to hold. Cloud feedback processes are among the most challenging problems facing the climate science community. Much of the difficulty stems from the fact that many cloud processes are at smaller or comparable scales to the grid spacing of our climate models. But clouds are also hooked to the larger scales of the general circulation of the atmosphere, for example, the mid-latitude storm belts that might be undergoing secular change with global warming.

---

8 Some authors prefer to start with preindustrial levels of 250 ppm.

2. Following the Chicago website, we adjusted the temperature "rigidly" up the whole column when we changed the surface temperature. This is probably not what actually happens in the air column. The lapse rate profile might change as the surface temperature is raised and this is likely to lower the response. This effect is called the *lapse-rate feedback* and it is probably weak but negative.
3. In increasing the temperature to compensate for the reduced outgoing radiation, we ignored the fact that the atmospheric column will now hold more water vapor because of the strong dependence of saturation vapor pressure on air temperature. The Clausius–Clapeyron equation suggests that a 1.0 K increase of temperature leads to about 7% increase in water vapor concentration. That relative humidity is roughly constant is supported by climatology and model simulations. This positive feedback is likely to be strong, possibly increasing the response by a factor of 2. Lapse rate and water vapor feedback are anti-correlated, but water vapor appears to be much stronger based on climate model simulation studies. See Held and Soden (2000, 2006).
4. Other known feedbacks such as those due to snow and ice cover are also ignored. These are positive feedbacks but are thought to be smaller than the combination of water vapor, lapse rate, and cloud feedbacks. Some studies suggest that transport of water vapor poleward from the tropics influences the sensitivity in the higher latitudes (Roe *et al.*, 2015).
5. As the CO_2 increases, changes in the temperature of the stratosphere will occur along with those of the troposphere. Generally, to maintain balance of air layers in the stratosphere, the temperature there will have to decrease during greenhouse warming below. This comes about because convection mixes the air in the troposphere, but the stratospheric layers are not coupled to the troposphere through convection. Basically, the stratosphere does not know the troposphere is warming as regards convective overturning, so the cooling in the stratosphere due to increased CO_2 leads to a lowering of stratospheric temperatures.
6. We ignored the rest of the planet. Middle and higher latitudes may have quite different sensitivities to GHG concentrations. If the tropics exhibit the largest sensitivity, this will have to be mixed with that of less-sensitive latitude belts.
7. There are likely to be slow feedbacks in the system that alter the composition of the atmosphere, including its GHGs. These feedbacks may take decades or even centuries to kick in, as permafrost is melted or GHGs are released from sources deep in the oceans and terrestrial biosphere.
8. A final consideration is that the differential of radiation as a function of wavenumber across the ditch for an increased concentration of CO_2 depends on the height of the tropopause on the year, month, or even the day of the occurrence. This is one source of variability of the sensitivity of climate. We could also ask whether a steadily increasing altitude of the tropopause due to global warming represents an additional feedback in the system.

While the exercises presented above are instructive, we cannot end the discussion at this point. The only way to accurately simulate the response to CO_2 is through a general circulation model (GCM) of the atmosphere. This is a formidable task given that many of the important processes are not yet well represented in the models.

4.4 Log Dependence of the CO_2 Forcing

This section provides a heuristic derivation of the log dependence of radiative forcing on CO_2 concentration.[9] From Figure 4.10, we see that the effect of doubling CO_2 is to lower the emission to space from the intervals (550, 640) and (700, 770) results in decreasing the radiation flux density to space, while the radiation flux density to space increases in the interval (640, 700). The eye tells us that the net result is a decrease in the emission to space. It appears that the emission elevation is raised to levels of lower temperatures for wavenumbers on the outer wings of the "CO_2 ditch," but the opposite holds in the center of the ditch, except for the sharp spike in the center. This suggests a very simple explanation for the concentration dependence.

First, we have to acknowledge the fact that the line widths actually depend on the local temperature and total pressure. We ignore this effect here. The justification might be found in Figure 4.3 which shows that in the wavenumber interval between 600 and 700 cm^{-1} (we call it the CO_2 ditch), there are more than 32 000 CO_2 lines, suggesting that there are more than 600 within the resolution of MODTRAN (2.0 cm^{-1}) or a spacing of the order of 0.003 cm^{-1}. Figure 3.18 of Goody and Yung (1989) shows that at 0.5 bar of pressure the half width is about 0.025 cm^{-1}, or about eight times the spacing between spectral lines. This suggests considerable overlap of the line widths and therefore we might be able to neglect the temperature dependence of the upward flux for a fixed elevation of emission. This is equivalent to saying the bandwidth of 2 cm^{-1} is effectively saturated. Our key assumption is that the elevation of emission (Chapman level[10]) is key to determining the changes due to changes in CO_2.[11]

Consider the (well-mixed) number density, $n_{CO_2}(z)$ of CO_2 at height z:

$$n_{CO_2}(z) = n_{CO_2}(0)e^{-z/H}. \tag{4.2}$$

Assume that the emission to space for a particular 1-cm^{-1} wavenumber band occurs at the level z_e which is (effectively) equivalent to optical depth $\tau_{CO_2} = 1$, defined by

$$1 = \int_{z_e}^{\infty} \varepsilon_{CO_2} n_{CO_2}(z)\, dz \;\Rightarrow\; z_e = -H \ln(\varepsilon_{CO_2} H n_{CO_2}(0)), \tag{4.3}$$

where ε_{CO_2} is the band-averaged emissivity/absorptivity ($\propto$ cross section) of a CO_2 molecule and H is the atmospheric scale height. The last equation leads to

$$\Delta z_e = -H\varepsilon_{CO_2} \ln\left[\frac{n_{CO_2}(\text{later})}{n_{CO_2}(\text{now})}\right]. \tag{4.4}$$

Now consider the change in upward radiative flux density for such a change in emission level corresponding to a lower temperature. Using $R = \sigma_{SB} T^4$,

$$\frac{\Delta R}{R} = 4\frac{\Delta T}{T}, \tag{4.5}$$

---

9 Hueristic might be a euphemism here. A similar derivation with more detailed considerations can be found in Huang and Shahabadi (2014).
10 This level is known as the *Chapman layer*. It is an estimate of the effective altitude where the radiation is emitted. See several passages in Goody and Yung (1989).
11 Key issues about the validity of our approximation are discussed in Chapter 4 of Pierrehumbert (2011).

where ΔT is the change of emission temperature at one optical depth from the TOA. The temperature in the troposphere is given by

$$T(z) = T(0) - \Gamma z, \qquad (4.6)$$

where $T(0)$ is the ground temperature. Then the change of the emission temperature is related to the change in emission level, $\Delta T_e = -\Gamma \Delta z_e$ and

$$\Delta T_e = -4\varepsilon_{CO_2} \Gamma H \frac{\Delta z_e}{T}. \qquad (4.7)$$

Finally,

$$\frac{\Delta R}{R} = -\frac{4\varepsilon_{CO_2} \Gamma H}{T} \ln\left[\frac{n_{CO_2}(\text{later})}{n_{CO_2}(\text{now})}\right]. \qquad (4.8)$$

This result would hold for the upward flux of a single 1-cm^{-1} width band (such as with MODTRAN), indexed as i. Each band will have a different R_i, T_i, and perhaps $(\varepsilon_{CO_2})_i$. But the factor containing i dependence lies to the left of the logarithmic factor, the latter having no i dependence. Hence, as we sum the contributions from individual bands, the log dependence on CO_2 concentration remains.

$$\Delta R = \left[\sum_i \left(\frac{4\varepsilon_{CO_2} \Gamma H R}{T}\right)_i\right] \ln\left[\frac{n_{CO_2}(\text{later})}{n_{CO_2}(\text{now})}\right]. \qquad (4.9)$$

More complicated LBL calculations such as those by Myhre *et al.* (1998) also lead to the same logarithmic behavior. As noted by the same authors, other gases might exhibit different behavior. The logarithmic behavior might result from the very heavy line density of CO_2 lines in the ditch area, but other gases may exhibit line spectra with more separation compared to their line widths. Then the dependence might be more due to the temperature and pressure dependence of widths of the lines (see many useful discussions by Goody and Yung, 1989).

4.5 Runaway Greenhouse Effect

Having devoted so much space to understanding radiation transfer in the atmosphere and the effects of various forcings on climate change, we would be remiss if we did not mention the "runaway greenhouse effect." The idea has been explored for many years and many studies have been included into the literature. The question is what happens as the planet is warmed continuously by some forcing which is often proxied by the outgoing IR flux (which, in steady state, is equal to the total solar absorbed flux density). As the planet warms up, the oceans evaporate more water into the atmosphere. This adds a strong positive feedback to the heating. Is there a point, as the heating is turned up, at which all the oceans simply evaporate leaving a very hot planet? Could the transition be catastrophic[12] as with the (modeled) runaway snowball Earth?

---

12 Here *catastrophic* is used in the mathematical sense that the solution jumps to another branch of the operating curve.

These questions are important in the search for life (especially higher life forms such as multicellular species)[13] outside our own solar system – the *exoplanets*. We have already discussed the ice-covered Earth in our discussions of ice-cap models in Chapter 2. Although crude, our modeling suggests that there is an outer radius of the Earth's orbit outside of which no advanced-life forms are likely to be possible because of the total ice coverage. As the planet is brought nearer to the Sun, we have to wonder if there might be a runaway greenhouse effect, which happens to be the plausible situation with Venus. The zone between these two limits is called the *habitable zone* (see Hart, 1979). There are many other factors that determine the habitable zone such as bounds on the obliquity that if too large can cause huge ice age cycles, perhaps even total ice-overs – very hazardous.

To get a crude idea how the runaway greenhouse works, consider the radiative-equilibrium solution to the gray atmosphere case that was discussed in the preceding chapter. The ground temperature in that model (see 3.48 in the preceding chapter) is given by

$$T_*^4 = \frac{Qa_p}{\sigma}\left(1 + \frac{3}{4}\tau_*\right), \tag{4.10}$$

where the subscript "$*$" indicates the ground level values, and σ is the Stefan–Boltzmann constant. If our gray atmosphere were a volatile one on a planet with liquid and water vapor in equilibrium (with no spectral windows because it is really gray in this simplified scheme), then the optical depth at the ground is a function of temperature. To see how this works qualitatively, rearrange the preceding equation:

$$\frac{\sigma T_*^4}{1 + \frac{3}{4}\tau_*(T_*)} = Qa_p. \tag{4.11}$$

Take the infrared absorbing (gray) gas to be water vapor. In addition, consider that the total optical thickness is proportional to the water vapor pressure at the surface. We can see qualitatively that the LHS of the last equation viewed as a curve would be hump shaped because for small T_* it rises from zero, and for large T_* the denominator will increase because of its strong (nearly) exponential dependence (Clausius–Clapeyron equation). To find the solution, simply plot the LHS versus T_* along with a given constant value of Qa_p as in Figure 4.13. The result is a root (in this example) below 300 K. There is a second root to the right, but this one surely represents an unstable climate solution, indicating a case where the steady-state ground temperature decreases as the solar brightness increases. Note also that a numerical solution found by the method of starting with an initial value and marching forward in time steps will never find the unstable solutions. Instead, the solutions found by this method will simply become very large as the bifurcation point (top of the hump) is passed. As the control parameter Qa_p is increased, we eventually reach the top of the hump and above that value there is no solution. Presumably, the temperature races to ∞ (actually, until the oceans run dry). This is the runaway greenhouse.

---

[13] Some experts on exoplanetary life claim that there may be microbes, but not likely multicellular forms, the latter being much more vulnerable to hazard.

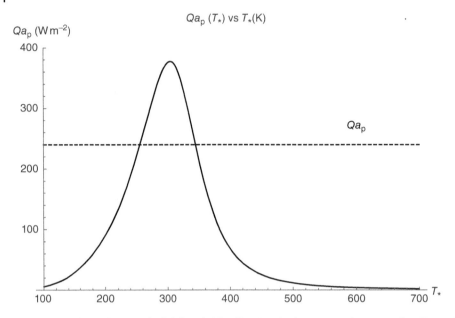

Figure 4.13 The ordinate is the left-hand side of Eq. (4.11) whose roots determine $Qa_p(T_*)$, and the abscissa is the equilibrium ground temperature T_* in kelvins. The flat line is for a particular value of Qa_p.

Weaver and Ramanathan (1995) find a way to incorporate the window of the water vapor spectrum and proceed to develop an analytical (approximate) solution to the window problem. Theirs is a simplified model commensurate with the spirit of this book. They begin by defining a fraction of the area under the Planck function:

$$\beta = \frac{\int_{\lambda_1}^{\lambda_2} B_\lambda(T)\, dT}{\int_0^\infty B_\lambda(T)\, dT}. \tag{4.12}$$

The derivation goes through similarly to the previous one except for the modification:

$$T_*^4 = \frac{Qa_p}{\sigma} \frac{\left(2 + \frac{3}{2}\tau_*\right)}{\left(2 + \beta\frac{3}{2}\tau_*\right)}. \tag{4.13}$$

Following the same procedure as the preceding, we find that if $\beta > 0$, for large τ_*, its dependence cancels out and there is no hump yielding multiple solutions or a bifurcation (leading to a catastrophe). The runaway greenhouse effect is eliminated by the presence of the (clear!) window.

In a paper following the same procedure, Pujol and North (2003) find analytical solutions to problems with several box-shaped or stepwise hazy windows. In this way, they find analytical solutions for cases that imitate H_2O (hazy window) and CO_2 (stepwise hazy window) in the outgoing radiation flux densities to space. Figure 4.14a shows the five infrared bands. Gray atmospheres in this work use mass absorption coefficients independent of wavelength in the individual hazy bands. The water vapor is assumed to be saturated in the atmospheric column. The solutions are more complicated because there have to be distinct βs for each of the hazy (nonzero absorbing)

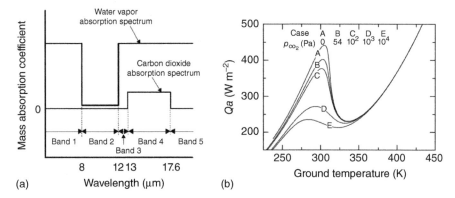

Figure 4.14 (a) The semi-gray model absorption coefficients indicating the step-function changes across the infrared spectrum. (b) The ordinate is the solar absorbed energy density (W m^{-2}) and the abscissa is the surface temperature (K). In this depiction, the ordinate is the independent variable. The curve labels the partial pressure (Pa) of CO_2 for the cases listed across the top of the figure. These solutions do show a catastrophic runaway greenhouse effect. (Figure modified from two figures in the paper by Pujol and North, 2003. Figures originally generated in the paper by coauthor GRN © *Tellus A*, permission not required.)

windows. Figure 4.14b shows the results. When both carbon dioxide and water vapor are included in this kind of model, the runaway greenhouse returns in dramatic form with catastrophic jumps in the surface temperature as the solar absorbed flux density is increased past a threshold that depends on the amount of CO_2.

Results from a model with more sophisticated radiative transfer ingredients are shown in Figure 4.15 based on the work of Kasting (1988). Kasting reviews a series of models evaluated by himself and others. Most of the models generally indicate a runaway greenhouse effect. However, the qualitative features of the results depend sensitively on the parameter values and the choice of processes used for the still-simple models. Whether a runaway greenhouse effect is possible for the Earth or is part of Venus's history seems to depend sensitively on whether CO_2 or other GHGs are included with the water vapor, whether the water vapor window is clear or hazy, as well as how the convection is treated. Fully three-dimensional models are surely called for, but simpler models can be useful in determining what is important in the problem.

4.6 Climate Feedbacks and Climate Sensitivity

In this section, we continue the discussion on global average models, but with emphasis on the important concept of feedbacks in the system. In Chapter 2, we introduced the semiempirical global average climate model based on the balance of fluxes (rate per unit surface area and per unit time) of incoming absorbed solar energy and the outgoing infrared energy. Empirical data were used to parameterize the outgoing radiation rate at the TOA, ($I = A + BT$) and the absorbed solar flux (Qa_p), where A (218 W m^{-2}), B (1.90 W m^{-2} °C^{-1}), and a_p are empirical constants based on satellite data, T is the surface temperature in °C, and Q is the total solar irradiance divided by 4. We found that the coefficients A and B differ dramatically from the blackbody values

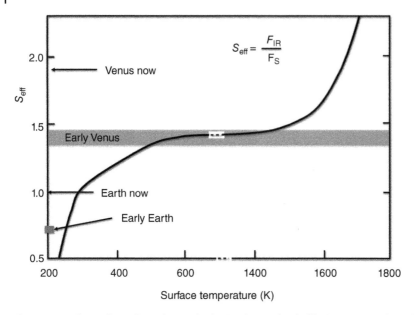

Figure 4.15 The ordinate S_{eff} is the total solar irradiance divided by its current value. The abscissa is the surface temperature in K. Depicted are the Early Earth with S_{eff} = 70% of its present value, the Earth now at the value unity, Early Venus at about 1.4, and Venus now at 1.9. T_{cr} denotes the temperature at which all the oceans have been evaporated dry. Note that in steady state $S_{eff} = F_{IR}/F_S$, where F_{IR} is the outgoing radiation flux density and F_S is the value of the net incident flux density. There is a break in the abscissa between 700 and 1300 K where the curve is flat. This schematic graphic was made from information in Kasting (1988). (Kasting (1988). Reproduced with permission of Elsevier.)

($A_0 = 314.9$ W m^{-2}, $B_0 = 4.61$ W m^{-2}K^{-1}). Because the sensitivity to increasing the TSI by 1% is given by $(A + BT_0)/(100B)$, where T_0 is the present global average temperature, the value of the *damping coefficient* B is a key parameter determining the sensitivity to external perturbations to the energy balance. Here we are assuming that the sensitivity to any external perturbation that is essentially of global scale will have the same response structure. We gained some insight in Chapter 3 that the introduction of a "gray" column

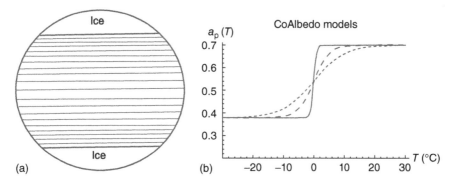

Figure 4.16 (a) Diagram of a planet with ice cap and (b) the global temperature dependence of the global coalbedo for different values of the smoothing parameter, γ. The solid line is for a case of a near-delta function shift as function of global temperature. The dashed curves are two different cases where the transition is smoother. See Section 2.4 for details. The coalbedo function is defined in (2.79).

of air would increase the global average surface temperature and in Chapter 4 we found that if the CO_2 concentration were doubled with all other variables held constant (water vapor, temperature lapse rate, snow/ice cover, etc.) the increase in temperature would be approximately 1.00 ± 0.10 K. But when the Earth is warmed by a change in CO_2 concentration or a change in total solar irradiance, such further influences as changes in water vapor or snow/ice cover area are invoked. Some of these features alter the energy balance along with the *direct forcing* (Figure 4.16). We call these invoked perturbations *climate feedbacks*. The techniques were first used in the climate change context by Hansen et al. (1984) and Schlesinger (1986). The review article by Roe (2009) presents a historical overview of feedback formalism, which comes originally from electrical engineering.

4.6.1 Equilibrium Feedback Formalism

The energy balance at the TOA can be (crudely) expressed as follows:

$$I(T, G, F_1^I, F_2^I, \ldots) = S(Q, F_1^S, F_2^S, \ldots), \tag{4.14}$$

where T is the surface temperature (in kelvin), G is the concentration of some test GHG, F_1^I or F_1^S is some measure of the water vapor in the column of air appropriate to either I or S, $F_2^{I,S}$ is the average vertical lapse rate of temperature in the air column, $F_3^{I,S}$ represents the parameters associated with clouds, and $F_4^{I,S}$ is the snow/ice area on the surface. If some control parameter such as Q or G is changed, we expect the system to re-equilibrate after some suitably long adjustment time; hence, the term "equilibrium sensitivity" is used in the perturbation. We consider only infinitesimal perturbations such as δQ or δG with the infinitesimal response δT.

Next consider a perturbation in which Q is held fixed and G is changed by a prescribed amount, say 100%.

$$B_{BB}\delta T + \frac{\partial I}{\partial G}\delta G + \sum_n \frac{\partial I}{\partial F_n^I}\frac{\partial F_n^I}{\partial T}\delta T = \sum_k \frac{\partial S}{\partial F_k^S}\frac{\partial F_k^S}{\partial T}\delta T, \tag{4.15}$$

where $B_{BB} = \frac{\partial I}{\partial T} = 4\sigma T^3|_{T=273\,K} = 4.61$ W m^{-2} K^{-1}. Gathering the coefficients of δT onto the left-hand side, and then dividing through by this factor, we find

$$\delta T = \frac{-\frac{\partial I}{\partial G}}{B_{BB} - \sum_k \frac{\partial S}{\partial F_k^S}\frac{\partial F_k^S}{\partial T} + \sum_n \frac{\partial I}{\partial F_n^I}\frac{\partial F_n^I}{\partial T}}\delta G, \tag{4.16}$$

or

$$\delta T = \frac{-\left(\frac{\partial I}{\partial G}\delta G\right)/B_{BB}}{1 - \sum_n \left(\frac{\partial S}{\partial F_n^S}\frac{\partial F_n^S}{\partial T} - \frac{\partial I}{\partial F_n^I}\frac{\partial F_n^I}{\partial T}\right)/B_{BB}}. \tag{4.17}$$

Finally,[14] in its simplest form,

$$\delta T = \frac{1}{1-f}(\delta T)^{\text{no feedback}}_{2\times CO_2}, \tag{4.18}$$

---

14 The formalism used in this section comes from the standard study of feedbacks in electrical circuitry. It was introduced in the climate sensitivity problem by Stone, see Stone & Risbey (1990). Our notation and normalization follow those of this reference.

where $f = f_1 + f_2 + \cdots$ with the individual terms being associated with the different feedback mechanisms. The factor $\frac{1}{1-f}$ is called the *gain*.

The numerator of the last expression is the amount of response to doubling CO_2 in the absence of any feedbacks. Its nominal value is 1.00 ± 0.1 K. The definitions of the f_n are implicit in the previous formula. Note that the f_n are dimensionless numbers whose magnitude is less than unity. f cannot be larger than or equal to unity because the quotient would diverge. A negative sign for the quotient would imply an unstable initial climate state (the slope-stability theorem of Chapter 2).

4.7 Water Vapor Feedback

We can use our empirical rules on outgoing radiation to estimate f_w the bulk water vapor feedback that includes lapse rate feedback. From Figure 2.1, a reasonable nominal value for the slope of $I(T) \approx A + B \cdot (T - 273)$ for cloud pixels eliminated (the so-called "clear sky") is $B = 2.26$ W m^{-2} K^{-1}. In this case, all the other feedbacks are held fixed. To get f_w, we need to divide 2.35 (=4.61 − 2.26) by 4.61, yielding

$$f_1 \approx 0.51, \tag{4.19}$$

in this very simple model. The water vapor feedback is then a *positive feedback* or an amplifier of the direct response by a factor of 2. Water vapor is probably the largest feedback term in the denominator of the feedback equation (4.18). Note that in this estimate of f_w, we neglected the contribution of water vapor to the solar absorbed part of the system. Since water vapor does absorb some sunlight, we might expect this positive addition to the sum of feedback factors would add even more to the magnitude of the positive feedback. The magnitude of the water vapor feedback is probably the best known of the climate feedback mechanisms. The interpretation is simply that as the temperature is increased by direct forcing, the air column will hold more watervapor from the Clausius–Clapeyron equation of thermodynamics. Moreover, it appears to be an established empirical fact that the relative humidity stays approximately constant as climate is forced to change. Simulations of forced climate change with large GCMs also indicate that relative humidity remains roughly constant. The upshot of this is that the water vapor feedback represents a doubling of the response to the direct forcing.

Lapse rate feedback refers to the possibility that the lapse rate of the temperature in the air column might change in response to a directly forced heating. The reason this matters is that the imbalance occurs primarily at a level very high in the tropical troposphere. If the column warms "rigidly," that is, the slope of the temperature versus height curve does not change as result of the warming, then there would be no lapse rate feedback. It seems likely that lapse rate feedback is negative, but the finding of the empirical curve in Figure 2.1b suggests that, when combined, the slope of the $I(T)$ versus T curve (2.26 W m^{-2}) for clear skies includes both water vapor as well as lapse rate feedbacks. Curves similar to that in Figure 2.1b can also be computed from GCM output and a similar slope is found (see Cess *et al.*, 1990).

Although we have been able to estimate the water vapor feedback to probably $\pm 10\%$, obtaining an accurate value is very important as it is the largest feedback, and because

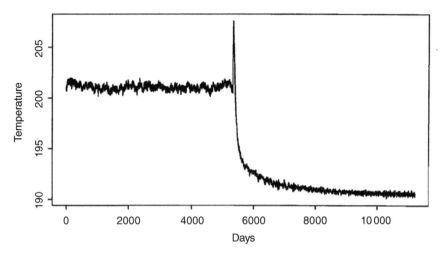

Figure 4.17 The daily global average temperature as simulated by Terra Blanda, a version of the CCM0 GCM but with no ocean, no topography, and no seasons. The simulation starts as with water vapor, precipitation, and cloud, then suddenly the soil moisture is no longer allowed to evaporate after year 15. A sudden spike (shock) is followed by decay to a cooler climate with no clouds. The resulting planet is about 15 K cooler, presumably because of the lack of water-related feedbacks. (North *et al.* (1993). Reproduced with permission of Springer.)

of its peculiar role in the feedback equation (4.18). Note the strong nonlinear amplification that would occur if the first feedback is large. For example, suppose the water vapor feedback where f_1 is 0.8 (sensitivity = 5.0 K) instead of 0.5 (sensitivity 2.0 K). Then, adding a small feedback $f_2 = 0.1$ leads to total response of 10 K as opposed to 2.5 K. In other words, the closer the first feedback f values are to unity, the more influence the later ones, although small, will have on the total response. See Held and Soden (2000, 2006).

As an example of how the climate adjusts from a moist surface planet to one with no evaporation permitted, Figure 4.17 shows a time series of daily global average temperatures for such a planet (North *et al.*, 1993).

4.8 Ice Feedback for the Global Model

In Chapter 2, we examined the global average climate problem in detail. In particular, we considered the problem of a temperature-dependent planetary coalbedo $a_p(T)$ where T_{eq} is the equilibrium global average temperature. The energy balance equation is written as follows:

$$A + BT_{eq} = Qa_p(T_{eq}). \quad (4.20)$$

Now imagine introducing a steady perturbation (forcing) ΔF to the energy balance, holding Q fixed. When this is done, the energy balance adjusts by a change in the temperature:

$$B\Delta T_{eq} = Qa'_p \Delta T_{eq} + \Delta F. \quad (4.21)$$

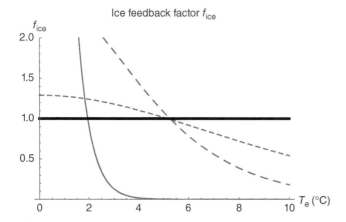

Figure 4.18 The variation of feedback parameter f_{ice} as the temperature is lowered in the simple icecap coalbedo models of Figure 2.8 of Chapter 2. The solid curve represents a model whose global coalbedo is nearly a step function as the global average temperatures goes through freezing point. The long-dashed curve is for a case where the transition to ice cover is milder in terms of the global coalbedo, and the short-dashed curve even more so. The solid black line denotes unity, the value for which the bifurcation in the solution occurs. These curves show how the feedback factor f_{ice} increases with increasing ice cap until the planet is iced over.

Recall that $B = B_{BB} - B_w$, where B_{BB} corresponds to the damping of outgoing radiation from a blackbody and B_w is the offsetting contribution (defined here as positive) from water vapor. Then,

$$\Delta T_{eq} = \frac{(\Delta T_{eq})^{\Delta F}_{f=0}}{1 - \frac{B_w}{B_{BB}} - Q\frac{a'_p}{B_{BB}}}, \qquad (4.22)$$

where the numerator is the change in temperature when there are no feedbacks and $f_w = \frac{B_w}{B_{BB}} = 0.51$ and $f_{ice} = Q\frac{a'_p}{B_{BB}}$.

We can find the value of $a'_p(T_{eq})$ from Chapter 2 by differentiating:

$$a'_p = \frac{\gamma}{2}(a_f - a_i)\text{sech}^2(\gamma T_{eq}). \qquad (4.23)$$

Figure 4.18 shows the variation of f_{ice} as a function of global average temperature for three ice-cap coalbedo models as discussed in Chapter 2; see also the three curves in Figure 4.16b. The sharpest transition is for $\gamma = 1$, shown by the solid curve. The next sharper curve occurs for $\gamma = 0.2$ shown by the long-dashed curves, and finally, the mildest transition (short dashes) is shown for $\gamma = 0.1$. Notice how rapidly the feedback factor rises as the planet's temperature is lowered toward the jump-off at 0 °C. The solid black line is included to indicate the point of divergence of the feedback (in the absence of any other feedback mechanisms).

4.9 Probability Density of Climate Sensitivity

Before considering the other feedbacks, let us look next at the effects on probability density of sensitivity of our uncertain knowledge of f. Suppose f has a probability density function (pdf)

$$P(f) \, df = \frac{1}{\sqrt{2\pi}\sigma} e^{-\frac{(f-\bar{f})^2}{2\sigma^2}} \, df, \tag{4.24}$$

where $\bar{f}$ is mean (nominally 0.5 if water vapor were the only contributor) of the density distribution of f. The distribution of f can be thought of as a lack of knowledge or it may be that f varies randomly over time with an unknown timescale but with a definite PDF whose standard deviation is σ. Given this uncertainty in our knowledge of the value of f, we would like to know how this translates into our lack of knowledge of sensitivity.

This is a standard problem in the theory of probability (e.g., Papoulis, 1984). In our case,

$$g = \frac{1}{1-f}. \tag{4.25}$$

Given a random variable f distributed as $P(f)$, what is the density distribution of a function $g = g(f)$? In the case of a univariate problem, the form is particularly simple:

$$g(f) = \frac{1}{1-f}; \quad f(g) = \frac{g-1}{g}, \tag{4.26}$$

$$P_g(g) \, dg = P_f(f(g)) \left| \frac{df}{dg} \right| dg = \frac{P_f(f(g))}{\left| \frac{dg}{df} \right|} \, dg, \tag{4.27}$$

or in our case,

$$P_g(g) \, dg = P_f(f(g))(1-f)^2 dg = \frac{P_f(f(g))}{g^2} \, dg. \tag{4.28}$$

Figure 4.19a shows two normal distributions with mean 0.5 and standard deviations 0.1 (solid curve) and 0.2 (dashed curve). Figure 4.19b shows how the two distributions in f become distorted into distributions of δT. Clearly, the broader distribution in f is more distorted. Two things are apparent, the transformed distributions are skewed, and

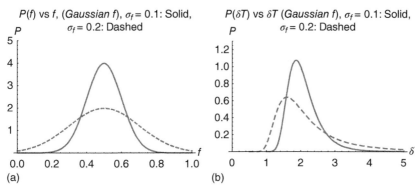

Figure 4.19 (a) Illustration of the normal probability density functions of the feedback coefficient f for two values of standard deviation, $\sigma_f = 0.1$ (solid) and $\sigma_f = 0.2$ (dashed). The mean value is $f = 0.5$. (b) Illustration of the probability density functions of the response given the probability density function of the feedback coefficient f (previous figure) for two values of standard deviation, $\sigma_f = 0.1$ (solid) and $\sigma_f = 0.2$ (dashed). The mean value is $f = 0.5$.

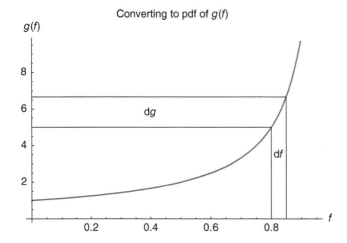

Figure 4.20 This plot shows how a small increment of f can lead to a large increment of $g(f) = \frac{1}{1-f}$. The mean of the normal distribution on the abscissa is 0.7 and the standard deviation is 0.1. Note that the portion intersected by df is smaller than the portion dg which projects onto the ordinate. The curvature of the function $g(f)$ leads to a distortion (positive skewness) of the density function portrayed on the ordinate.

the mode is lowered from 2.26 to about 1.86 in the narrower distribution and 1.59 in the wider distribution. Because the distribution in f spreads beyond $f = 1$, the computation of the mean and higher moments diverges. Besides the mode being reduced, the tail becomes fatter, indicating that extremely large sensitivities are possible because of our uncertainty in the value of f and this has led to some concern because of our lack of precise knowledge of the feedbacks in the climate system. Further insight can be gained by viewing Figure 4.20, where it is shown directly how an infinitesimal portion of the probability in f, $P_f(f)\,df$ is mapped into a corresponding probability in g, $P_g(g)\,dg$. See Andronova and Schlesinger (2001) and Roe and Baker (2007).

4.10 Middle Atmosphere Temperature Profile

As seen in the previous section and from Figure 1.4, the stratospheric profile is quite different from the tropospheric one. The linear tropospheric thermal profile is caused by a combination of radiative and convective adjustments to a radiative–convective equilibrium. The troposphere is to a large extent well mixed. Above the tropopause, the temperature (Figure 1.4) up to about 45 km has a negative lapse rate (inversion). The giant thermal maximum occurring above that is due mainly to the dissociation of O_3 by ultraviolet solar incoming radiation. When the loose O atom rapidly recombines with the abundant O_2 molecules nearby, the energy release is by collisions with neighboring molecules resulting in local increases of temperature. Eventually, the profile comes to a radiative equilibrium balancing this heating with the cooling due to the net locally emitted (infrared) radiation. The lower stratosphere is then nearly uncoupled from the tropospheric mixing due to convection ending at the tropopause. Above this level, for many kilometers, the profile is stable and in radiative equilibrium. There are two important consequences for us to consider in climate change scenarios.

4.10.1 Middle Atmosphere Responses to Forcings

One might ask why the observed cooling in the stratosphere occurs while the surface temperature is increasing as CO_2 concentration is increasing. The answer lies in Figure 4.21a, which shows the cooling rate (K per day) due mainly from CO_2 in the range centered at 50 km. Likewise, the heating rate in the same range of heights is dominated by O_3 dissociation and recombination. Areas left of the heating must equal areas right of the cooling to maintain equilibrium. While increasing concentration of CO_2 induces warming at the Earth's surface, the heating in Figure 4.21b does not change. But CO_2 concentrations are increasing. To counter the increased imbalance of the increased cooling, the temperature must decrease in the lower stratosphere (due to excess cooling). A key element in this argument is that the air above the tropopause is dynamically uncoupled from that below it (Figure 4.22).

Another important change in the stratosphere occurs as CO_2 is increased gradually over time. The result of the slight cooling in the stratosphere has an effect on the surface temperature. This is called the *stratospheric adjustment process*. It is neatly explained in the IPCC Third Assessment Report (TAR) appendix which can be found at http://www.ipcc.ch/ipccreports/tar/wg1/258.htm. It is also discussed in the paper by Ramanathan and Coakley (1978) and many papers following. The sequence that occurs goes as follows. First the stratosphere cools rapidly because of the elevated effective level of emission of CO_2 in the stratosphere. Remember there is little or no coupling between the troposphere and the stratosphere. We note that the adjustment time of the tropospheric profile (see Figure 3.11) is of the order of a month over land or years over ocean. Most studies consider the surface to be ocean and it is therefore held fixed as the ocean surface does not change much over a few weeks or even months. The slight cooling of the

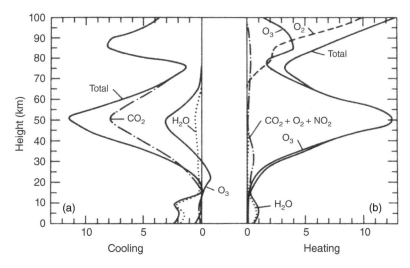

Figure 4.21 Heating and cooling rates up to 100 km. (a) Cooling rates (K per day) due to net infrared emissions. (b) Heating rates due mostly to solar absorption processes. Note the huge effects of cooling by ozone in the 50 km range (a) and the heating in the same altitudes by O_3 (b). During greenhouse-induced climate change the magnitude of cooling increases as CO_2 concentration increases, while the heating rates are essentially unchanged. (London (1980). US Department of Transportation.)

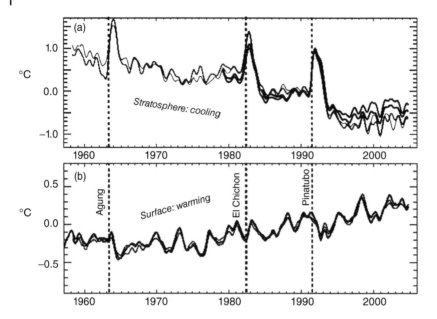

Figure 4.22 Time series of measurements of decreasing temperatures in the lower stratosphere (the separate curves indicate an envelope of measurements from different data sources) and rising temperatures at the Earth's surface. (a) The dashed curves influencing the peaks in the stratospheric curve are volcanic eruptions by (left) Agung, (middle) El Chichon, and (right) Pinatubo. (b) The volcanic influence is to cool the surface. (Solomon and Dahe (2007). Reproduced with permission of Nature.)

stratosphere changes the energy budget of the surface. In the natural adjustment time of the troposphere, a new equilibrium is established. The steps are spelled out in cartoon form in Figure 4.23. The exact alteration of climate sensitivity due to the stratospheric adjustment process is difficult to pin down and is currently a subject of study in the modeling community. We presume in this book, given our level of approximation, that

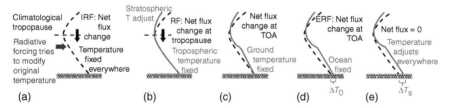

Figure 4.23 Steps to incorporate the stratospheric adjustment to a forcing change due to GHG increase. Schematically, the vertical is height and the abscissa is temperature. In these steps, the dashed line is the original profile before perturbation. Step a: The initial state, wherein an imbalance is induced at the tropopause due to the increase of infrared flux downward. The tropospheric temperature is held fixed at this stage, as the adjustment in the stratosphere is fast compared with that below. Step b: The lower stratosphere cools because of the increase in GHG there. Steps c and d: The surface and troposphere are now warm and equilibrium is restored. Steps d and e: The surface–troposphere equilibrate to end the process. (Stocker (2013). Reproduced with permission of Cambridge University Press.)

it is negligible, but this may prove to be erroneous as future research may reveal. Using the Chicago website referred to in this chapter, we find that the change at the TOA for doubling CO_2 from 400 ppm, we find a lowering of upward flux by 3.23 W m^{-2}, whereas at 18 km height (taken to be the tropopause in the tropics), we find a decrease of upward flux to be 4.58 W m^{-2}. In other words, the change at the TOA where we usually define "forcing," is different from that at the tropopause. The stratosphere does have a role of the order of 1.3 W m^{-2}. Our use of the online calculator is likely to be deficient for such an estimate. Realistic research focusing on these adjustments and their effects have to be conducted in the interior works of a GCM.

4.11 Conclusion

The Earth's surface temperature has been rising steadily over the last century. Most of the potential drivers of climate change (volcanic activity, solar brightness variability, scattering and absorbing atmospheric aerosols, and GHG concentration increases) have been examined in great detail in recent years. We now have moderately good estimates of the strengths and time dependencies of these drivers and although much needs to be done in substantiating these assertions, virtually all have been eliminated except for the increasing influence of GHGs (although there is a renewed effort to estimate the size of the warming influence of black carbon aerosols, see Bond et al., 2013) as well as the role of aerosols in general including their effect on clouds. The importance of the GHG driver is also consistent with paleoclimate evidence. We have shown a series of pedagogical computer experiments that provide estimates of the response in the tropics to doubling the CO_2 concentration while excluding all feedbacks. These experiments can be repeated by the reader by going to the Chicago website. While virtually all experts on the radiative aspects of climate science would agree with the values we have obtained, we have to acknowledge that many additional physical effects will come into play along with intensification of any of the primary drivers. Some of these additional physical effects (feedbacks) are likely (at the time of this writing) to amplify the response to CO_2 doubling to a value perhaps as much as four times.

One more important effect not considered here is the time dependence of the response. The illustrations used here were from one equilibrium climate state to another equilibrium state. The Earth system has a number of effective heat-storage components that have varying effective heat capacities. Among these are the atmospheric column with a response time of about a month, the oceanic mixed layer with a response time of a few years, continental glaciers with centuries of response time, and finally the deep ocean which communicates with the surface waters through small passageways that limit the flow of heat toward the deeper parts. Many parcels of the deep ocean water have not touched the surface in 800 years. The upshot of this is that while climate sensitivity is an important index for comparing one atmospheric model with another, the global system has many interlocking parts that cause the response of the surface to be delayed in its full response by perhaps hundreds of years. This means that even if we were to stop or reverse the rate of GHGs entering the atmosphere, the reversal of the response is likely to be delayed by these sluggish and nearly inaccessible components.

Notes for Further Reading

Many books cover infrared spectra, but Goody and Yung (1989) and Pierrehumbert (2011) are especially recommended.

The history of the runaway greenhouse problem is too long to relate in detail here, but a few important references include Simpson (1927a, 1927b), Ingersoll (1969), Lindzen *et al.* (1982), Nakajima *et al.* (1992), Weaver and Ramanathan (1995), and Kasting (1988) paper.

The habitable zone is discussed by Kasting (2010), Ward and Brownlee (2003) as well as Hart (1978, 1979), who gave it the name. Another book about the history of life on our planet is by Ward and Kirschvink (2015).

The feedback calculus introduced in this chapter comes from electrical engineering. It seems to come first into climate science by way of Schlesinger (1986), who gives a thorough description and applies it to climate forcing and feedback. The probability distribution of climate sensitivity is introduced by Andronova and Schlesinger (2001), and nicely described by Roe and Baker (2007).

Exercises

Use the Chicago website to answer the following questions.

4.1 Use the default values for CO_2, O_3, and water vapor scales and tropical atmosphere with no clouds or rain. Maintain the view from 70 km looking down. Describe qualitatively the outgoing spectrum of infrared radiation. What is the total flux density of radiation energy to space? Double the concentration of CO_2 and determine the new value of outgoing radiation flux density. Restore the radiative balance by adjusting the Ground T offset upward (by trial and error). How much is the change in tropical ground temperature to achieve the balance?

4.2 Continue the experiment of 4.1, but with a change in cloudiness. First, use the Altostratus Cloud Base 2.4 km Top 3.0 km setting. By what amount does the upward flux at 70 km change? What happens in the water vapor window? What happens to the "ditch"? Double the CO_2 and compare the change in equilibrium temperature in this cloudy case.

4.3 Return to the clear tropical atmosphere with present CO_2. Now change the Trop. Ozone (pp.) to 0. What happens to the outgoing radiation flux? Restore the tropospheric ozone and change the Strat. Ozone Scale from 1.0 to 0. What happens to the total outgoing radiation flux density? Now restore the equilibrium by adjusting the surface temperature. How much warming is due to the stratospheric ozone?

4.4 Repeat these experiments but at middle latitudes in summer.

4.5 How accurate are these procedures for estimating climate sensitivity?

4.6 How is water vapor treated in this calculator? You can check this by changing the ground temperature and looking at Show Raw Model Output. Scroll down to the table that shows the profiles of various gases. How might you estimate water vapor feedback by adjusting water vapor according to keeping the relative humidity constant? You can expect specific humidity to increase by 7% for each kelvin of increase.

5

Latitude Dependence

We turn now to the latitude dependence of the surface temperature in the one-horizontal-dimension energy balance model (EBM) and the main factors that govern its shape. The satellite data indicate that there is far more net solar radiation absorbed combined with cooling to space by infrared radiation in tropical latitudes than in the polar regions. In order to maintain steady state, there must be some mechanism in the atmosphere/ocean system that transports heat poleward. In the 1960s, there were pioneering papers written independently by Budyko (1968, 1969) and Sellers (1969) that considered EBMs that employed simplified mechanisms for the transport depending only on the mean-annual zonally averaged surface temperature. These papers drew the attention of a number of investigators who proceeded to build simple EBMs for climate studies (see the suggested readings at the end of this chapter). In the spirit of these earlier studies, this chapter introduces a diffusive heat transport device that lends itself to study by the same kinds of analytical methods already familiar from Chapter 2. As in every chapter in this book, we stress the importance of analytical solutions, not only because of their aesthetic appeal but also for the additional insight provided in such procedures. Another reason is that the phenomenological parameters introduced are explicit. The unnecessary resort to numerical solutions to simplified climate models can result in the inadvertent inclusion of hidden adjustments to make model solutions conform to the investigator's wishes. Of course, it is often the case that the numerical solution is the only resort, especially with more complex models. The problem of communicating the procedures in numerical modeling is receiving attention throughout the community because it is very difficult to compare or reproduce calculations from different research groups (see, e.g., Irving, 2016; Stevens, 2015).

The first step is to introduce the spherical coordinate system in preparation for the simplest one-parameter model for transport of heat across latitude circles: macroscopic-thermal conductivity or diffusion. The same techniques as in Chapter 2 are used in reconciling the balance of heat energy fluxes, but this time, the rate of energy deposition per unit area in a latitude belt of infinitesimal width (the divergence of horizontal heat flux) must be balanced with the difference between solar absorbed and infrared terrestrial radiation to space for each latitude belt simultaneously. The resulting energy balance can be expressed as a second-order ordinary differential equation for surface temperature with latitude as the dependent variable. The system is constrained by a zero horizontal flux condition at the poles (boundary conditions). These two conditions fix the arbitrary constants inherent in the solution of a second-order differential equation. The homogeneous part of this linear system is readily identified

Energy Balance Climate Models, First Edition. Gerald R. North and Kwang-Yul Kim.
© 2017 Wiley-VCH Verlag GmbH & Co. KGaA. Published 2017 by Wiley-VCH Verlag GmbH & Co. KGaA.

with Legendre's equation and the complete system can be solved using Legendre polynomials (LPs), which, for completeness, are described in the chapter. Retention of only a few Legendre modes suffices for many applications. Once the solutions are obtained, it is possible to enumerate all the properties of such climate systems and compare them with observational data.

The chapter proceeds with a simple introduction to the latitude dependence of the solar insolation. Some extreme transport conditions (none and infinite) are studied and compared to the actual observed zonally and annually averaged surface temperatures. Diffusive heat transport is introduced in spherical coordinates, followed by a derivation and discussion of the LPs. This allows a complete solution to this class of problems when the coefficients are constants. Some attention is devoted to a solution with just two Legendre modes. This solution appears to be a very good approximation for latitude-dependent models when forced at very large scales, such as solar irradiance changes or long-lived greenhouse gas (GHG) changes. The two-mode model is shown to provide a very good fit to the poleward transport of heat as derived from satellite data. After a few applications to specific forcings,[1] such as a ring of heating, the problem of polar amplification due to ice cap feedback is discussed in its linear form.

5.1 Spherical Coordinates

Since most of the models in this book endeavor to describe the surface temperature of the Earth (Chapters 3, 4 and 10, being the exceptions), we will be continually required to work on the surface of a sphere. We adopt the conventional spherical coordinates:

$$x = r \sin \vartheta \cos \varphi; \tag{5.1}$$

$$y = r \sin \vartheta \sin \varphi; \tag{5.2}$$

$$z = r \cos \vartheta. \tag{5.3}$$

In our case, $r = R_e$. The angle ϑ is the polar angle (same as colatitude) and φ is the longitude. Figure 5.1 shows a diagram defining the spherical coordinates. It is convenient to use $\mu \equiv \cos \vartheta = \sin(\text{latitude})$ instead of ϑ to designate latitude. An area element on the sphere is $d^{2A} = r^2 \sin \vartheta \, d\vartheta \, d\varphi = -r^2 \, d\varphi \, d(\cos \vartheta)$. Then ignoring the sign, $d^{2A} = r^2 \, d\varphi \, d\mu$. The variable μ is proportional to z (see 5.3) in the Cartesian system; it is also the projection of $\hat{\mathbf{r}}$ onto the polar axis for the Earth, where $\hat{\mathbf{r}}$ is the unit vector from the center of the sphere to the point on the surface in question. In this chapter, we are concerned with zonally symmetric models; that is, there is no φ dependence to the problems. In this case, the area element in question is a zonal strip or belt around the Earth (cf. Figure 5.2). The area of the strip is $d^{2A}_{\text{strip}} = 2\pi R_e^2 \, d\mu$. The use of the coordinate μ is particularly useful as area averages over zonal belts are easily computed:

$$\bar{f} = \frac{1}{\Delta \mu} \int_{\Delta \mu} f(\mu) \, d\mu. \tag{5.4}$$

Functions of latitude will be labeled as functions of μ, that is, $f(\mu)$. Note that at the North pole $\mu = 1$ and at the equator, $\mu = 0$ and at the South pole, $\mu = -1$. At latitude 30°N, $\mu = 1/2$, which incidentally means that half the Earth's area lies between ±30°.

---
1 We remind the reader that the term "forcing" indicates an imposed imbalance on the radiation budget at the top of the atmosphere.

Figure 5.1 The spherical coordinate system. The polar angle is ϑ, with its complement the latitude θ (= 90° − ϑ). The longitude is given by φ.

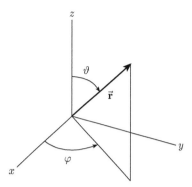

Figure 5.2 A zonal strip or narrow latitude belt on the sphere. The small shaded box has area $r^2 \sin\vartheta \, d\vartheta \, d\varphi$. The corresponding zonal strip has area $d^2A = -2\pi r^2 \, d(\cos\vartheta)$.

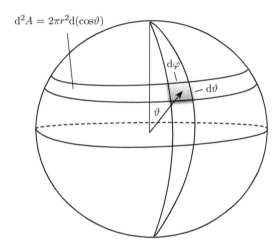

5.2 Incoming Solar Radiation

Since the Earth's surface is not perpendicular at all latitudes to the incoming solar beam, the amount of radiation power per unit local surface area at an instant in time is not simply $\sigma_\odot$; rather, it is reduced by a factor of the cosine of the local instantaneous zenith angle that depends upon the time of day, time of year, and latitude. In this chapter we are interested in the *mean annual* energy/area/time reaching the top of the atmosphere averaged through the local day at a given latitude, $\sin^{-1}\mu$. We postpone the derivation of this function so that we can get right to work with the climate problem, presenting here the approximate result:

$$S_{\text{m.a.}}(\mu) \approx \frac{1}{4}(5 - 3\mu^2). \tag{5.5}$$

We will also occasionally be interested in the distribution at equinox,

$$S_{\text{eqnox}} = \frac{4}{\pi}\sqrt{1 - \mu^2}. \tag{5.6}$$

These functions, by convention, are normalized such that

$$\frac{1}{2}\int_{-1}^{1} S(\mu) \, d\mu = 1. \tag{5.7}$$

The radiation flux (W m^{-2}) reaching the strip of the Earth–atmosphere system (after averaging through the day) is $QS_{\text{m.a.}}(\mu)\,d\mu$ (the factor common to all terms in the energy budget, πR_e^2, has been omitted). The total energy *absorbed* by the strip is this factor multiplied by the local coalbedo of the Earth–atmosphere system $a_{\text{m.a.}}(\mu)$. Now we can compute the planetary coalbedo used in the last two chapters:

$$a_p \equiv \frac{1}{2}\int_{-1}^{1} a_{\text{m.a.}}(\mu) S_{\text{m.a.}}(\mu)\,d\mu. \tag{5.8}$$

From satellite data, we find that $a_{\text{m.a.}}(\mu) \approx 0.68 - 0.20(3\mu^2 - 1)/2$ (North and Coakley, 1979; the most up-to-date values come from Graves *et al.*, 1993). The reason for this peculiar way of grouping the terms will be evident in the next few sections. Strictly speaking, the coalbedo and the sunlight (insolation) functions should be multiplied together and then averaged through the annual cycle instead of the way it has been done here.

5.3 Extreme Heat Transport Cases

Before proceeding to specific parametric forms for the horizontal heat transport in the system, consider the case of a planet with no horizontal heat transport. We simply balance the heat absorbed with the heat radiated outward by the Earth–atmosphere system for each strip

$$A + BT(\mu) = QS(\mu)a(\mu) \quad \text{(no cross-latitude heat transport).} \tag{5.9}$$

Note that we have applied the Budyko radiation rule in every latitude belt; that is, we have used the same empirical coefficients A and B as in Chapter 2. Figure 5.3 shows the steep temperature profile one gets for the no-heat transport case (the heavy solid-gray curve in the figure) compared with Northern Hemisphere (NH) data (short-dashed curve in the figure). Consider next the case of infinitely efficient heat transport (analogous to the electrostatic potential of a perfect conductor). This extreme makes the planet isothermal. The temperature will be the same as the global average but uniform over the globe as indicated by the long-dashed horizontal line in the figure. Also shown in Figure 5.3 is the approximately parabolic form of the true Earth's zonally averaged temperature that lies in between these two cases. Note that, in all cases, the global average temperature is the same. We will see later that this is because the heat-transport term in the energy balance is a *divergence*, and the area integral $(d\mu)$ from pole to pole (combined with the boundary conditions) of such a term always vanishes.

5.4 Heat Transport Across Latitude Circles

Next, we seek a modification of the last EBM that allows for the divergence of heat from the strip due to finite horizontal transport. This must be of the form

$$\text{divergence} = (\text{heat energy leaving the strip per unit time})/(\text{area of the strip}). \tag{5.10}$$

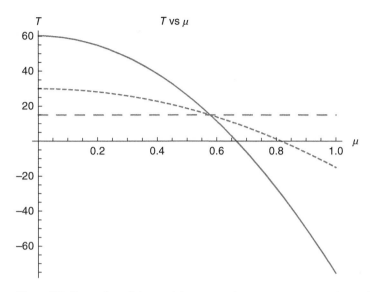

Figure 5.3 Illustration of the model-computed temperature versus the cosine of the polar angle μ (note: $\mu = 0$ at the equator, unity at the pole) for two extremes of heat transport: none (solid line) and infinite (long-dashed line). In between is a curve (short dashed) similar to the Earth's actual temperature. Note that infinite heat transport is analogous to a perfect conductor in electrostatics, while the zero transport case is simply proportional to the solar heating rate.

Note that, in this formula, we canceled out the factor "per unit longitudinal length of the strip" in the numerator and denominator. By this last step, we have effectively assumed that the planet is uniform around latitude belts. For a flux vector in the ϑ direction, $\vec{F} = F_\vartheta(\vartheta)\hat{\vartheta}$, and the divergence in spherical coordinates is given by

$$\nabla \cdot \vec{F} = \frac{1}{R_e \sin\vartheta}\left(\frac{\partial}{\partial \vartheta}\sin\vartheta F_\vartheta\right) \tag{5.11}$$

$$= -\frac{1}{R_e}\left(\frac{\partial}{\partial \mu}\sqrt{1-\mu^2}F_\vartheta\right). \tag{5.12}$$

In the last equation, we used the chain rule,

$$\frac{d}{d\vartheta} = \frac{d\mu}{d\vartheta}\frac{d}{d\mu} = -\sqrt{1-\mu^2}\frac{d}{d\mu}. \tag{5.13}$$

In order to proceed, we need an expression or rule for the flux of poleward heat transport. The simplest such form is taken up in the following section.

5.5 Diffusive Heat Transport

In a paper published in 1969, Sellers proposed a model for the heat transport that was diffusive. The diffusion coefficient contained the sum of three terms: (i) oceanic heat transport, (ii) sensible heat transport, and (iii) latent heat transport. The last included a nonlinear form because of the strong temperature dependence of evaporation rate. In the treatment here, we lump all three mechanisms into one with a constant diffusion coefficient whose value is chosen to best fit the latitude dependence of the observed zonally averaged temperatures. A good analog would be the conduction of heat on a spherical shell with a given thermal conductivity, or the current density of electrical charge on a spherical shell with a constant resistivity.

Consider the form of heat *conduction* on the surface of the sphere. Such an expression would have the flux density of heat across a plane interface proportional to the local gradient of temperature (often referred to as *diffusive heat transport*):

$$\vec{q}_{\text{heat}} = -D\nabla T, \tag{5.14}$$

where D is a kind of macroturbulent thermal conductivity. Note that $\vec{q}_{\text{heat}}$ is the amount of heat energy per unit time passing across a unit length of longitudinal distance. To get the heat transported across a full latitude circle one must multiply by the circumference of the latitude circle. We cancel the 2π factor that is common to all terms. The divergence of the heat flux is

$$\nabla \cdot \vec{q}_{\text{heat}} = -\nabla \cdot (D\nabla T), \tag{5.15}$$

or in our zonally symmetric case,

$$\text{heat flux divergence} = \frac{d}{d\mu}\left(D \cdot (1-\mu^2)\frac{dT}{d\mu}\right), \tag{5.16}$$

where we have used the expression for the gradient

$$(\nabla)_\vartheta T = -\frac{\sqrt{1-\mu^2}}{R_e}\frac{dT}{d\mu}. \tag{5.17}$$

In the last equation, we have introduced a thermal conductivity D that has absorbed into it two factors of R_e. In general, D might be a function of latitude $D(\mu)$, but for now we take it to be a constant independent of μ. Now we can write our complete (time-independent) mean annual EBM with diffusive heat transport:

$$-\frac{d}{d\mu}\left(D \cdot (1-\mu^2)\frac{dT(\mu)}{d\mu}\right) + A + BT(\mu) = QS(\mu)a(\mu). \tag{5.18}$$

This is a second-order differential equation in μ and to specify its solution uniquely we must satisfy certain conditions at the two endpoints of the interval $-1 < \mu < 1$. These constitute the boundary conditions for the problem.[2] In the present case, the boundary conditions are to be that no heat flux enters the poles:

$$D\sqrt{1-\mu^2}\frac{dT}{d\mu}\bigg|_{\mu\to\pm1} = 0. \tag{5.19}$$

Often we will deal with a planet that is symmetric between the two hemispheres, in which case, we can take the interval on μ to be (0, 1) with the boundary conditions

$$D\sqrt{1-\mu^2}\frac{dT}{d\mu}\bigg|_{\mu=0,1} = 0, \quad \text{North–South symmetric planet.} \tag{5.20}$$

---

2 An alternative approach is to note that the homogeneous solution of the energy balance equation (5.18) is the Legendre function (Kamke, 1959), $T^{(h)}(\mu) = E\,P_\nu(\mu) + F\,Q_\nu(\mu)$ with constant coefficients E and F to be determined by boundary conditions and the index ν is a function of parameters B and D and not necessarily an integer. North (1975a) solves the problem using this method, but we do not pursue it in this book, because the approach using LPs is simpler and they lead to a modal interpretation.

Before proceeding, we need to look at a set of special functions that will be useful throughout the text.

5.6 Deriving the Legendre Polynomials

In this section,[3] we introduce the LPs that form a very simple example of *eigenfunctions* that can be used as a basis set in series expansions. The mathematics problem is the following: consider the nontrivial (i.e., not identically zero) functions $P_n(\mu)$ that satisfy

$$-\frac{d}{d\mu}\left((1-\mu^2)\frac{dP_n(\mu)}{d\mu}\right) = \lambda_n P_n(\mu), \tag{5.21}$$

together with the boundary conditions

$$\sqrt{1-\mu^2}\frac{dP_n(\mu)}{d\mu}\bigg|_{\mu\to\pm 1} = 0, \tag{5.22}$$

for some λ_n not equal to zero. We have introduced the index $n = 0, 1, \ldots$ to indicate that there may be countably infinitely many functions $P_n(\mu)$ that satisfy the above conditions. The floating parameter λ_n corresponding to each function $P_n(\mu)$ is called its *eigenvalue*.

The above formulation is a special case of a much more general class of relations known as *eigenvalue problems*; for example,

$$\mathcal{L}f_n(\mu) = \lambda_n r(\mu) f_n(\mu), \tag{5.23}$$

subject to boundary conditions $\alpha f_n(\mu = a, b) + \beta f_n'(\mu = a, b) = 0$. As can be seen in (5.23), $r(\mu) = 1$ in our case. The object $\mathcal{L}$ is to be a member of a class of differential or integral operators. Later we will return to a broader class of transport operators. In particular,

$$\mathcal{L} = \frac{d}{d\mu}\left\{p(\mu)\frac{d}{d\mu}\right\} + q(\mu), \tag{5.24}$$

leads to a class of eigenvalue problems known as the *Stürm-Liouville* problems. They have the following remarkable property:

$$\int_a^b f_n^*(\mu) f_m(\mu) r(\mu)\, d\mu = \frac{\delta_{nm}}{N_n}, \tag{5.25}$$

where $\sqrt{N_n}$ is the normalization factor for $f_n(\mu)$, and if the functions $q(\mu)$ and $r(\mu)$ are positive definite, it can be proved that the eigenvalues are all positive. The more general form of the Stürm–Liouville problem comes up again in Chapter 8. To find out if such functions $P_n(\mu)$ exist, let us try the series expansion

$$P_n(\mu) = \sum_{m=0}^{\infty} c_m^{(n)} \mu^m. \tag{5.26}$$

---

3 On first reading, some readers may wish to jump to the next section. The technique used here is a common approach to solving such equations as can be found in many reference books on mathematical methods for engineers and physicists.

The superscript indicates the series of coefficients $c_m^{(n)}$ will depend upon the index n as well as m. Substituting, we find

$$\sum_{m=1} \frac{d}{d\mu}(1-\mu^2)c_m^{(n)} m\mu^{m-1} + \sum_{m=0} \lambda_n c_m^{(n)} \mu^m = 0. \tag{5.27}$$

After differentiating and changing the dummy summation index, we can rearrange the terms to obtain

$$\sum_{k=0} \{[(k+1)(k+2)]c_{k+2}^{(n)} - [k(k+1) - \lambda_n]c_k^{(n)}\}\mu^k = 0. \tag{5.28}$$

If this is to be true for any value of μ, the coefficient must vanish term by term, hence we have the following *recursion relation*:

$$c_{m+2}^{(n)} = \left[\frac{m(m+1) - \lambda_n}{(m+1)(m+2)}\right] c_m^{(n)}. \tag{5.29}$$

This means that if we know $c_0^{(n)}$, we can compute all the even coefficients in the series. In fact, the entire even part of the series is proportional to this number. The same goes for $c_1^{(n)}$ and the entire odd part of the series. We have two independent solutions, one for the even-termed entries and one for the odd:

$$P_n(\mu) = \sum_{m \text{ even}} c_m^{(n)} \mu^m + \sum_{m \text{ odd}} c_m^{(n)} \mu^m. \tag{5.30}$$

The $c_0^{(n)}$ and the $c_1^{(n)}$ are arbitrary constants at our disposal in satisfying all the constraints. Now consider the convergence of the series (5.26). For large m, $\{m(m+1) \gg \lambda_n\}$, we have

$$\left|\frac{c_{m+2}^{(n)}}{c_m^{(n)}}\right| \sim 1. \tag{5.31}$$

This means that the series will diverge unless something drastic happens. The way to prevent the divergence catastrophe is first to set $c_0^{(n)} = 0$ for n odd and set $c_1^{(n)} = 0$ for n even. This will allow us to set λ_n such that the series cuts off and becomes a finite degree polynomial. The prescription is $\lambda_n = n(n+1)$; the series will continue up to the term $m = n$, after which $c_{n+2}^{(n)}$ and all succeeding terms will vanish (whether n is even or odd).

We can now generate the first few LPs. First, $P_0(\mu) = c_0^{(0)}$ that we take to be unity. Then

$$P_2(\mu) = c_0^{(2)} + c_2^{(2)} \mu^2, \tag{5.32}$$

and $c_2^{(2)} = -3c_0^{(2)}$ from the recursion relation. The overall coefficient $c_0^{(2)}$ is arbitrary as it will be for every $P_n(\mu)$. The convention is to fix this normalization by requiring that

$$P_n(1) = 1. \tag{5.33}$$

This means that

$$P_2(\mu) = \frac{1}{2}(3\mu^2 - 1). \tag{5.34}$$

5.6.1 Properties of Legendre Polynomials

The LPs[4] are a set of polynomials in μ that are mutually orthogonal:

$$\int_{-1}^{1} P_n(\mu) P_m(\mu)\, d\mu = \frac{2}{2n+1} \delta_{nm}, \tag{5.35}$$

where δ_{mn} is the Kronecker delta that is unity if $n = m$ and zero otherwise. As noted in the previous section, $P_n(1) = 1$ and $P_n(-1) = (-1)^n$. $P_n(\mu)$ with n odd (even) are odd (even) functions consisting only of odd (even) powers of μ. In fact, they can be derived by defining $P_0(\mu) \equiv 1$, then finding a a first-order polynomial that is orthogonal to it with $P_1(\mu = 1) = 1$; this leads to

$$P_1(\mu) = \mu. \tag{5.36}$$

Using this rule, we can generate the entire family of LPs.

Following is a list of the first few LPs:

$$P_0(\mu) = 1, \tag{5.37}$$
$$P_1(\mu) = \mu, \tag{5.38}$$
$$P_2(\mu) = \frac{1}{2}(3\mu^2 - 1), \tag{5.39}$$
$$P_3(\mu) = \frac{1}{2}(5\mu^3 - 3\mu), \tag{5.40}$$
$$P_4(\mu) = \frac{1}{8}(35\mu^4 - 30\mu^2 + 3). \tag{5.41}$$

Graphs of the first few LPs are shown in Figure 5.4a.

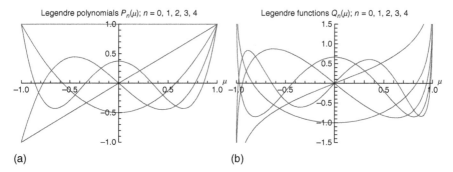

Figure 5.4 (a) Plots of the first five Legendre polynomials, $P_0(\mu) = 1, P_1(\mu) = \mu, \ldots, P_4(\mu)$. The order of the polynomial n is equal to its number of zeros on the interval, $-1 \leq \mu \leq 1$. (b) Plots of the first five Legendre functions of the second kind $Q_0(\mu), Q_1(\mu), Q_2(\mu), Q_3(\mu)$, and $Q_4(\mu)$. Each function has one more zero than its index. Note that the Legendre functions of the second kind, $Q_n(\mu)$, have logarithmic singularities at the poles.

---

4 In this section we list only a few of the properties of these special functions. Many more properties and relations can be found in various books on mathematical methods in physics and/or engineering.

5.6.2 Fourier–Legendre Series

We can make use of the orthogonality property of the LPs (5.35) to expand a function $f(\mu)$ into a Fourier–Legendre series:

$$f(\mu) = \sum_{n=0}^{\infty} f_n P_n(\mu), \tag{5.42}$$

and the coefficients f_n can be calculated by multiplying each side by $P_m(\mu)$ and integrating

$$\int_{-1}^{1} f(\mu) P_m(\mu) \,\mathrm{d}\mu = f_m \frac{2}{2m+1}, \tag{5.43}$$

or

$$f_n = \frac{2n+1}{2} \int_{-1}^{1} f(\mu) P_n(\mu) \,\mathrm{d}\mu. \tag{5.44}$$

This is very similar to an ordinary Fourier series except that instead of the basis functions being $\sin(n\pi\mu)$ and $\cos(n\pi\mu)$, they are $P_n(\mu)$. Similar to the Fourier case, the LPs are oscillatory. We are now in a position to expand our temperature field $T(\mu)$ into a series of LPs[5]

$$T(\mu) = \sum_{n=0}^{\infty} T_n P_n(\mu). \tag{5.45}$$

Also similarly to the Fourier series case, for functions that have discontinuities or discontinuous derivatives, the series can be expected to converge very slowly.

5.6.3 Irregular Solutions

Occasionally, one encounters a need for the irregular solutions to Legendre's equation. These solutions, the Legendre functions of the second kind, are divergent at the poles and can normally be excluded in diffusion problems on the whole sphere. But in certain problems, such as an expansion on a finite-width latitude belt, there is no reason to exclude them and a solution might then be of the form

$$T(\mu) = A_n P_n(\mu) + B_n Q_n(\mu), \tag{5.46}$$

where $P_n(\mu)$ are the LPs discussed in previous sections and the $Q_n(\mu)$ are the Legendre functions of the second kind. While we do not wish to derive their explicit forms here, we list the first few to get an idea how they behave:

$$Q_0(\mu) = \frac{1}{2} \ln \frac{1+\mu}{1-\mu}, \tag{5.47}$$

$$Q_1(\mu) = \frac{1}{2} \mu P_2(\mu) \ln \frac{1+\mu}{1-\mu} - 1, \tag{5.48}$$

$$Q_2(\mu) = \frac{1}{2} \ln \frac{1+\mu}{1-\mu} - \frac{3}{2} P_1(\mu), \tag{5.49}$$

$$Q_3(\mu) = \frac{1}{6}\left(\frac{3}{2}\mu(3-5\mu^2)\ln\frac{1+\mu}{1-\mu} + 4 - 15\mu^2\right). \tag{5.50}$$

---

5 The convergence and completeness properties of these infinite series representations on a finite interval are nearly identical to those of the Fourier series.

In numerical solutions to equations such as Legendre's on the whole interval $[-1, 1]$, one must be careful to apply the boundary conditions such that a small contribution from the irregular solution does not enter, leading to a divergence at the poles. The first few $Q_n(\mu)$ are shown in Figure 5.4b.

5.7 Solution of the Linear Model with Constant Coefficients

First note that the series representation of $T(\mu)$ satisfies the polar boundary conditions term by term. The polar boundary conditions eliminate the $Q_n(\mu)$, which otherwise would have to be included.[6] This is important as it means that we can truncate the series at a finite level without fear of our approximation failing to satisfy the boundary conditions. Now insert the series into the energy balance equation to obtain

$$\sum_n \left[(Dn(n+1) + B)T_n P_n(\mu) + A\delta_{n0}\right] = QS(\mu)a(\mu). \tag{5.51}$$

Now multiply through by $P_m(\mu)$ and integrate from -1 to 1. After making use of the orthogonality relation, we obtain

$$\sum_n \left[(Dn(n+1) + B)T_n \frac{2\delta_{mn}}{2n+1} + 2A\delta_{n0}\delta_{m0}\right] = \frac{2}{2m+1} QH_m, \tag{5.52}$$

where

$$H_m = \frac{2m+1}{2} \int_{-1}^{1} S(\mu)a(\mu)P_m(\mu)d\mu, \tag{5.53}$$

are called the *heating components*. Because of the Kronecker deltas in the sum, we can find

$$T_n = \frac{QH_n - A\delta_{n0}}{n(n+1)D + B}. \tag{5.54}$$

This represents an analytical solution to the problem, as $T(\mu)$ can now be recomposed from the coefficients T_n using (5.45). Note that T_0 is the global average temperature as H_0 is the integral of $S(\mu)a(\mu)$. Figure 5.5 indicates symbolically the terms in the Fourier–Legendre expansion.

5.8 The Two-Mode Approximation

We can get a rough idea how well the model works by comparing with the Northern hemisphere zonally and annually averaged surface air temperature taken from Hartmann (1994). From the data, we find that

$$T_0 = 288 \text{ K (nominally } 15°\text{C)}, \tag{5.55}$$

$$\text{and } T_2 = -30 \text{ K}. \tag{5.56}$$

---

[6] For example, if only a broad latitude belt is to be considered, the irregular solutions $Q_n(\mu)$ would have to be included for completeness.

$T(\mu) = T_0($ 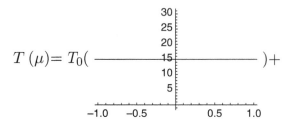 $) +$

$T_2($ 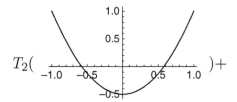 $) +$

$T_4($ 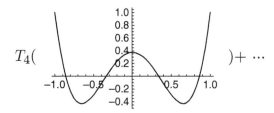 $) + \cdots$

Figure 5.5 The first three terms of an expansion of $T(\mu)$ into Legendre polynomials. This is a case where there is North–South symmetry.

We take the coalbedo from a fit to satellite data (Graves *et al.*, 1993): $a_0 = 0.68, a_2 = -0.24$; similarly, $A = 219$ W m^{-2} and $B = 2.00$ W m^{-2} K^{-1}. We can compute the mean annual solar insolation for the NH: $H_0 = 0.70, H_2 = -0.53, H_4 = 0.061$.

$$T(\mu) \approx T_0 + T_2 P_2(\mu), \tag{5.57}$$

with

$$T_0 = \frac{QH_0 - A}{B}, \tag{5.58}$$

$$T_2 = \frac{QH_2}{6D + B}, \tag{5.59}$$

and for comparison

$$T_4 = \frac{QH_4}{20D + B}. \tag{5.60}$$

T_0 is set by the global energy balance. We can compute D by solving (5.59), as we are given T_2 from the data. We find

$$D = \frac{1}{6}\left(\frac{QH_2}{T_2} - B\right) \approx 0.67 \text{ W m}^{-2} \, ^\circ\text{C}^{-1}. \tag{5.61}$$

This gives the desired parabolic dependence of $T(\mu)$ on μ.

We can calculate the dimensionless quantity

$$D/B \approx 0.34; \tag{5.62}$$

hence, using $R_e = 6368$ km,

$$\ell = \sqrt{\frac{D}{B}} R_e = 0.58 R_e \approx 3700 \text{ km} \tag{5.63}$$

is a fundamental length scale in the problem.

Figure 5.6 shows a graph of the two-mode solution just calculated:

$$T(\mu) \approx T_0 + T_2 P_2(\mu), \tag{5.64}$$

and on the same graph, the data are plotted as the dashed line. Including model-calculated $T_4 = 1.30$ K does not improve the fit. In fact, a best fit comes from a value closer to -4 K. Nevertheless, the fit in Figure 5.6 is quite good, considering the simplicity of the model, especially considering that the same value of $D = 0.67$ W m^{-2} K^{-1} is used in each hemisphere. The origins of the error at the T_4 level are ambiguous, as it could be attributed to a latitude dependence to either or several of D, A, or B. Introducing such dependencies would involve more adjustable parameters, and the choice of which to include or exclude would not be unique (for a discussion of this, see North, 1975b). For many purposes, it is probably best to leave the truncation at $n = 2$. A heuristic perusal of Figure 5.6 suggests that the errors in the tropics might be attributable to tropical clouds along the Intertropical Convergence Zone (ITCZ), or, less likely, bright subtropical deserts. The cold temperatures at the South pole might be attributable to the very high albedo and/or elevation of the surface.

The small values of T_4 from both the data and the constant-diffusion-coefficient model calculation indicate that the Fourier–Legendre series is converging rapidly. This makes

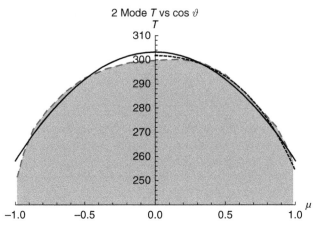

Figure 5.6 Illustration of the level of agreement at large spatial scales of the two-mode EBM with the observations. The solid line denotes the pole-to-pole solution of the two-mode model-computed temperature (K) versus μ, where $\mu \equiv \cos\vartheta = \sin(\text{latitude})$. The dashed curve indicates zonally averaged Northern Hemisphere surface air temperature taken from data in a graph in Hartmann (1994). The black-dashed curve shows the temperature curve with the T_4 (from Hartmann's data) mode added. Values for the temperature amplitudes are $T_0 = 288.3$ K, $T_2 = -30$ K, and $T_4 = -4$ K. The diffusive model yields a much smaller value of $T_4 = +1.30$ K. (Data from Hartmann (1994).)

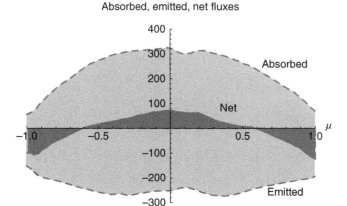

Figure 5.7 Radiation components for the zonal averages of the year-long average 2010 data (Loeb and Wielicki, 2014; Loeb et al., 2009).

sense considering the smooth parabolic-like zonal averages of the heating and cooling components.

The zonal average estimates for the year-long averages for the year 2010 of the radiation components: absorbed solar radiation, terrestrial infrared radiation flux to space, and the net radiation are shown in Figure 5.7. Note the dip in absorbed solar radiation at about $\mu \approx 0.1$–0.2 in the NH. This is attributable to the deep, bright convective clouds in the NH tropics where the ITCZ resides just north of the equator in the Pacific most of the year. There is a corresponding reduction in the outgoing infrared radiation flux density due to emission from the high cold cloud tops in the ITCZ. This is only for 1 year of observational data and other years may show slight differences.

As a further check on the two-mode approximation, consider the model's representation of the components just discussed: outgoing terrestrial radiation and solar absorbed radiation as a function of μ. These are shown in Figure 5.8. In Figure 5.8a, the absorbed solar radiation data (zonally and annually averaged for the year 2010) from (Loeb and Wielicki, personal communication, 2013) are shown as the dashed line. The solid curve is the two-mode approximation for the solar absorbed radiation energy, the latter makes use of the mean annual solar insolation $QS_{m.a.}(\mu)$ and the coalbedo $a(\mu)$ with the product expanded into LPs, retaining only terms for $n = 0, 2$. As already discussed, we can expect the simple model to fail in the neighborhood of $\mu \approx 0.1$ in the NH where the bright, deep convection of the ITCZ is seated north of the equator most of the year. From Figure 5.8, the absorbed flux fits well at the South pole, but the error discussed regarding Figure 5.6 might be more likely due to the infrared error (see Figure 5.8b).

Lindzen and Farrell (1977) argue that the flattened temperature curve (refer to Figure 5.6, long-dashed curve) seen in the tropics is because of increased heat transport in the tropics, basically an infinite diffusion coefficient leading to isothermal tropics. Such a model does fit the temperature profile better, but at the expense of adding an additional parameter, the latitude of the edge of the tropics. The latter is a piece of information we know rather well, so this approximation is attractive. In the presence of an interactive ice cap, this model has solutions that lower the value of solar brightness for which the ice-covered Earth occurs. We will return to this feature in Chapter 7.

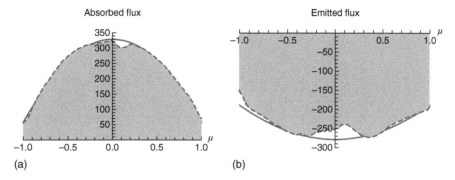

Figure 5.8 Fits of the two-mode model to the absorbed (a) and emitted (b) by the Earth–atmosphere system. The curved or dashed line are data from the year 2010 and the parabolic curves are based on the two-mode climate model. The vertical axes are energy fluxes in units of W m^{-2} and the abscissa (dimensionless) is cos ϑ = sin (latitude), where ϑ is the polar angle. Disagreement in tropical latitudes in (a, b) are clearly associated with clouds of the Intertropical Convergence Zone (ITCZ).

5.9 Poleward Transport of Heat

As suggested in the previous sections, the Earth–atmosphere system transports heat poleward because of the excess net radiation heating in the tropics and the deficiency in the poleward regions. In the simple diffusive model of this chapter, this is accomplished by the down-gradient flow of heat toward the poles. In the real world, the heat transport is accomplished by several factors. In the atmosphere, there is the overturning of the Hadley cell in the tropics; in the mid-latitudes, the strong equatorward thermal gradient leads to strong vertical shear (via the thermal wind condition), and hence large eddies (mid-latitude weather systems) generated by instabilities that we take to be simulated by the diffusive mechanism (at least in some kind of ensemble average). Poleward heat transport is more complicated in the oceans where the land–sea geography and bottom topography combine to make basin-wide gyres with swift currents capable of heat transport and that generate eddies that can move heat energy. In addition, the ocean currents are driven not only by diffusive eddies but by the sinking of cold saline surface waters in the North Atlantic sector. Sinking also occurs off the coasts of Antarctica (see Huang, 2009). Figure 5.9 shows estimates of the net annualized radiation at the top of the atmosphere (Trenberth and Caron, 2001). From the data in Figure 5.9, one can estimate the poleward flux (per unit of longitude) of heat by noting from the energy balance that

$$\text{div } F_\vartheta = R_{\text{net}} = F^\uparrow - F^\downarrow \quad \text{(data).} \tag{5.65}$$

The total poleward flux per unit length of circumference is

$$\frac{1}{\sin \vartheta} \frac{\text{d}}{\text{d}\vartheta}(\sin \vartheta F_\vartheta) = R_{\text{net}}; \tag{5.66}$$

$$\sin \vartheta F_\vartheta \big|_0^\vartheta = \int_0^\vartheta \sin \vartheta R_{\text{net}} \, \text{d}\vartheta; \tag{5.67}$$

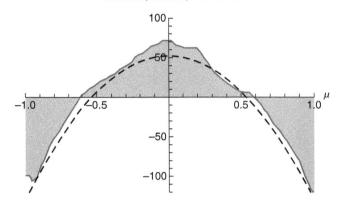

Figure 5.9 Estimates of net annualized radiation at the top of the atmosphere from satellite data 1984–1989. The two-mode model is represented by the dashed line while the solid gray line is from observational estimates. (Data from Trenberth and Caron (2001).)

$$F_\vartheta(\vartheta) = \frac{1}{\sin\vartheta} \int_0^\vartheta \sin\vartheta R_{net} \, d\vartheta. \tag{5.68}$$

The total poleward heat transport, F_H, crossing the entire latitude circle at ϑ is found by integrating around the latitude circle or, in the zonally symmetric case, by multiplying by the circumference of the latitude belt $2\pi R_e \sin\vartheta$:

$$F_H(\vartheta) = 2\pi R_e \int_0^\vartheta \sin\vartheta R_{net} \, d\vartheta. \tag{5.69}$$

Observed data for R_{net} as shown in Figure 5.7 can be inserted into the last formula (5.69). The result of the calculation is shown as the dashed curve in Figure 5.10. The solid line is the two-mode curve for the heat flux given by

$$F_{2mH} = 2\pi R_e D \cdot \left(\sqrt{1-\mu^2}\right)^2 \frac{dT}{d\mu} = 6\pi R_e D\mu \cdot (1-\mu^2)T_2, \tag{5.70}$$

where, in the last equation, we used $T(\mu) = T_0 + T_2 P_2(\mu)$, with $dT/d\mu = -3T_2\mu$.

Figure 5.10 suggests that the two-mode approximation provides a very good representation of the data for poleward heat transport in both hemispheres. This is particularly interesting considering that the heat is transported by three mechanisms: (i) transport by the world oceans, (ii) transport of sensible heat exchanged with the atmosphere and the surface, and (iii) transport of latent heat from evaporation at the surface and condensation at roughly the atmospheric boundary layer or above.

5.10 Budyko's Transport Model

In 1968, Budyko proposed an interesting form for the heat flux divergence. He assumed that heat would leave a latitude belt (per unit area per unit time) by an amount

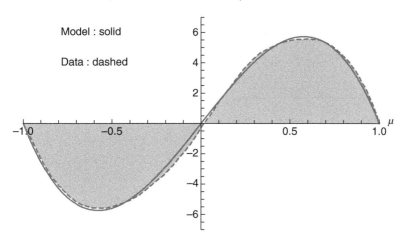

Figure 5.10 Illustration of the level of agreement in large-scale total poleward transport of heat from pole to pole for the two-mode EBM despite the huge differences in the geography of the two hemispheres. Poleward transport of heat derived from radiation budget data of Trenberth and Caron (2001) based on 4 years of ERBE data (1984–1988). The dashed curve represents the transport derived from the satellite observations and the solid line is based on the two-mode approximation to the surface temperature field with diffusive transport. The thermal diffusion coefficient is $D = 0.67\text{W m}^{-2}\,(°C)^{-1}$. (Trenberth and Caron (2001). ©American Meteorological Society. Used with permission.)

proportional to the difference between the local temperature of the strip and the global average

$$\gamma(T(\mu) - T_0), \tag{5.71}$$

instead of the diffusion form. If we compare Budyko's model in two-mode form with the two-mode approximation to the diffusive model, we find that they are identical as North (1975b) pointed out:

$$-D\frac{d}{d\mu}(1-\mu^2)\frac{d}{d\mu}[T_0 + T_2 P_2(\mu)] = 6DT_2 P_2 = 6D[T(\mu) - T_0]. \tag{5.72}$$

In other words, $\gamma = 6D$ in the two-mode approximation, and the two transport parameterizations are equivalent. It is also possible to show that if D has a latitude dependence, it is equivalent to the constant D case in the two-mode approximation (North, 1975b).

Budyko's model is interesting because it leads to an algebraic EBM as compared to the differential equation generated by the diffusive model. This makes the model very simple to solve (algebraically!). It can be stated that

$$\gamma(T(\mu) - T_0) + A + BT(\mu) = QS(\mu)a(\mu). \tag{5.73}$$

Integrating on μ from -1 to 1 leads to the same equation as in Chapter 2 for the global average temperature T_0. Then one can easily solve for $T(\mu)$, giving us the equivalent of the two-mode approximation for the latitude dependence. Differences arise as one allows $a(\mu)$ to have a discontinuity at say an ice sheet edge. Diffusion handles such situations nicely by a smoothing effect at such a discontinuity, but the Budyko model gives a discontinuity in the temperature field proportional to the ice sheet discontinuity in $a(\mu)$. The next section shows how the diffusive model handles a ring-of-heat source around a latitude circle. The Budyko model would handle it quite differently.

5.11 Ring Heat Source

If a ring of heat source is distributed around a latitude circle, the temperature will be elevated in the latitudinal band covering the line source. This is an easy problem to solve using LPs (Salmun et al., 1980). The anomaly in temperature is governed by the energy balance:

$$-D\frac{d}{d\mu}(1-\mu^2)\frac{d}{d\mu}G(\mu,\mu_0) + BG(\mu,\mu_0) = \delta(\mu-\mu_0). \tag{5.74}$$

We can expand

$$\delta(\mu-\mu_0) = \sum_{n=0}^{\infty} \frac{2n+1}{2} P_n(\mu)P_n(\mu_0), \tag{5.75}$$

and we quickly arrive at

$$G(\mu,\mu_0) = \sum_{n=0}^{\infty} \left(\frac{2n+1}{2}\right) \frac{P_n(\mu)P_n(\mu_0)}{n(n+1)D+B}. \tag{5.76}$$

We can extend our analysis to a general heat source distribution, $h(\mu)$, by expanding as follows:

$$h(\mu) = \sum_{n=0}^{\infty} h_n P_n(\mu), \tag{5.77}$$

so that

$$\Delta T_h(\mu) = \sum_{n=0}^{\infty} \frac{h_n P_n(\mu)}{n(n+1)D+B} \tag{5.78}$$

$$= \sum_{n=0}^{\infty} \int_{-1}^{1} d\mu' h(\mu')P_n(\mu') \left(\frac{2n+1}{2}\right) \frac{P_n(\mu)}{n(n+1)D+B} \tag{5.79}$$

$$= \int_{-1}^{1} G(\mu,\mu')h(\mu')d\mu'. \tag{5.80}$$

Hence, the response to a steady general latitudinal distribution of heat source is the integral of the function $h(\mu')$ weighted by the function $G(\mu,\mu')$. It is therefore of interest to know $G(\mu,\mu')$, which is an example of a *Green's function*. An example of the response to a ring heat source is shown in Figure 5.11. The solid gray line is for the series representing $G(\mu,\mu')$ to be truncated at $n=300$, while the gray-dashed curve represents a more severe truncation at $n=30$. This is typical of Fourier as well as Fourier–Legendre series where many terms are required to fit a curve snugly in the neighborhood of a point where there is a discontinuity or a discontinuous derivative (as in this case). Green's functions

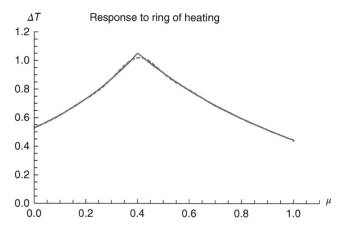

Figure 5.11 Temperature responses to a heat source consisting of a latitudinal ring whose strength is 1 W m^{-2} (in the Northern Hemisphere only; odd terms are retained). Shown are responses to a ring located at $\mu = 0.4$. The solid line is the solution for $n = 300$ for which a cusp at the heat source is shown at the delta function heat source, $\mu_0 = 0.4$. The solution for $n = 30$ is also shown as the dashed line indicating that the convergence for this perturbation is very slow.

are discussed in almost any book on mathematical physics or engineering (e.g., Arfken and Weber, 2005).

5.12 Advanced Topic: Formal Solution for More General Transports

The simple case of constant diffusion is not likely to hold, as we will see later, because transport mechanisms vary by latitude on the Earth. Consider the general EBM defined by

$$\mathcal{L}T(\mu) + A = QS(\mu)a(\mu), \tag{5.81}$$

where $\mathcal{L}$ is a linear operator with certain desirable properties that will come out below. If the operator is a differential operator, some boundary conditions may be required as well. A linear operator is one that has the following property:

$$\mathcal{L}(\alpha h(\mu) + \beta g(\mu)) = \alpha \mathcal{L}h(\mu) + \beta \mathcal{L}g(\mu), \tag{5.82}$$

where α and β are arbitrary constants. We can ask about the eigenfunctions and eigenvalues of $\mathcal{L}$, which are defined by

$$\mathcal{L}f_n = \lambda_n f_n, \tag{5.83}$$

plus boundary conditions if necessary. We will require that the eigenvalues λ_n be discrete, real, and positive.[7] These requirements impose constraints on the type of

---

[7] This will be true if the operator $\mathcal{L}$ is symmetric (or, in the complex case, Hermitian). Symmetric here means that $(f, \mathcal{L}g) = (g, \mathcal{L}f)$ with f, g, and $\mathcal{L}$ real.

operators $\mathcal{L}$ that qualify. When these conditions are met, we can enjoy the use of expansions such as

$$T(\mu) = \sum_n T_n f_n(\mu), \tag{5.84}$$

as it can be shown that the $f_n(\mu)$ are orthogonal

$$\int f_n(\mu) f_m(\mu) \mathrm{d}\mu = 0, \quad \text{if} \quad m \neq n. \tag{5.85}$$

In fact, the functions can be normalized so that they are also *orthonormal*.

$$\int f_n(\mu) f_m(\mu) \mathrm{d}\mu = \delta_{mn}. \tag{5.86}$$

The model will be solved by inserting the series into the energy balance equation and finding

$$T_n = \frac{Q h_n - A_n}{\lambda_n}, \tag{5.87}$$

where

$$h_n = \int S(\mu) a(\mu) f_n(\mu) \, \mathrm{d}\mu, \tag{5.88}$$

and similarly for A_n.

Generally, it is expected that the modes will not look very different from the LPs. For example, a simple case such as the smoothly varying diffusion coefficient is expected to produce eigenfunctions that merely distort the locations of zeros from the usual locations for LPs.

5.13 Ice Feedback in the Two-Mode Model

Ice feedback is especially important in paleoclimate problems where large ice sheets exist and their growth has a substantial effect on sensitivity as already noted in the 0-D model. In this section, we consider the two-mode diffusive model. We do this by maintaining zonal symmetry and retaining only two Legendre modes. The edge of the zonally symmetric ice cap is denoted by μ_s and it is hooked to the $-10°$ (mean annual) isotherm. In other words,

$$T(\mu_s) = T_s = -10° \text{ C}; \quad \text{ice cap edge condition.} \tag{5.89}$$

In the two-mode model this constraint is expressed as

$$T_s \equiv T_0 + T_2 P_2(\mu_s), \tag{5.90}$$

where

$$T_0 = \frac{Q h_0(\mu_s) - A}{B}, \tag{5.91}$$

and

$$T_2 = \frac{Q h_2(\mu_s)}{6D + B}. \tag{5.92}$$

Since we are considering the sensitivity of the climate to GHG concentrations, we take the control variable to be A in the outgoing radiation rule: $I = A + BT$. In the case of carbon dioxide, the relation is

$$\Delta A = -5.3 \text{ (W m}^{-2})\Delta(\ln [CO_2]). \tag{5.93}$$

If we change the concentration of GHGs, we change the latitude of the ice cap edge in response, owing to the change in the latitude distribution of temperature. Thus we can think of μ_s to be a function of the control variable A. In terms of solving the problem, it is better yet to reverse the independent versus dependent variables and think of $A(\mu_s)$ as the dependent variable, letting μ_s be the control variable in the algebra to follow, then re-inverting to $\mu_s(A)$.

In changing the GHG concentration, we hold the total solar irradiance Q, constant. We can write the latitude dependence of $T(\mu)$ in the two-mode model as

$$T(\mu) = \frac{Qh_0(\mu_s)}{B} - \frac{A(\mu_s)}{B} + \frac{Qh_2(\mu_s)P_2(\mu)}{6D + B}. \tag{5.94}$$

By evaluating $T(\mu)$ at $\mu = \mu_s$ and using (5.89), we obtain a formula for $A(\mu_s)$ explicitly.

$$A(\mu_s) = Qh_0 - BT_s + \frac{B}{6D + B}Qh_2P_2, \tag{5.95}$$

where we have suppressed the arguments of h_0, h_2, and P_2, each of which is to be evaluated at μ_s. The sensitivity can be obtained from

$$\frac{dT_0}{dA} = -\frac{1}{B}\left(1 - Q\frac{h'_0}{A'}\right), \tag{5.96}$$

where the primes indicate derivative with respect to μ_s.

For doubling CO_2 with no feedbacks, the change in T_0 is

$$(\Delta T_0)_{f=0} = -\frac{\Delta A}{B_0}, \tag{5.97}$$

where $B_0 = 4.61\text{W m}^{-2}\text{K}^{-1}$, the blackbody damping coefficient. Using this in (5.96) and rearranging, we have

$$(\Delta T_0)_{2\times} = (\Delta T_0)_{f=0}\left(1 - Q\frac{h'_0}{A'}\right). \tag{5.98}$$

If the second term in parentheses is small, we have

$$(\Delta T_0)_{2\times} \approx \frac{(\Delta T_0)_{f=0}}{1 + Q\frac{h'_0}{A'}}, \tag{5.99}$$

which means $f_{\text{ice}} \approx -Q\frac{h'_0}{A'}$. Figure 5.12 shows plots of $A(\mu_s)$ (a) and f_{ice} (b) for the case of $B = B_0 = 4.61 \text{ W m}^{-2}\text{K}^{-1}$. This latter gives us the value of f_{ice} in the presence of no other feedbacks. As f_{ice} crosses the dashed horizontal line denoting $f_{\text{ice}} = 1$ in Figure 5.12b and the maximum in Figure 5.12a, there is a transition to an ice-covered planet, as discussed for the 0-D model. Note that, as we would expect, the feedback factor for ice caps increases as the ice cap gets larger. The value of f_{ice} in the range of $0.8 \leq \mu_s \leq 1.0$ is from 0.26 to 0.15.

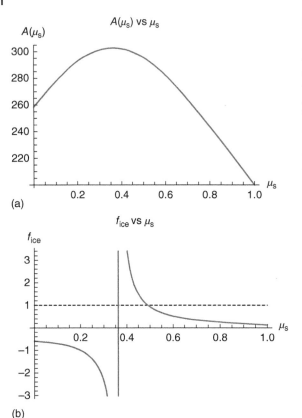

Figure 5.12 Illustration of the extreme behavior of climate sensitivity to the location of the ice cap edge, μ_s. In fact, the feedback factor f_{ice} exceeds unity (catastrophe) when the ice cap crosses the half-covered threshold. (a) A plot of $A(\mu_s)$ and (b) a plot of the ice cap feedback factor as a function of μ_s, the sine of the latitude of the ice cap edge. Both plots are for the radiation damping coefficient corresponding to a blackbody, $B_0 = 4.61$ W m^{-2} K^{-1}.

5.14 Polar Amplification through Ice Cap Feedback

Climate data over the last half century indicate that the northern polar regions are warming faster than the middle latitudes as well as the tropics. We explore here what happens in the two-mode energy balance climate model (EBCM). First, it is clear that if the warming is caused by a perturbation of A in the outgoing radiation formula, and, if A and D, the thermal diffusion, have no dependence on latitude, the response of the temperature is confined to that of the forcing, namely, the global average mode. This latter result comes from examination of (5.91) and (5.92) with μ_s held fixed and noting that A does not occur in (5.92). This means there is no polar amplification in this linear EBCM. Polar enhancement would be indicated in the second Legendre mode amplitude T_2. Recall that $P_2(\mu) = \frac{1}{2}(3\mu^2 - 1)$ determines the equator-to-pole temperature difference as well as the curvature of the zonally averaged temperature as a function of sine of the latitude.

But the T_2 mode can be affected by the ice cap feedback via the change in μ_s in (5.92). From that equation, we find

$$\delta T_2 = \frac{Q}{6D + B} \frac{dh_2}{d\mu_s} \frac{d\mu_s}{dA} \delta A. \tag{5.100}$$

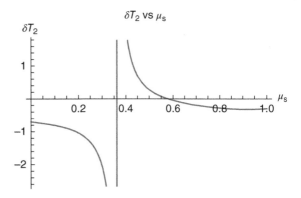

Figure 5.13 Change in mode two amplitude T_2 for a doubling of CO_2 ($\delta A = 4.0\,\text{W m}^{-2}$) for a blackbody Earth (i.e., the damping coefficient of the outgoing radiation is $B_0 = 4.61\,\text{W m}^{-2}\,°\text{C}^{-1}$. Note that $\mu_s = 1$, corresponds to no ice cap. If water vapor feedback is used (the value of $B = 2.00\,\text{W m}^{-2}\,°\text{C}^{-1}$), the magnitude of the value of δT_2 increases about a factor of 3 to about $-1.0\,\text{K}$.

The nominal value of T_2 for equilibrium is of the order of $-30\,°\text{C}$. Hence, for the chosen parameters for coalbedo over ice and ice-free areas, the perturbation in T_2 is rather small ($\sim -0.32\,\text{K}$) according to Figure 5.13. On the other hand, if we include water vapor feedback (i.e., using $B = 2.00\,\text{W m}^{-2}\,\text{K}^{-1}$), this perturbation is about a factor of 3 larger, $\delta T_2 \approx -1.0\,\text{K}$. Hence, if water vapor feedback is engaged simultaneously the effect of ice feedback is three times as large. As discussed in Chapter 4, the inclusion of more feedbacks is not strictly additive but nonlinear through the following formula:

$$(\Delta T)_{\text{w/feedback}} = \frac{(\Delta T)_{\text{no feedback}}}{1 - f_{\text{wv}} - f_{\text{ice}}}. \tag{5.101}$$

We will return to ice cap feedback in Chapter 8, where we will solve the full nonlinear 1-D model analytically.

5.15 Chapter Summary

The zonally symmetric, steady-state, latitude-dependent EBCM can be expressed mathematically as a second-order ordinary differential equation subject to boundary values of zero heat flux density into the poles. The single dependent variable is the zonally averaged surface temperature field $T(\mu)$ where μ is the sine of the latitude or cosine of the polar angle. On the range of latitudes, the surface temperature is assumed to be governed by a damping term $(BT(\mu))$ due to outgoing radiation flux density, a diffusive transport mechanism, and a source or forcing term on the right-hand side, all of which amount to a nonhomogeneous, ordinary second-order differential equation. In addition, the source term is given by the mean annual solar insolation function of the latitude, which is modulated by the coalbedo (or energy flux density of solar absorption).

We were able to find a general solution to the problem for $T(\mu)$ as posed in the form of an infinite series of LPs, $P_n(\mu)$. But it turns out that a truncation of the series including only the $P_0(\mu) = 1$ and $P_2(\mu) = \frac{1}{2}(3\mu^2 - 1)$ modes provides a remarkably good fit to the zonally averaged mean annual data. The relevant coefficients turned out to be $T_0 = 288\,\text{K}$ ($15\,°\text{C}$) and $T_2 = -30\,°\text{C}$, which lead to satisfactory fits to the data – at least good enough considering the level of our model's realism regarding transport of heat, and so on. In order to obtain T_2, we needed to adjust the diffusion parameter to a value of

0.67 W m^{-2} °C^{-1}. This last operation amounts to tuning the curvature of the parabolic fit of the model's temperature to that of the observations as a function of μ.

Once we have the two-mode model, we can calculate various quantities that serve as a partial check on our assumptions. For example, it is no surprise that the modeled outgoing radiation fits rather well except in the ITCZ, where a narrow band of clouds departs from parabolicity. Likewise, the parabola fits the absorbed solar radiation tolerably well. It is very interesting that the poleward heat fluxes in both hemispheres can be derived from the satellite data and fit almost perfectly to the two-mode model with the same value of D in both hemispheres.

Because the LPs are the eigenfunctions of the damped diffusion operator (left-hand side of the energy balance equation), inhomogeneous terms (those on the right-hand side) will only excite a response in the same modes (eigenfunctions) as the modes exhibited in the forcing terms. We found that with a ring of heat around the planet, many higher modes (values of the index n) are needed to represent the response because of the large number of modes required to represent such latitudinally narrow features in the heating function.

5.15.1 Parameter Count

As we increase the dimension of our climate model, we always run into the need for new parameters. In this chapter, we encountered the thermal diffusion coefficient, D. This allowed us to fit the surface temperature versus latitude rather well, except possibly in the tropics, where we argue the discrepancy is with clouds in the ITCZ. With this single new parameter, we were able to predict the latitudinal form of the poleward heat transport in each hemisphere.

Notes for Further Reading

Besides the articles by Budyko (1969) and Sellers (1969), the review article by Schneider and Dickinson (1974) and the paper by Chýlek and Coakley (1975) induced one of the authors of this book (GRN) into climate science with North (1975a, 1975b). Virtually simultaneously and independently came the papers by Held and Suarez (1974) and Ghil (1976). The review article by North *et al.* (1981) covers much of the early science of EBMs. The articles in the volume edited by Archer and Pierrehumbert (2010) provide further historical material.

Exercises

5.1 Assuming that the solar constant does not change in time and is given by $\sigma_\odot = 1360$ W m^{-2}, calculate the total amount of insolation reaching the zonal band of width $d\vartheta$ centered at latitude ϑ at any given time t. Assume the equinox condition with the Sun directly above the equator. On the basis of your answer, show that the total amount of insolation received by the Earth is $\pi R_E^2 \sigma_\odot$.

5.2 On the basis of your calculation in Exercise 5.1, calculate the equinox distribution of solar radiation as a function of $\mu = \cos \vartheta$.

5.3 Modify Exercise 5.2 to address the insolation distribution function when the Sun's declination angle is δ.

5.4 Suggest how the mean annual insolation distribution function can be determined by using the insolation distribution function obtained in Exercise 5.3. Calculate the mean annual insolation distribution numerically by assuming the declination angle varies uniformly between $\delta \in [-23.5°, 23.5°]$ throughout the year.

5.5 Show that the mean-annual and equinox insolation distribution functions as given by
$$S_{\text{m.a.}}(\mu) = \frac{1}{4}(5 - 3\mu^2) \quad \text{and} \quad S_{\text{eqnox}} = \frac{4}{\pi}\sqrt{1 - \mu^2}$$
yield 1 (unity) when averaged over $\mu \in [-1, 1]$.

5.6 Given the annual mean albedo in the form
$$a_{\text{m.a.}}(\mu) = 0.68 - 0.10(3\mu^2 - 1),$$
calculate the planetary albedo a_{p}.

5.7 The LPs can be determined by using the recursion formula
$$(n + 1)P_{n+1}(x) = (2n + 1)xP_n(x) - nP_{n-1}(x).$$
Starting from $P_0(x) = 1$ and $P_1(x) = x$, determine the LPs up to order $n = 5$. Determine the general form for the eigenvalues λ_n.

5.8 Let us assume that the LP of order n can be defined as
$$P_n(\mu) = \sum_{m=0}^{\infty} c_m^{(n)} \mu^m,$$
where the LP satisfies the second-order differential equation
$$-\frac{d}{d\mu}\left((1 - \mu^2)\frac{dP_n(\mu)}{d\mu}\right) = \lambda_n P_n(\mu)$$
together with the boundary condition
$$\sqrt{1 - \mu^2}\frac{dP_n(\mu)}{d\mu}\bigg|_{\mu=\pm 1} = 0.$$
Show that the expansion coefficients satisfy
$$[(k + 1)(k + 2)]c_{k+2}^{(n)} - [k(k + 1) - \lambda_n]c_k^{(n)} = 0.$$

5.9 Using the result in Exercise 5.7, verify that the expansion coefficients $c_m^{(n)}$ in
$$P_n(\mu) = \sum_{m=0}^{\infty} c_m^{(n)} \mu^m$$
satisfy the following:
$$[(m + 1)(m + 2)]c_{m+2}^{(n)} - [m(m + 1) - \lambda_n]c_m^{(n)}.$$

5 Latitude Dependence

5.10 Show that for $n = 0, \ldots, 4$ that
$$\int_{-1}^{1} P_n^2(x)\, dx = \frac{2}{2n+1}.$$

5.11 Let us assume that the mean-annual insolation distribution and coalbedo are given as functions of μ (sine of latitude) as
$$S_{\text{m.a.}}(\mu) = \frac{1}{4}(5 - 3\mu^2), \quad \text{and} \quad a_{\text{m.a.}}(\mu) = 0.68 - 0.12(3\mu^2 - 1).$$
Write the annually averaged coalbedo-weighted insolation distribution function in terms of the Legendre polynomials, i.e.,
$$S_{\text{m.a.}}(\mu)\, a_{\text{m.a.}}(\mu) = \sum_n c_n P_n(\mu).$$

5.12 Using the results in Problem 11, solve the 1-D EBM with a constant diffusion coefficient ($D = 0.67\ \text{W m}^{-2}\ \text{K}^{-1}$):
$$A + BT(\mu) - \frac{d}{d\mu}\left(D(1 - \mu^2)\frac{dT(\mu)}{d\mu}\right) = QS(\mu)a(\mu), \quad Q = \frac{\sigma_\odot}{4}.$$

5.13 Let us consider the ring heat source problem
$$-D\frac{d}{d\mu}\left((1 - \mu^2)\frac{d}{d\mu}\right)T'(\mu + BT'(\mu)) = h(\mu),$$
where $h(\mu)$ is the heat source as a function of μ (sine of latitude) and the prime indicates that $T'(\mu)$ is departure from the equilibrium. This problem is often solved by using the so-called *Green's function technique*. Green's function $G(\mu, v)$ satisfies the differential equation
$$-\frac{d}{d\mu}\left(D(1 - \mu^2)\frac{dG(\mu, v)}{d\mu}\right) + BG(\mu, v) = \delta(\mu - v).$$
Show then that the solution of the heat source problem is given by
$$T'(\mu) = \int_{-1}^{1} G(\mu, v) h(v)\, dv.$$

5.14 The ice-feedback model can be written as
$$A(\mu_s) = Qh_0(\mu_s) - B(T_0 + T_2 P_2(\mu_s)) + \frac{B}{6D + B} Qh_2(\mu_s) P_2(\mu_s),$$
where h_0 is the latitude of the ice boundary, $h_0(\mu_s)$ and $h_2(\mu_s)$ are the expansion coefficients for $S_{\text{m.a.}}(\mu_s)$ evaluated at the latitude μ_s, and $A(\mu_s)$ is the temperature-independent component of the longwave radiation to space. Show that
$$\frac{dT_0}{dA} = -\frac{1}{B}\left(1 - Q\frac{h'_0}{A'}\right), \quad h'_0 = dh_0/d\mu_s, \quad A' = dA/d\mu_s.$$

6

Time Dependence in the 1-D Models

In Chapter 5, we found that time-independent, zonally symmetric models can be cast into the form of a linear differential equation in the latitude variable with zero-flux boundary conditions at the poles. In the case of a latitude-independent thermal diffusion coefficient, the system can be solved with Legendre polynomials. In this chapter, we extend this procedure to the case of time dependence. Such an extension allows for idealized seasonal-cycle simulations for the zonally symmetric planet. Introducing time dependence necessarily requires some kind of effective heat capacity for the column of air–land or air–ocean medium. When we take the planet to have a homogeneous surface and constant diffusion coefficient, we also specify that the heat capacity is not spatially dependent, that is, it is a constant. An all-land planet is one for which the heat capacity is characteristic of a land surface. This means it takes a value calculated from a fraction (nominally half) of the mass of a column of air. This leads to a time constant C/B of about 1 month. This is our new phenomenological coefficient for Chapter 6.

The time-dependent models are amenable to analytical solution for uniform planets (constant D and C). Model solutions are in the form again of Legendre polynomial modes, each having a characteristic decay time with larger scales (low Legendre index) having longer relaxation times, and smaller scales, shorter times. The seasonal cycle of latitude-dependent insolation has a very simple form when truncated at the few-mode level. Legendre polynomials of index 0, 1, and 2 describe the latitude dependence of the seasonal cycle of insolation, except, of course, for the discontinuous derivative near the poles, which is necessary to describe the onset of perpetual night or day and the appropriate latitudes and times of year. This insolation formula allows us to obtain very simple expressions for the dependence of insolation on obliquity, precession of the equinoxes, and eccentricity. The seasonal cycle simulations also give us an idea about the lag of seasonal response to the driving seasonal heating cycle.

Later in the chapter, we return to the stochastic nature of climate forcing (presumably due to weather instabilities) by showing how the random walk of heated parcels is approximately imitated by diffusion in an ensemble average. This analysis relates the mean square of the parcel's velocity and relaxation time to the diffusion coefficient of the ensemble.

A short section is added to show how one might go about solving one-dimensional problems via finite difference methods. The appendices to the chapter give derivations of the insolation functions.

6.1 Differential Equation for Time Dependence

Consider the problem of a time-dependent one-dimensional climate system for a planet whose surface is uniformly land or ocean. The energy balance model for constant coefficients is as follows:

$$C\frac{\partial T}{\partial t} - D\frac{\partial}{\partial \mu}(1-\mu^2)\frac{\partial T}{\partial \mu} + A + BT = QS(\mu, t)a(\mu, t), \tag{6.1}$$

where μ is the sin(latitude), t is time, $T(\mu, t)$ is the latitude and time-dependent temperature field, $S(\mu, t)$ is the insolation function possibly allowing for a seasonal cycle, $a(\mu, t)$ is the coalbedo, which is a function of latitude and possibly time dependent; Q is the total solar irradiance divided by $4 = \sigma_\odot/4$, A, B, and D are phenomenological coefficients as defined in Chapter 5, and C is an effective heat capacity. As discussed in Chapter 2, this means a timescale of about 30 days ($\tau_0 = C/B$) for an all-land planet and a few years for a planet covered with a mixed-layer ocean. This amounts to using the heat capacity of half a column of air (at constant pressure) for the all-land case and that of 50–100 m of water for the mixed-layer ocean case. In the latter, we assume the ocean below is uncoupled with the mixed layer.

We expand the time-dependent temperature field into Legendre polynomials:

$$T(\mu, t) = \sum_n T_n(t) P_n(\mu). \tag{6.2}$$

Note that each coefficient $T_n(t)$ is a function of t. After inserting the series, multiplying through by $P_m(\mu)$ and integrating as before we obtain an ordinary differential equation for the $T_n(t)$:

$$C\frac{dT_n(t)}{dt} + \delta_{n0} A + (n(n+1)D + B)T_n(t) = QH_n, \tag{6.3}$$

where

$$H_n(t) = \frac{2n+1}{2} \int_{-1}^{1} S(\mu, t) a(\mu, t) P_n(\mu)\, d\mu. \tag{6.4}$$

It is important to note that none of the terms in the array indicated by (6.3) is coupled to any of the others. In other words, for the equation dependent on mode number n there is no other mode index, say $m \neq n$, that is referred to in the equation indexed by n.

6.2 Decay of Anomalies

Consider first the case where $S(\mu, t)$ and $a(\mu, t)$ are set at their mean annual values as in Chapter 5. We can examine the behavior of the climate as it is perturbed away from

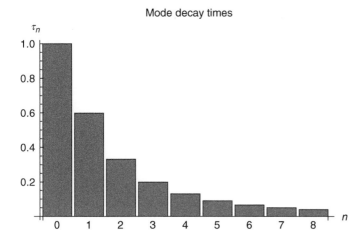

Figure 6.1 This figure illustrates that large spatial scales (low Legendre index) have longer time constants than for small spatial scales. Shown is a bar chart with relaxation times for various Legendre modes for the homogeneous planet with n in terms of $\tau_0 = C/B = 1$ month. In this case, $D = 0.67\,\text{W\,m}^{-2}\,°\text{C}$, $B = 2.00\,\text{W\,m}^{-2}\,°\text{C}$, so that $D/B = 0.34$ (dimensionless).

steady state. We have the solution to the initial-value problem:

$$T_n(t) = T_n^{(\text{ss})} + (T_n(0) - T_n^{(\text{ss})})e^{-t/\tau_n}, \tag{6.5}$$

where $T_n^{(\text{ss})}$ is the solution to the time-independent (steady-state) problem given above and

$$\tau_n = \frac{C}{n(n+1)D + B} = \frac{\tau_0}{n(n+1)(D/B) + 1}. \tag{6.6}$$

Each mode $T_n(t)$ has its own characteristic time τ_n. Note that the characteristic times fall off rapidly as a function of n as shown in Figure 6.1. This agrees with our intuition that larger spatial scales have longer characteristic (adjustment) times. We will establish a similar result for the autocorrelation times later in noise-driven models.

6.2.1 Decay of an Arbitrary Anomaly

An arbitrary distribution of thermal anomaly will decay in time according to a superposition of the modes just discussed. For example, an initial distribution of anomaly (departure from equilibrium) whose shape is $T'(\mu, 0)$ will decay according to

$$T'(\mu, t) = \sum_{n=0}^{\infty} T'_n(0) P_n(\mu) e^{-t/\tau_n}. \tag{6.7}$$

For example, consider an initial shape rendered by a steady ring of heat source (Chapter 5) at a specific latitude as given by (5.11) and shown in Figure 5.11. The solid line in Figure 6.2 shows such a steady-state anomaly located at $\mu = 0.6$ (37°N). If suddenly the heat source is switched off, the distribution will decay according to (6.7). The figure shows stages of the decay at $0.25\tau_0$, $0.5\tau_0$, $0.75\tau_0$, and $1.00\tau_0$. We see that the higher modes, those rendering the cusp-like peak, are lost quickly leaving behind only the smoother long-lived modes.

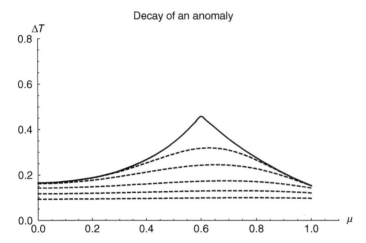

Figure 6.2 Decay of a symmetric (between the hemispheres) initial distribution (solid line) after times $0.25\tau_0$, $0.5\tau_0$, $0.75\tau_0$ and τ_0 (dashed lines). The model parameters are as earlier and the sum is truncated at $n = 200$. The steady-state response has a discontinuous derivative at the ring's latitude. Its Fourier–Legendre series is very slowly converging, and this indicates that much power resides in higher-mode indices. These modes decay quickly leaving behind a smoother latitudinal profile as the temperature relaxes toward the old steady-state solution (zero here).

6.3 Seasonal Cycle on a Homogeneous Planet

We can begin the study of the seasonal cycle by placing the solar heating distribution as a forcing on the right-hand side of the energy-balance equation. The heating distribution can be written for a circular orbit as follows:

$$S(\mu, t) \approx 1 + S_{11} \cos 2\pi t P_1(\mu) + (S_{20} + S_{22} \cos 4\pi t) P_2(\mu), \tag{6.8}$$

where $S_{11} = -0.797$, $S_{20} = -0.477$, and $S_{22} = 0.147$ (see Figure 6.3). The subscripts denote the Legendre index first, then seasonal time harmonic second. A derivation for this heating distribution function is sketched in the appendix to this chapter. But for now we can observe some important properties. The mean annual version can be quickly recovered by averaging from 0 to 1 year in t,

$$S_{\text{ma}} = 1 + S_{20} P_2(\mu), \tag{6.9}$$

which is the formula used earlier in Chapter 5. The main part of the seasonal cycle is carried by the coefficient of the $P_1(\mu)$ term. Note that it is antisymmetric between the hemispheres (recall the identity $P_1(-\mu) = -P_1(\mu)$). The seasonal forcing is largest near the poles ($P_1(\pm 1) = \pm 1$). Another interesting feature captured by this representation is that along the equator ($\mu = 0$) there are two maxima (see Figure 6.3). This is the effect of the Sun crossing the equator twice each year at the equinoxes. Table A.6.1 shows the coefficients for some higher terms in the expansion for the present elliptical orbit of the Earth. Figure 6.4 shows the seasonal insolation when only modes 0, 1, and 2 are included. This figure indicates that retention of only these three modes captures the most important features of the insolation. Figure 6.5 shows how in the tropics there are two peaks as one passes through the year. The figure shows that for $-0.1 \le \mu \le 0.1$ the

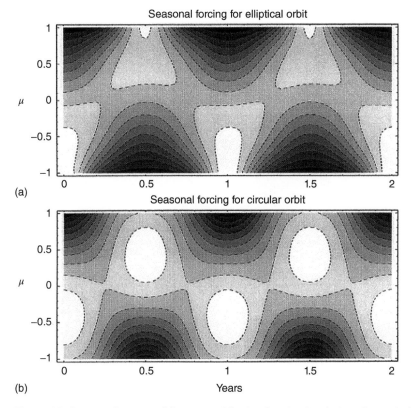

Figure 6.3 Contour diagrams of the seasonal forcing $S(\mu, t)$ or insolation. The vertical axis is cosine of colatitude, μ; the horizontal axis is time, t in years and $t = 0$ corresponds to summer solstice for the Northern Hemisphere. The units of the contour plot is for the present elliptical orbit forcing through the Legendre mode 4 and time harmonic 2. The seasonal cycle of $S(\mu, t)$ for the present elliptical orbit with eccentricity 0.016 (a) and the seasonal cycle for the same heating, but for a circular orbit (b). The value of time equal to zero in the Northern Hemisphere winter solstice.

semiannual harmonic is strong. These three modes are able even to capture the passage of the Sun over the equator twice per year. Away from the equator, the seasonal harmonic dominates. Figure 6.6 shows latitudinal time sections of the forcing. The dashed lines show the forcing for zero eccentricity.

We can express the seasonal dependence of insolation as

$$S(\mu, t) = \sum_{n=0}^{N} \sum_{k=0}^{K} [a_{nk} \cos 2\pi kt + b_{nk} \sin 2\pi kt] P_n(\mu). \tag{6.10}$$

If we take the coalbedo, diffusion coefficient, and heat capacity to be constants independent of season and latitude, we can write equations for the mode responses:

$$C\frac{d}{dt}T_n(t) + [Dn(n+1) + B]T_n(t)$$
$$= Qa_0 \sum_{k=0}^{2} [a_{nk} \cos 2\pi kt + b_{nk} \sin 2\pi kt], \tag{6.11}$$

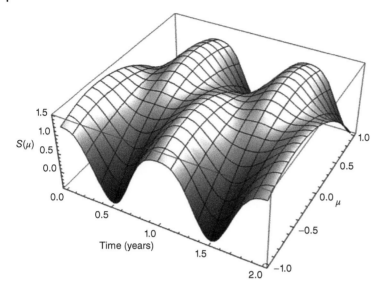

Figure 6.4 Seasonal cycle of $S(\mu, t)$ for the present orbital obliquity δ when only the Legendre modes 0, 1, and 2 are retained along with mean annual, annual harmonic, and semiannual harmonic. The time span is over 2 years in units of years with the origin at northern winter solstice. The latitudinal coordinate μ runs from the South pole ($\mu = -1$) to the North Pole ($\mu = +1$). The semiannual harmonic captures the passage of the Sun over the equator twice a year in this image.

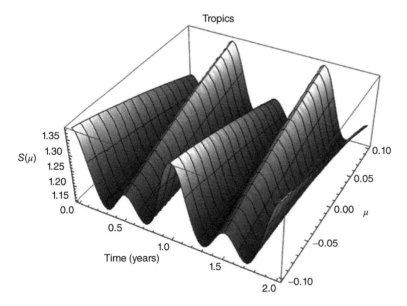

Figure 6.5 Seasonal cycle of $S(\mu, t)$ in the tropics ($\pm 5.7°$) for the present orbital obliquity δ when only the Legendre modes 0, 1, and 2 are retained along with mean annual, annual harmonic, and semiannual harmonic. This graphic shows the tropical variation with two maxima per year caused by the S_{22} term. Note the change in vertical scale from the previous figure. The time span is over 2 years in units of years with the origin at northern winter solstice. Note that the details of polar day and night are missed in this truncation, but the semiannual harmonic captures the passage of the Sun over the equator twice a year.

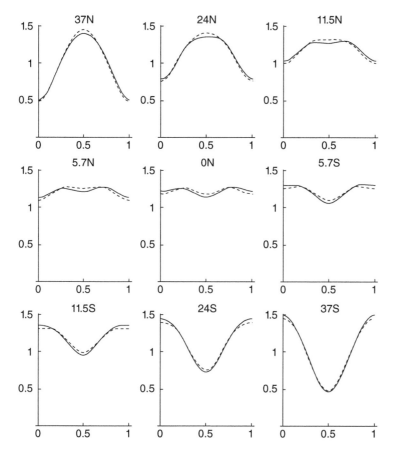

Figure 6.6 This figure illustrates the effect of eccentricity on the seasonal cycle of zonal average temperatures for an all-land planet. Shown is the modeled seasonal cycle (retaining Legendre modes 0, 1, and 2) of temperatures for the bare planet for present orbital parameters (solid curves) at selected latitudes: $\mu = \pm 0.6$ (36.9N/S), ± 0.4 (23.6N/S), ± 0.2 (11.5N/S), ± 0.1 (5.7N/S), and 0 (Equator). The dashed curves are for the corresponding circular orbit ($e = 0$).

where a_0 is the constant coalbedo. To simplify the algebra. we employ complex notation (superscript c indicates a complex variable) and the complex Fourier series:

$$T_n(t) = \sum_{k=0,1,2} T_{nk}^c \, e^{2\pi i k t}. \tag{6.12}$$

Inserting this into the governing mode equations:

$$C\frac{d}{dt}\left(\sum_{k=0}^{2} T_{nk}^c \, e^{2\pi i k t}\right) + [Dn(n+1) + B]\left(\sum_{k=0}^{2} T_{nk}^c \, e^{2\pi i k t}\right)$$

$$= Q a_0 \sum_{k=0}^{2} S_{nk}^{c*} \, e^{2\pi i k t}. \tag{6.13}$$

To recover the physical mode amplitudes,

$$|T_{nk}| = \sqrt{(T_{nk}^c)^* T_{nk}^c}, \quad (6.14)$$

whereas the phase lag behind the forcing is

$$\phi_{nk} = \arctan \frac{\Im(T_n^c)}{\Re(T_n^c)}. \quad (6.15)$$

The complex S_{nk}^c are related to the real Fourier coefficients as follows:

$$S_{nk}^c = a_{nk} + ib_{nk}. \quad (6.16)$$

After a time ($\gg \tau_0 = C/B$), the transients die out and we are left with the repeating steady-state solutions. First, consider the time-independent parts:

$$T_{00}^c = \frac{Qa_0 - A}{B}; \quad (6.17)$$

$$T_{20}^c = \frac{Qa_0 S_{20}^c}{6D + B}. \quad (6.18)$$

Next, consider the seasonal harmonic terms:

$$T_{01}^c = \frac{Qa_0 S_{01}^{c*}}{2\pi iC + B}; \quad (6.19)$$

$$T_{11}^c = \frac{Qa_0 S_{11}^{c*}}{2\pi iC + 2D + B}; \quad (6.20)$$

$$T_{21}^c = \frac{Qa_0 S_{21}^{c*}}{2\pi iC + 6D + B}; \quad (6.21)$$

and finally, the semiannual terms:

$$T_{02}^c = \frac{Qa_0 S_{02}^{c*}}{4\pi iC + B}; \quad (6.22)$$

$$T_{12}^c = \frac{Qa_0 S_{12}^{c*}}{4\pi iC + 2D + B}; \quad (6.23)$$

$$T_{22}^c = \frac{Qa_0 S_{22}^{c*}}{4\pi iC + 6D + B}. \quad (6.24)$$

In each case, we can compute the amplitude

$$|T_{nk}^c| = \frac{Qa_0 |S_{nk}^c|}{\sqrt{[n(n+1)D + B]^2 + 4\pi^2 k^2 C^2}}, \quad (6.25)$$

with phase lag

$$\phi_{nk} = \arctan \left(\frac{2\pi kC}{n(n+1)D + B} \right). \quad (6.26)$$

It is interesting to evaluate these quantities for the all-land planet. Using typical values of the parameters, $A = 210$ W m^{-2}, $B = 1.90$ W (m^{-2} K^{-1}), $D = 0.15\,B$, $a_0 = 0.70$, we find $T_0 = 14.7°$ C, $|T_1| = 69.1$ K, $\phi_1 = 25$ days, $T_{20} = -31$ K, $|T_{22}| = 8.1$ K, $C = 30$ days. While the phase lag is approximately correct for an all-land planet, the amplitude of both harmonics is larger than observed values for the Northern Hemisphere by a factor

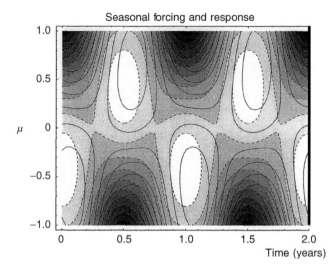

Figure 6.7 Illustration of the lag of zonal-average seasonal temperatures behind the forcing for the all-land planet. Shown is a contour plot (solid contours) of seasonal temperature response for the circular orbit insolation in dashed contours superimposed on the forcing to show the lag between heating and response. For a mixed-layer all-ocean planet where the response time is several years, the lag will be close to $\pi/2$ radians or 0.25 year.

of about 4 (Northern Hemisphere zonal averages). These large amplitudes are due to the absence of ocean surface in the zonal averages. In the Northern Hemisphere, the land fraction is about 60% and in the Southern Hemisphere, the fraction is 80%. We will attempt to remedy this situation in Chapter 8 by introducing land–sea geography. Figure 6.7 shows the forcing for a circular orbit and the response superimposed in solid line contours. Note the lag of about a tenth of a year in the response. There is also a displacement poleward of the maximum response from the heating. Figure 6.8 indicates the response through time at some selected latitudes, this time including the effects of the eccentric orbit.

6.4 Spread of Diffused Heat

Since EBMs make use of diffusion as a mechanism to transport heat poleward in the atmosphere/ocean system, it is useful to see how diffusion is related to random walk processes. Random walk means that the progress (root mean square average distance from the point of origination) of a passive scalar is the sum of a large number of steps. Let Y_N be the random variable denoting the displacement after N steps. We can write

$$Y_N = \sum_{n=1}^{N} X_n. \tag{6.27}$$

The mean over an ensemble of such random walks is 0, because we assume the individual steps have mean 0, that is,

$$\langle Y_N \rangle = \sum_{n=1}^{N} \langle X_n \rangle = 0. \tag{6.28}$$

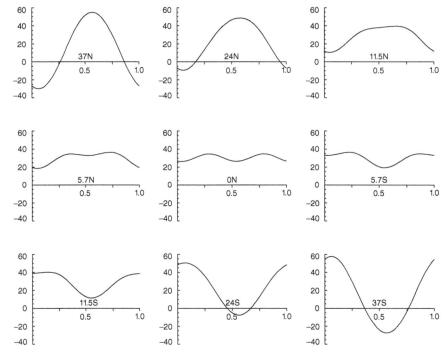

Figure 6.8 EBM solutions for the seasonal cycle of $T(\mu, t)$ for the all-land planet with present orbital parameters at selected latitudes: $\mu = \pm 0.6$ (36.9N/S), ± 0.4 (23.6N/S), ± 0.2 (11.5N/S), ± 0.1 (5.7N/S) and 0 (equator).

The variance of Y_N can be calculated if all the X_n are uncorrelated and have equal variance:

$$\sigma_{Y_N}^2 = N\sigma_X^2. \tag{6.29}$$

The standard deviation of Y_N is proportional to the square root of the number of steps.
Now consider the one-dimensional damped diffusion equation

$$\frac{\partial T}{\partial t} = \frac{D}{C}\frac{\partial^2 T}{\partial x^2} - \frac{T}{\tau_0}; \quad \tau_0 = \frac{C}{B}. \tag{6.30}$$

It is convenient to solve this equation with Fourier transforms. $T(x, t)$ can be represented as follows:

$$T(x, t) = \frac{1}{2\pi}\int_{-\infty}^{\infty} T_k(t)e^{ikx}\, dk. \tag{6.31}$$

We can write the second spatial partial derivative as follows:

$$\frac{\partial^2 T}{\partial x^2} = -\frac{1}{2\pi}\int_{-\infty}^{\infty} k^2 T_k(t)e^{ikx}\, dk. \tag{6.32}$$

The inverse Fourier transform is

$$T_k(t) = \int_{-\infty}^{\infty} T(x, t)e^{-ikx}\, dx. \tag{6.33}$$

We can insert the other terms.

$$\frac{1}{2\pi}\int_{-\infty}^{\infty}\left(\dot{T}_k + \frac{D}{C}k^2 T_k + \frac{T_k}{\tau_0}\right) e^{ikx}\, dk = 0. \tag{6.34}$$

The integrand must vanish for every wave number.

$$\dot{T}_k = \left(-k^2\frac{D}{C} - \frac{1}{\tau_0}\right) T_k. \tag{6.35}$$

The solution to this first-order homogeneous linear equation for wave number k is

$$T_k(t) = T_k(0) e^{-(1/\tau_0 + k^2 D/C)t}. \tag{6.36}$$

If the initial distribution is peaked at $x = 0$, that is, $T(x, 0) = T_0 \delta(x)$, then $T_k(0) = T_0$, a constant. The inverse Fourier transformation

$$T(x, t) = \frac{T_0 e^{-t/\tau_0}}{2\pi}\int_{-\infty}^{\infty} e^{-k^2 Dt/C}\, e^{ikx}\, dk \tag{6.37}$$

gives (from tables or MATHEMATICA)

$$T(x, t) = \frac{T_0 e^{-t/\tau_0}}{2\pi}\sqrt{\frac{\pi C}{Dt}}\, e^{-Cx^2/4Dt},$$

$$= \frac{T_0}{2\lambda_{dd}}\sqrt{\frac{\tau_0}{\pi t}} \exp\left(-\frac{x^2}{4\lambda_{dd}^2\, t/\tau_0}\right) e^{-t/\tau_0}, \tag{6.38}$$

where we have used the now familiar length scale $\lambda_{dd} = \sqrt{D/B}$ and $\tau_0 = C/B$ to make the notation more compact and to see the variables x and t proportional to their natural scales, λ_{dd} and τ_0. We see that except for the overall damping factor e^{-t/τ_0}, the solution spreads like a bell-shaped curve with standard deviation

$$\sigma(t) = \lambda_{dd}\sqrt{2t/\tau_0}. \tag{6.39}$$

The interpretation is that heat spreads symmetrically away from a point-concentrated initial anomaly a distance that is comparable to the damped diffusive length scale λ_{dd} in about one characteristic time τ_0. This is shown in Figure 6.9, wherein the spatial units are proportional to λ_{dd} and temporal units are proportional to τ_0.

6.4.1 Evolution on a Plane

This section generalizes the treatment to two horizontal dimensions on the infinite x–y plane. We begin by including diffusion in two Cartesian dimensions:

$$D\left(\frac{\partial^2}{\partial x^2} + \frac{\partial^2}{\partial y^2}\right) T - BT = C\frac{\partial T}{\partial t}. \tag{6.40}$$

This time we use the two-dimensional Fourier transform pair (in the two spatial dimensions):

$$T(\mathbf{r}, t) = \frac{1}{(2\pi)^2}\int_{-\infty}^{\infty} T_{\mathbf{k}}(t) e^{i\mathbf{k}\cdot\mathbf{r}}\, d^2\mathbf{k}; \quad T_{\mathbf{k}}(t) = \int_{-\infty}^{\infty} T(\mathbf{r}, t) e^{-i\mathbf{k}\cdot\mathbf{r}}\, d^2\mathbf{r}, \tag{6.41}$$

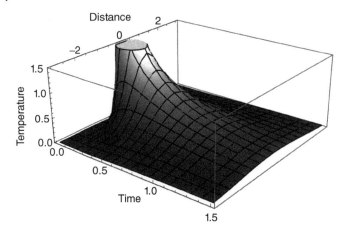

Figure 6.9 Temporal evolution of a point pulse of heat at the origin at time equal to 0. Horizontal distance in units of λ_{dd}, temporal span in units of τ_0.

where the two-dimensional vectors $\mathbf{r} = (x, y)$ and $\mathbf{k} = (k_x, k_y)$ have been introduced. Following the steps from the subsection 6.4, we find

$$T(\mathbf{r}, t) = \frac{T_0 e^{-t/\tau_0}}{(2\pi)^2} \int \int e^{-k^2 Dt/C} e^{i\mathbf{k}\cdot\mathbf{r}} \, d^2\mathbf{k}, \tag{6.42}$$

where $k^2 = k_x^2 + k_y^2$, and the integration limits $(-\infty, \infty)$ have been suppressed. We proceed by using polar coordinates in the $\mathbf{k} = (k_x, k_y)$ plane. We write $\mathbf{k}\cdot\mathbf{r} = k\, r \cos\theta$ where θ is the angle between the vectors $\mathbf{r}$ and $\mathbf{k}$, and $d^2\mathbf{k} = k\, dk\, d\theta$.

$$T(\mathbf{r}, t) = \frac{T_0\, e^{-t/\tau_0}}{(2\pi)^2} \int_0^\infty e^{-k^2 Dt/C} \left(\int_0^{2\pi} e^{ikr\cos\theta} \, d\theta \right) k\, dk. \tag{6.43}$$

We proceed by considering first the portion of the double integral into its θ part (enclosed in parentheses): This angular integral is well known (it is a special case, $n = 0$, of Bessel's integral (see Whittacker and Watson, 1962 p. 362; or other books on mathematical physics)):

$$\int_0^{2\pi} e^{ikr\cos\theta} \, d\theta = 2\pi J_0(kr), \tag{6.44}$$

where $J_0(\cdot)$ is the Bessel function. The resulting integral can also be solved (e.g., MATHEMATICA), yielding the rather unsurprising answer:

$$T(\mathbf{r}, t) = \frac{T_0\, e^{-(t/\tau_0)}}{4\pi \lambda_{dd}^2 (t/\tau_0)} \exp\left(\frac{-(r/\lambda_{dd})^2}{4(t/\tau_0)} \right). \tag{6.45}$$

Note that in this compact form r is expressed proportional to the damped diffusion length scale λ_{dd} and t is always found proportional to the global timescale τ_0. We have found exactly what we found in one dimension: a delta function point expanding into a 2-D Gaussian shape, further expanding its disk width in a time τ_0 that is about $\sqrt{2}\lambda_{dd}$ in radius. The volume under the Gaussian surface also diminishes by the factor e^{-t/τ_0} owing to the damping from infrared radiation to space.

6.5 Random Winds and Diffusion

Next consider the one-dimensional energy-balance equation in which heat is advected by a wind field that for simplicity has no x-dependence.

$$\frac{\partial T}{\partial t} + v(t)\frac{\partial T}{\partial x} + \frac{T}{\tau_0} = 0. \tag{6.46}$$

Once again, look at the component of the wave number k

$$\dot{T}_k = \left(-ikv(t) - \frac{1}{\tau_0}\right)T_k. \tag{6.47}$$

This is a first-order linear equation and can be solved using the usual integrating factor:

$$T_k(t) = T_k(0)e^{-t/\tau_0}\exp\left(-\int_0^t ikv(t')dt'\right). \tag{6.48}$$

Let $v(t)$ be a stationary random function of time. The ensemble average of a random quantity is denoted by $\langle v(t) \rangle$. Let the ensemble average vanish; stationarity demands that $\langle v(t_1)v(t_2)\rangle = v_0^2 \rho_v(|t_1 - t_2|)$ with $\rho_v(0) = 1$. This means that $T_k(t)$ will have mean zero and be stationary as well. Finally, we assume that $v(t)$ is a Gaussian random field. In other words, our wind field is similar to a random eddy field that carries heat one way or another depending on the vagaries of such a wind field. Note, however, that we have not permitted the wind to be a function of position. This makes our wind field a rather peculiar one: at a point in time it is everywhere pointed in the same direction, until the next instant, when it suddenly switches direction and magnitude the same everywhere. The case of the wind field that is variable in space as well as time is more difficult to solve. Rather than going into the complexities involved in that, we proceed with what we have.

Consider the series expansion of the second exponential factor, $\exp\left(-\int_0^t ikv(t')dt'\right)$. The first three terms are

$$1 - ik\int_0^t \langle v(t')\rangle dt' - \frac{k^2}{2!}\int_0^t\int_0^t \langle v(t')v(t'')\rangle dt'dt'' + \cdots. \tag{6.49}$$

The second term vanishes because $\langle v(t)\rangle = 0$. In fact, all the odd powered terms vanish.[1] We turn to the third term after running the integrals from time $-\Omega$ to Ω instead of 0 to t. This symmetry helps in analyzing the result.

$$1 - \frac{k^2}{2!}\int_{-\Omega}^{\Omega}\int_{-\Omega}^{\Omega}\langle v(t')v(t'')\rangle dt' \, dt'' = 1 - \frac{k^2}{2!}v_0^2\int_{-\Omega}^{\Omega}\int_{-\Omega}^{\Omega}\rho_v(|t_1 - t_2|)dt_1 \, dt_2. \tag{6.50}$$

The double integral can be computed by a transformation of variables from $dt_1 \, dt_2$ to a one-dimensional integral. Referring to Figure 6.10, as the integrand only depends on τ, the integration is best done by summing slabs that are at a 45° angle (along which $|\tau| = |t_1 - t_2|$) to the horizontal. The length of the slab can be found from the hypotenuse of the right isosceles triangle whose lower edge has length $2\Omega - \tau$. We can show this

---

[1] It is a property of stationary Gaussian random fields that ensemble averages of lagged odd powers vanish.

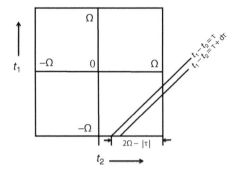

Figure 6.10 Geometry for a two-dimensional integral over a lag covariance function for a stationary process: $\langle v(t_1)v(t_2)\rangle = v_0^2 \rho_v(|t_1 - t_2|)$. Since ρ_v only depends on $|\tau|$, we can find the integral by summing the diagonal infinitesimal strip whose width is $d\tau/\sqrt{2}$. The value of τ in the lower-right corner is 2Ω and its value in the upper-left corner is -2Ω. The quantity to be summed is the length of the diagonal strip ($\sqrt{2}(2\Omega - |\tau|)$) times $\rho_v(\tau)$. The integral becomes $\int_{-2\Omega}^{2\Omega}(2\Omega - |\tau|)\rho_v(\tau)d\tau$. (Redrawn from original figures in Papoulis (1984).)

last by noting that the same side has length $\Omega - t_1$ where t_1 is evaluated at the point the hypotenuse intersects the line $t_2 = -\Omega$. At that point, $t_1 - t_2 = \tau = t_1 + \Omega$, or $t_1 = \tau - \Omega$. Using $\Omega - t_1 = \Omega - \tau + \Omega$ we finally arrive at the width of the lower side of the triangle to be $2\Omega - \tau$. The length of the slab is $\sqrt{2}(2\Omega - |\tau|)$. The width of the slab is $dt_1/\sqrt{2} = d\tau/\sqrt{2}$; the two square roots cancel to leave us with

$$\int_{-\Omega}^{\Omega}\int_{-\Omega}^{\Omega} \rho_v(|t_1 - t_2|)\, dt_1\, dt_2 = \int_{-2\Omega}^{2\Omega}(2\Omega - |\tau|)\rho_v(\tau)\, d\tau, \tag{6.51}$$

where $\rho_v(\tau)$ is the autocorrelation function for the magnitude of the lag $\tau = |t_1 - t_2|$. Note that $\rho_v(0) = 1$ and v_0^2 is the variance of the velocity field. We specify that the integration time 2Ω is very long compared to the autocorrelation time τ_v that we take to be a in Figure 6.11. This means that the kernel (nearly an isosceles triangle of base width $2a$ and height unity) of the integral is concentrated near $\tau = 0$ and the integral is $\sim 2\Omega a$, which is

$$\approx 1 - v_0^2 \tau_v \frac{k^2}{2!}(2\Omega) \approx e^{-\frac{v_0^2 \tau_v}{2}k^2 t}. \tag{6.52}$$

After replacing 2Ω by t to regain the original notation, we can now identify

$$D \equiv v_0^2 \tau_v \sim (\text{Length})^2\, (\text{Time})^{-1}, \tag{6.53}$$

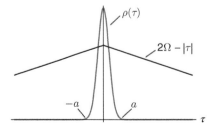

Figure 6.11 Depiction of how a short autocorrelation time leads to an estimate of a double integral related to second moments. (Redrawn from original figures in Papoulis (1984).)

which gives us a handy formula for the diffusion coefficient in terms of the variance of the wind speed and its autocorrelation time. The last expression is reminiscent of the form of (6.36). Note that the units of D are (length)^2(time)$^{-1}$, when the diffusion equation is set with the coefficient of $\frac{\partial T}{\partial t}$ set to unity. When the approximations above are valid, we can think of diffusion equations as being the equations describing the ensemble averages of (damped) random walk equations. Basically, the approximation is that the winds are white noise ($\tau_v \sim 3$ days) compared to our averaging times. All timescales, including relaxation time of a column of air (~ 30 days), and forcing functions such as the seasonal cycle (few months), must be long compared to the weather–noise timescale (few days). This suggests the following:

$$\left\langle v(t)\frac{\partial T}{\partial x}\right\rangle \approx D\frac{\partial^2}{\partial x^2}\langle T\rangle. \tag{6.54}$$

When this approximation is valid, we can safely think of the EBCM equation as an equation for the ensemble average of $T(\mathbf{r}, t)$. Note that there is a residual whose ensemble average is zero. Later, we will use this residual as a noise driver of climate fluctuations.

6.6 Numerical Methods

6.6.1 Explicit Finite Difference Method

We start with the energy-balance equation:

$$\frac{\partial T}{\partial t} = \frac{\partial}{\partial \mu}D(\mu)\frac{\partial}{\partial \mu}T - \frac{T}{\tau_0} + q(\mu), \tag{6.55}$$

where

$$D(\mu) = D_0 \frac{1}{C}(1-\mu^2) = \frac{\lambda_{dd}^2}{\tau_0}(1-\mu^2), \tag{6.56}$$

$$q(\mu) = \frac{Q_0}{C}S(\mu)a - \frac{A}{C}. \tag{6.57}$$

The coefficient in (6.56) was constructed by multiplying numerator and denominator by B, then using $\lambda_{dd}^2 = D/B$ and $\tau_0 = C/B$. We start with a grid from -1 to 1, divided into J equal segments. More details on the grid are presented in the following. The centered finite difference form for the derivative is

$$\frac{\partial T}{\partial \mu} \approx \frac{1}{\Delta \mu}\left(T_{j+\frac{1}{2}} - T_{j-\frac{1}{2}}\right). \tag{6.58}$$

The derivative of $D(\mu)\frac{\partial T}{\partial \mu}$ is then

$$\frac{\partial}{\partial \mu}\left(D(\mu)\frac{\partial T}{\partial \mu}\right) \approx \frac{1}{(\Delta\mu)^2}(D_j^+ T_{j+1} - (D_j^+ + D_j^-)T_j + D_j^- T_{j-1}), \tag{6.59}$$

where

$$D_j^\pm = D\left(\mu_j \pm \frac{1}{2}\Delta\mu\right). \tag{6.60}$$

To step forward in time, we have the following algorithm (in the explicit finite difference case):

$$T_j^{n+1} = T_j^n + s(D_j^+ T_{j+1}^n - (D_j^+ + D_j^-)T_j^n + D_j^- T_{j-1}^n)$$
$$- \frac{\Delta t}{\tau_0} T_j^n + \Delta t\, q(\mu_j), \qquad (6.61)$$

where the important parameter s is given by

$$s = \frac{\Delta t}{(\Delta \mu)^2}. \qquad (6.62)$$

Next, we must make sure our solution enforces the Neumann boundary conditions, implying that no net heat flux enters the infinitesimal latitude circles surrounding the poles:

$$\sqrt{1-\mu^2}\,\frac{\partial T}{\partial \mu}\bigg|_{\mu=-1,1} = 0. \qquad (6.63)$$

The fact that the time-dependent version of the equation has a regular singular point at the pole poses a problem because as noted earlier, small numerical errors will lead to erroneous encroachment by the irregular solutions ($Q_n(\mu)$). This does not turn out to be a serious problem as we will see. The most straightforward way of representing the equation in finite difference form is to break the interval $-1 \leq \mu \leq 1$ into J intervals, $\mu_1 = -1, \ldots, \mu_{J+1} = 1$. Similarly, the time is discretized in steps of Δt.

Deferring discussion of the boundary points, we see that the term on the left is the value of T at time $n+1$, while all the terms on the right are to be evaluated at the previous time n. This is very convenient because we presumably have knowledge of the field at time n. For example, if we start at time $n = 0$, we know the initial profile of the field. This process allows us to evaluate it at $n = 1$ and so on. Such an algorithm is very appealing intuitively as it feels like we are solving the equation just as nature does it. Note that there is a problem at the end points for the reason that when $j = 1$, the RHS contains the value T_0^n and when $j = J$, it contains the value T_{J+1}^n. Hence, we must use the aforementioned algorithm only for $2 \leq j \leq J-1$. Had the boundary conditions been of the *Dirichlet* type where the values of the field are specified at the boundaries, we could merely specify the values of T_1^n and T_J^n.

Enforcing the Neumann boundary conditions is a little tricky as the straightforward implementation of the thinking above leads to $T_1 = T_2$ and $T_J = T_{J-1}$. We can get around this by using a centered difference of width $2\Delta t$ for these points. We introduce fictitious points just outside the range, T_0, and T_{J+1}. Then the boundary conditions read as follows:

$$\frac{T_2^n - T_0^n}{2\Delta t} = 0; \quad \frac{T_{J+1}^n - T_{J-1}^n}{2\Delta t} = 0. \qquad (6.64)$$

We compute the outside points using the explicit algorithm, but then force T_2 to equal the lower outside value T_0 and likewise T_{J-1} is forced to be T_{J+1}.

The curves in Figure 6.12 show an example of implementation of this kind of algorithm for $\tau_0 = C/B = 1, N = 20, \Delta \mu = 1/10, \Delta t = \tau_0/100 = 0.01, s_D = D(\mu)\Delta t/(\Delta \mu)^2 = 0.32$, where we have modified the definition of the parameter s to include $D(\mu)$, that is, $s_D = D(\mu)s$, thereby allowing the stability parameter to be a function of latitude. The forcing is $q(\mu) = -A + QS(\mu)a(\mu)$. The initial condition for the integration is $T(\mu, 0) = 10.0\ °C$.

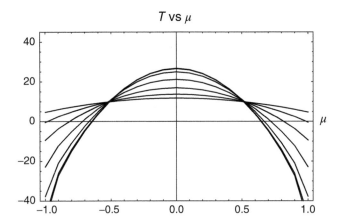

Figure 6.12 Sequence of the evolution toward steady state where the forcing is the usual $-A + QS(\mu)a(\mu)$. The initial condition is flat with $T(\mu) = 10\,°C$. The time step is 0.05, which is 1/20 of the relaxation time, τ_0. The individual graphs are at $t = 10, 20, 40, 80, 160, 320$, and 640 steps. In this case, $\tau_0 = C/B = 1$ month, $D_{max} = 0.67\,W\,m^{-2}\,(°C)^{-1}$, $N = 20$, $\Delta\mu = 1/20$, $s = D_{max}\Delta t/(\Delta\mu)^2 = 3.2$. An additional run (not shown) of length 3000 steps showed no change.

There are 20 intervals in μ from pole to pole, and the time step is $\frac{1}{100}$ of the relaxation time τ_0. The figure shows a sequence of profiles after 10, 20, 40, 80, 160, 320, and 640 time steps. The last is nearly indistinguishable from the 320 time-step case. The solutions in this case are smooth and stable after integration to 3.20 relaxation times. Advancement to 3000 steps (30 relaxation times) indicates no change. We infer stability of the solution.

If we increase the time step by a factor of 10 to $0.1\tau_0$, we run into trouble. This case is shown in Figure 6.13 starting from the same initial condition. The stability parameter s_D

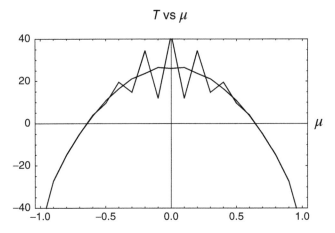

Figure 6.13 The smooth line represents the approximate solution after 300 time steps where the approximate solution seems to have settled down to the correct answer. But if the integration continues to 425 steps, one encounters the spiky line indicating numerical instability. In this case, $\tau_0 = C/B = 1$ month, $D_{max} = 0.67\,W\,m^{-2}\,(°C)^{-1}$, $N = 20$, $\Delta\mu = 1/10$, $\Delta t = \tau_0/10 = 0.1$, $s = D_{max}\Delta t/(\Delta\mu)^2 = 3.2$.

increases to 3.20. In the integration, everything goes well until time step 400 in which a small ripple occurs at the equator. By time step 425, the instability is full blown and propagating from the equator toward the poles. The reason for this peculiar behavior is that the explicit method is unstable when the time steps are too large compared to the spatial increment. For the diffusion equation, the condition can be made quantitative: the solution will be unstable if $s_D = D_{\max}\Delta t/(\Delta \mu)^2 \geq 0.5$. Since $D(\mu)$ is largest at the equator, the instability breaks out there first as indicated in the figure. The physical interpretation of the instability is that for diffusion or random walk processes, information propagates an rms distance $\sigma_\mu = \sqrt{2D_0 \Delta t}$. Turning this around, we have $\Delta t \sim \sigma_\mu^2/2D_0$; if Δt in the numerical integration is larger than the time for propagation of information in the continuous exact solution, we can expect instability of the numerical procedure. A similar limitation occurs in the integration of wavelike (hyperbolic) equations with wave motion entering as the propagation mechanism ($\Delta x = c\Delta t$, $c =$ wave speed).

The finite difference algorithm as noted above in (6.61) is not the only one that is accurate with errors of order $(\Delta t, \Delta \mu^2)$. The next subsection considers some alternatives that have different numerical stability properties.

6.6.2 Semi-Implicit Method

An intermediate method is often used in practice. It is simply the weighted average of the explicit and implicit algorithms:

$$T_j^{n+1} = T_j^n + \lambda \left[s(D_j^+ T_{j+1}^n - (D_j^+ + D_j^-)T_j^n + D_j^- T_{j-1}^n) - \frac{\Delta t}{\tau_0} T_j^n \right]$$
$$+ (1-\lambda) \left[s(D_j^+ T_{j+1}^{n+1} - (D_j^+ + D_j^-)T_j^{n+1} + D_j^- T_{j-1}^{n+1}) - \frac{\Delta t}{\tau_0} T_j^{n+1} \right]$$
$$+ \Delta t \, q_j^n, \tag{6.65}$$

with $0 \leq \lambda \leq 1$.

One has to gather the coefficients of T^{n+1} and place them onto the left-hand side. The coefficient matrix has to be inverted to obtain T^{n+1}. The result is

$$\sum_k \mathcal{M}_{jk} T_k^{n+1} = -\lambda s(D_j^+ T_{j+1}^n - (D_j^+ + D_j^-)T_j^n + D_j^- T_{j-1}^n) + \left(1 - \lambda \frac{\Delta t}{\tau_0}\right) + \Delta t \, q_j^n,$$

where $\mathcal{M}$ is a large matrix with j and k spatial indices. The matrix has to be inverted to find the next time-step values as a function latitude as indicated by the index j: T_j^{n+1}. Fortunately, the matrix $\mathcal{M}$ is usually sparse (most entries are zero), and very fast algorithms are available. A special case that is commonly used is for $\lambda = 0.5$, which is called the *Crank–Nicolson scheme*. Stability properties can be found in books on numerical analysis, but experience shows that use of this method allows larger time steps before instability sets in. Semi-implicit algorithms are commonly used in numerical solution of general circulation models.

In the finite difference solution to more complicated models treating dynamics of the flows as well as radiation, and so on, it is found to be advantageous to treat some terms on the RHS of the equation with one semi-implicit weighting and other terms with different weightings. These are called *splitting* methods. Again, they are commonly used in numerical solutions of climate models.

6.7 Spectral Methods

6.7.1 Galerkin or Spectral Method

An alternative to the finite difference methods is to write an approximate solution to the field as a series of (usually) orthogonal functions

$$T(\mu, t) \approx \sum_{n=0}^{N} T_n(t)\phi_n(\mu), \tag{6.66}$$

where N is some finite cutoff to the series. A rather obvious choice for the ϕ_n for this class of problems is the Legendre polynomials, $\phi_n(\mu) \equiv P_n(\mu)$. In the present case, this renders the problem trivial as the $P_n(\mu)$ are the eigenfunctions of $\frac{d}{d\mu}(1-\mu^2)\frac{d}{d\mu}$. The problem becomes nontrivial if the diffusion coefficient depends on μ. Such an EBM may be written as

$$\frac{\partial T}{\partial t} = \frac{\partial}{\partial \mu} D(\mu) \frac{\partial}{\partial \mu} T - \frac{T}{\tau_0} + q(\mu), \tag{6.67}$$

with the usual Neumann boundary conditions at the poles. If $T = \sum T_n(t) P_n(\mu)$ is inserted and each side is multiplied by $P_m(\mu)$ and integrated from pole to pole with respect to μ, we obtain

$$\dot{T}_m = \sum_{n=0}^{N} D_{mn} T_n - \frac{2T_n}{(2n+1)\tau_0} \delta_{mn} + q_n, \tag{6.68}$$

where

$$D_{mn} = -\frac{2m+1}{2} \int_{-1}^{1} P_m(\mu) \frac{d}{d\mu}\left(D(\mu)\frac{dP_n(\mu)}{d\mu}\right) d\mu$$

$$= \frac{2m+1}{2} \int_{-1}^{1} P'_m(\mu) D(\mu) P'_n(\mu) \, d\mu. \tag{6.69}$$

Now the problem has been reduced to a set of N first-order coupled ordinary differential equations for the time-dependent coefficients $T_n(t)$. The coupling matrix is symmetric, hence in this case we can even find its eigenvectors and conduct a stability analysis. From discussions earlier in this chapter, it is easy to relate the eigenvalues to relaxation times for the eigenmodes of the problem. The numerical integration will be unstable if the time constant for the highest eigenmode retained (Nth) is shorter than the time step employed in the time-stepping algorithm. In other words, the criterion encountered in the explicit scheme comes up again unless special precautions are taken.

Other basis sets are possible besides the Legendre polynomials. For example, one might try a Fourier series

$$T(\mu = \cos\vartheta, t) = \sum_{n=0}^{N} a_{n(t)} \cos n\vartheta, \quad 0 \le \vartheta \le \pi. \tag{6.70}$$

This choice is equivalent to the use of Chebyshev polynomials. This method is successful in the present one-dimensional case but requires considerable work and extra discussion when applied to the case of two horizontal dimensions. One advantage is that transformations from grid points to components a_n and *vice-versa* can be accomplished via fast Fourier transform.

6.7.2 Pseudospectral Method

A complication arises in the spectral method if one of the coefficients is space dependent, for example, $B = B(\mu)$. Then,

$$\dot{T}_m = \sum_{n=0}^{N} D_{mn} T_n - \sum_{n=0}^{N} B_{mn} T_n + q_n \tag{6.71}$$

with

$$D_{mn} = \frac{2m+1}{2} \int_{-1}^{1} P_m(\mu) \frac{d}{d\mu}\left(D(\mu)\frac{dP_n(\mu)}{d\mu}\right) d\mu$$

$$= -\frac{2m+1}{2} \int_{-1}^{1} P'_m(\mu) D(\mu) P'_n(\mu) d\mu, \tag{6.72}$$

and

$$B_{mn} \equiv \frac{2m+1}{2C} \int_{-1}^{1} P_m(\mu) B(\mu) P_n(\mu) d\mu$$

$$= \frac{2m+1}{2C} \sum_{l=0}^{N} B_l \int_{-1}^{1} P_m(\mu) P_l(\mu) P_n(\mu) d\mu$$

$$= \sum_{l=0}^{N} B_l C_{lmn}, \tag{6.73}$$

where

$$C_{lmn} = \int_{-1}^{1} P_l(\mu) P_m(\mu) P_n(\mu) d\mu, \tag{6.74}$$

and the C_{lmn} are known as interaction coefficients. They can be tabulated or generated from recurrence relations. It is important to notice how many are necessary in a high-resolution simulation. For example, in a GCM simulation the truncation level might be 15 or 42 typically. This means there are 3375 or 74 088 coefficients that have to be stored in a lookup table. Some storage savings can be gained by noting that most of the interaction coefficients are 0. However, when the problem is elevated to two dimensions on the sphere, it becomes truly formidable. One way to get around it is to use a so-called *pseudospectral* method.

In the pseudospectral method, we transform back and forth between a grid point representation and the spectral representation. The grid point representation for the sphere involves use of the Gaussian quadrature method of numerically estimating integrals. An integral may be estimated by the sum

$$\int_{-1}^{1} f(x) dx \approx \sum_{i=1}^{N} w_i f(x_i), \tag{6.75}$$

where the w_i are weights associated with each term and the x_i are unequally spaced but specified points along the axis. The x_i and w_i are optimal for a given level N. Gaussian integration[2] of order N has the property that it integrates a $(2N-1)$ degree

---

[2] Gaussian integration is discussed in most books on numerical analysis, for example, Stoer and Bulirsch (2002). In addition, many derivations can be found online.

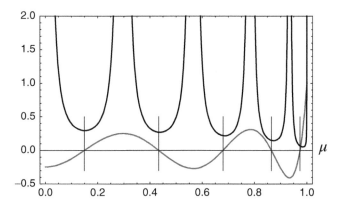

Figure 6.14 Plot of $P_{10}(x)$ (solid gray curve) with zeros at x_i. Also a plot of $2/((1 - x_i^2)[P_N'(x_i)]^2)$ (heavy black U-shaped curve) with vertical thin black lines corresponding to the value of the weights at the roots x_i of $P_{10}(x_i) = 0$. The ordinate values of the intersections are the weights. The graph is only shown for $\mu > 0$, as all the functions are even.

polynomial exactly. The abscissas x_i are the ith zero of $P_N(x)$ and the weights w_i are $2/((1 - x_i^2)[P_N'(x_i)]^2)$. As an example, Figure 6.14 shows a plot of $P_{10}(x)$ (solid gray line) along with a plot of $2/(1 - x_i^2)[P_N'(x_i)]^2$ (black line) with heavy points on the dashed line corresponding to the value of the weights at the roots x_i of $P_{10}(x_i) = 0$. Next, the quantities D_{mn} and B_{mn} are calculated by

$$D_{mn} = -\frac{2m+1}{2} \sum_{i=1}^{N} w_i P_m'(x_i) D(x_i) P_n'(x_i); \tag{6.76}$$

$$B_{mn} = \frac{2m+1}{2C} \sum_{i=1}^{N} w_i P_m(x_i) B(x_i) P_n(x_i). \tag{6.77}$$

The next time step may now be evaluated:

$$\frac{\Delta T_n}{\Delta t} = \sum_{m=1}^{N} (D_{nm} - B_{mn}) T_n + q_n, \tag{6.78}$$

where the ratio on the LHS stands for time advancement by some ODE algorithm such as Runge–Kutta. When the new set T_n are computed, we can reevaluate $T(x_i, t + \Delta t)$ by

$$T(x_i, t + \Delta t) = \sum_{n=0}^{N} T_n(t + \Delta t) P_n(x_i). \tag{6.79}$$

The procedure delineated in (6.76)–(6.79) can be repeated as many times as necessary as long as the employed ODE algorithm is stable.

Studies have shown that the pseudo-spectral method pays over the interaction coefficient method when the number of stored coefficients exceeds a few tens of thousands. Since this is rather quickly exceeded in GCM calculations, the pseudospectral method is commonly used in numerical simulations.

6.8 Summary

This chapter has introduced time into the energy balance models. The first result is that for the linear zonally symmetric models, we find that if a climate is perturbed from equilibrium and then released, it relaxes back to its steady-state solution. The relaxation can be characterized as a sum of contributions from relaxation modes, each having an exponential decay with large scales having longer relaxation times than smaller ones. The zonally symmetric planet can also be solved for its seasonal cycle. It is remarkable that the solar insolation can be represented for many purposes as a sum of terms involving only the Legendre modes 0, 1, and 2. The largest part of the circular orbital insolation is controlled by Legendre mode of index 1, which is antisymmetric on the globe and driven by a single annual cycle sinusoid in time. The Legendre mode of index 2 also has a non-negligible contribution in the semiannual cycle representing the passage of the Sun over the equator twice per year. These simple facts allow the seasonal cycle to be solved for this configuration. We can readily study the major effects on insolation of changing orbital parameters. The amplitude of the annual cycle is much too large if the zonally symmetric planet is considered to be all-land and too small if all-ocean. We postpone the study of land–sea distribution until Chapter 8.

Next in the chapter was a study of how a patch of heat deposited in a small area spreads laterally to larger circular areas in a time roughly that of the characteristic time of the global model, $\tau_0 = C/B$ to a radius of the characteristic length $\lambda_{dd} = \sqrt{D/B}$. The spread has an exponentially damped (time constant τ_0) Gaussian shape with standard deviation proportional to $\sqrt{\text{time}}$. It was also shown that if the diffusion term is replaced by a random-wind advection term, the ensemble result will be the same as diffusion. If the autocorrelation time of the velocity field is short compared to other times, such as τ_0, in the problem, then the ensemble average of this term is essentially the diffusion term. This is very close to the Brownian motion problem solved by Einstein in 1904.

The chapter concludes with a study of how 1-D EBMs can be solved by finite difference methods. In particular, an explicit problem is worked through in some detail with an illustration of how the procedure becomes unstable if the time step exceeds a certain threshold depending on the characteristic time and length scales of the EBM. Basically, the time step must be shorter than the spreading time across a grid box in the horizontal direction.

The appendix to the chapter provides some derivations of the orbital dependence of the insolation function.

6.8.1 Parameter Count

We have introduced the time dimension, which means we must now have an effective heat capacity that cannot be calculated from first principles. For the all-land planet and at frequencies around 1/year, our guess is to take it to be 1/2 the heat capacity of the air column's mass at constant pressure. This neglects the heat capacity of soil. We are also neglecting the topography, where atop mountains, the effective heat capacity might be less. For an all-ocean planet we take the effective heat capacity to be that of the ocean's mixed layer. It depends on position, especially latitude, but we take it to be constant. So for the latitude-only model we have introduced one new parameter, but we have learned a lot about how the seasonal cycle works for a uniform planet.

Notes for Further Reading

Papoulis (1984) is an excellent book on random processes and spectral analysis written mainly for electrical engineers. It is rather compact, but comprehensive. Numerical methods for partial differential equations are covered in Ames (1992) and Fletcher (1991). Multigrid methods for solving partial differential equations are covered in Hackbusch (1980) and Briggs *et al.* (2000) and applied to EBMs by Bowman and Huang (1991), Huang and Bowman (1992), and Stevens and North (1996). Mars has no ocean, so a constant heat capacity can be used on the (approximately) homogeneous planet. A one-dimensional model with a zonally symmetric seasonal cycle was used successfully by James and North (1982). Much more can be found about the Martian climate as well as the other planets in Ingersoll's recent book on planetary climates Ingersoll (2015). Planetary climates are covered at a higher level of physics by Pierrehumbert (2011).

Exercises

6.1 Given the time-dependent, one-dimensional energy balance model

$$C\frac{\partial T}{\partial t} - D\frac{\partial}{\partial \mu}\left((1-\mu^2)\frac{\partial T}{\partial \mu}\right) + A + BT = QS(\mu, t)a,$$

where $A, B, C, D,$ and coalbedo a are constants, derive the basic timescale and spatial scale of the model by nondimensionalizing the model.

6.2 Consider the time-dependent, one-dimensional model of the previous exercise, where the parameters, $A, B, C, D,$ and a are constants. Let the solution of the energy balance model be denoted as follows: $T(\mu, t) = T_s$. An external forcing is applied at time $t = t_0$ so that the temperature field is perturbed into the form $T(\mu, t_0) + T'(\mu, t_0)$. The external forcing disappears instantaneously (delta function in time). (a) Derive the governing equation for $T'(t)$. (b) Solve the governing equation to obtain the solution for $T'(t)$.

6.3 Consider next the same one-dimensional model as before with the same constant parameters. Let the solar distribution function be decomposed as follows:

$$S(\mu, t) = \sum_n S_n(t)P_n(\mu) = \text{Re}\left(\sum_{n=0}^{N}\sum_{m=0}^{M} S_{mn} e^{i2\pi mt} P_n(\mu)\right),$$

where Re($\cdot$) denotes the real part of its argument. Solve the energy balance model for $T(\mu, t)$.

6.4 Show that the solution of the one-dimensional damped diffusion equation

$$\frac{\partial T}{\partial t} = \frac{D}{C}\frac{\partial^2 T}{\partial x^2} - \frac{T}{\tau_0},$$

with the initial condition $T(x,0) = T_0\delta(x)$ is given by

$$T(x,t) = \frac{T_0}{2\lambda_d}\sqrt{\frac{\tau_0}{\pi t}}\exp\left(-\frac{x^2}{4\lambda_d^2 t/\tau_0}\right)e^{-t/\tau_0},$$

where $\tau_0 = C/B$ and $\lambda_d = \sqrt{D/B}$.

6.5 A semi-implicit algorithm for a time-dependent, one-dimensional energy balance model is given by

$$T_j^{n+1} = T_j^n + w\left[s(D_j^+ T_{j+1}^n - (D_j^+ + D_j^-)T_j^n + D_j^- T_{j-1}^n) - \frac{\Delta t}{\tau_0}T_j^n\right]$$

$$+ (1-w)\left[s(D_j^+ T_{j+1}^{n+1} - (D_j^+ + D_j^-)T_j^{n+1} + D_j^- T_{j-1}^{n+1}) - \frac{\Delta t}{\tau_0}T_j^n\right],$$

where

$$D_j^{\pm} = D(\mu_j \pm \Delta\mu/2), s = \Delta t/(\Delta\mu)^2$$

and j and n denote a point in space and time, respectively. This equation can be rewritten as

$$\sum_k M_{jk}T_k^{n+1} = \sum_k L_{jk}n + \Delta t q_j^n.$$

Determine the matrices $\mathbf{M}_{jk}$ and $\mathbf{L}_{jk}$ explicitly.

6.6 Consider a time-dependent, one-dimensional energy balance model in the form

$$C\frac{\partial T}{\partial t} - \frac{\partial}{\partial \mu}\left[D(\mu)\frac{\partial T}{\partial \mu}\right] + B(\mu)T = q(\mu),$$

where C represents the constant heat capacity and $D(\mu) = D_0(1-\mu^2)$. Set up the solution procedure using the spectral method. Use the Legendre polynomials as basis functions.

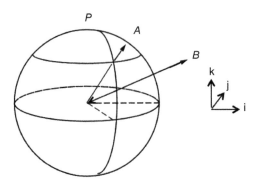

Figure 6.15 Diagram for Exercise 6.7.

6.7 In Chapter 5, the insolation distribution function was determined by using spherical geometry. Consider the situation depicted in Figure 6.15. (a) Determine the normal vector at point $A(\hat{\mathbf{r}})$ and at point $B(\hat{\mathbf{r}}_s)$ in terms of the longitude, latitude, and declination angle, and unit vectors $\mathbf{i}$, $\mathbf{j}$, and $\mathbf{k}$. Then, calculate the cosine of the angle between the two vectors. (b) Determine the range of longitude for which the Sun is visible. (c) Based on your answers for (a) and (b), determine the total amount of insolation received at the surface when the solar constant is $\sigma_\odot$.

6.9 Appendix to Chapter 6: Solar Heating Distribution

This chapter made use of the solar heating (insolation) function $S(\mu, t)$, which is the amount of radiant energy per unit time per unit surface area averaged through the day reaching the top of the atmosphere. In this appendix, we present a short derivation of this function, as its development into Legendre functions in latitude and sinusoids in time helps to understand the excitation of these modes in the response field under different orbital conditions. The appendix is organized as follows. First the derivation of $S(\mu, t)$ is given, followed by a discussion of how the orbital elements enter the mode amplitudes. Figure 6.16 shows an octant of the Earth's surface with a given point $\hat{\mathbf{r}}$ singled out for our attention. The North Pole lies in the z direction, and hence the point $\hat{\mathbf{r}}$

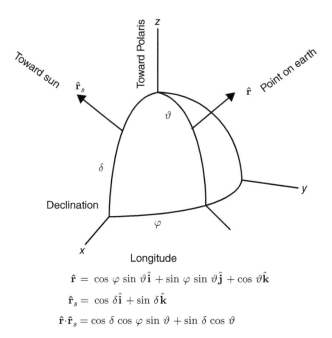

$$\hat{\mathbf{r}} = \cos\varphi \sin\vartheta \hat{\mathbf{i}} + \sin\varphi \sin\vartheta \hat{\mathbf{j}} + \cos\vartheta \hat{\mathbf{k}}$$

$$\hat{\mathbf{r}}_s = \cos\delta \hat{\mathbf{i}} + \sin\delta \hat{\mathbf{k}}$$

$$\hat{\mathbf{r}} \cdot \hat{\mathbf{r}}_s = \cos\delta \cos\varphi \sin\vartheta + \sin\delta \cos\vartheta$$

Figure 6.16 An octant of the Earth's surface showing a unit vector toward the Sun $\hat{\mathbf{r}}_s$, lying in the z–x plane. A given point on the Earth's surface is designated $\hat{\mathbf{r}}$. The North Pole is along the z-axis. The local time of day at point $\hat{\mathbf{r}}$ is proportional to φ. The declination δ is the angle the Sun makes with a perpendicular to the equatorial plane (the x–y plane here) for a given day of the year, $\delta = \cos^{-1}\hat{\mathbf{k}} \cdot \hat{\mathbf{r}}_s$. The declination depends on the tilt of the Earth's axis with respect to a perpendicular to the plane of the ecliptic (the obliquity) and the time of the year.

will rotate uniformly around the z-axis in the course of a day. As the day passes, φ will vary linearly from 0 to 2π. At a given time of day, $\varphi(t_0)$, the amount of solar power per unit area deposited at $\hat{\mathbf{r}}$ is given by the solar constant $\sigma_\odot$ times the cosine of the angle between zenith direction $\hat{\mathbf{r}}$ and the line joining the Sun and the Earth $\hat{\mathbf{r}}_s$. In other words, $\hat{\mathbf{r}} \cdot \hat{\mathbf{r}}_s \sigma_\odot$. We define the solar vector $\hat{\mathbf{r}}_s$ to lie in the $x-z$ plane. Using the Cartesian unit vectors $\hat{\mathbf{i}}, \hat{\mathbf{j}}$, and $\hat{\mathbf{k}}$, we may write

$$\hat{\mathbf{r}} = \cos\varphi \sin\vartheta \hat{\mathbf{i}} + \sin\varphi \sin\vartheta \hat{\mathbf{j}} + \cos\vartheta \hat{\mathbf{k}}, \tag{6.80}$$

$$\hat{\mathbf{r}}_s = \cos\delta \hat{\mathbf{i}} + \sin\delta \hat{\mathbf{k}}, \tag{6.81}$$

where δ is the angle the solar vector makes with the equatorial plane (x–y plane). Clearly, δ depends on the time of the year. We readily compute

$$\hat{\mathbf{r}} \cdot \hat{\mathbf{r}}_s = \cos\delta \cos\varphi \sin\vartheta + \sin\delta \cos\vartheta. \tag{6.82}$$

To obtain the diurnal average of the solar power at $\hat{\mathbf{r}}$ per unit area, we must integrate $\hat{\mathbf{r}} \cdot \hat{\mathbf{r}}_s$ through the daylight hours (the range of φ for which $\hat{\mathbf{r}} \cdot \hat{\mathbf{r}}_s \geq 0$) and divide by the length of the *whole* day (2π). Dawn and dusk can be defined as $\pm H$, the roots of $\hat{\mathbf{r}} \cdot \hat{\mathbf{r}}_s = 0$.

$$\cos H = -\tan\delta \cot\vartheta. \tag{6.83}$$

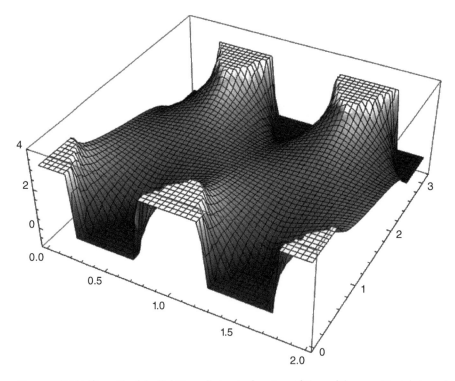

Figure 6.17 Half length of daylight in radians as a function of time of the year (t) and the polar angle ϑ for obliquity $\delta_0 = 35°$ and eccentricity $e = 0$. When the half day $H = \pi$, it indicates perpetual daylight. The obliquity is chosen to be 35° (as opposed to its present value of 23.47°) for illustration.

Figure 6.18 Graphic of the seasonal cycle of heating energy flux at the top of the atmosphere for a circular orbit and obliquity $\delta_0 = 23.47°$ as a function. The function is normalized by the total solar irradiance Q. (a) The three mode (T_0, T_1, T_2) insolation function. (b) The exact solution as described in Appendix A. A larger value of obliquity (~ 25°) is used to illustrate the polar day and night characteristics in the figure.

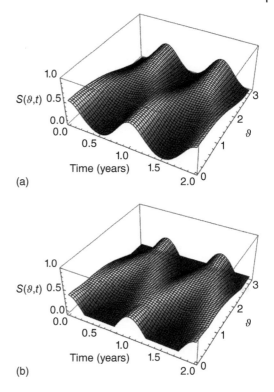

The half day is shown in Figure 6.17. The formula can now be written for the solar power per unit area averaged through the day,

$$\bar{\sigma}(\vartheta, \delta) = \frac{\sigma}{2\pi}(2\cos\delta\sin\vartheta\sin H + 2H\sin\delta\cos\vartheta), \tag{6.84}$$

where σ differs slightly from the total solar irradiance (because it is not annualized). Because of the elliptical orbit,

$$\sigma = \sigma_\odot \frac{r_0^2}{r^2}, \tag{6.85}$$

where r is the Earth–Sun distance at the given time of year and r_0^{-2} is the annual average of r^{-2}. Figure 6.18 shows the seasonal cycle of insolation function for a circular orbit.

6.9.1 The Elliptical Orbit of the Earth

The Sun lies at the focus of an ellipse that constitutes the Earth's trajectory over the year (Figure 6.19). Using the Sun as the origin of a polar coordinate system with λ as polar angle in the ecliptic plane (celestial longitude), we can write

$$r = \frac{ke}{1 - e\cos(\lambda - \lambda_0)}, \tag{6.86}$$

where e is the eccentricity (presently $e \approx 0.0167$),

$$k = \frac{a}{e}(1 - e^2), \tag{6.87}$$

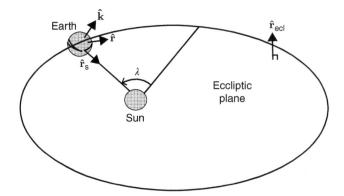

Figure 6.19 Diagram of the Earth's orbit illustrating the orientation of various unit vectors. The vector $\hat{\mathbf{k}}$ points out along the North polar axis. The vector $\hat{\mathbf{r}}$ is the position vector of a given point on the Earth's surface. The declination on a given day is the angle between $\hat{\mathbf{k}}$ and $\hat{\mathbf{r}}_s$, that is, $\cos\delta = \hat{\mathbf{k}} \cdot \hat{\mathbf{r}}_s$. The unit vector $\hat{\mathbf{r}}_{ecl}$ is perpendicular to the plane of the Earth's orbit (the ecliptic plane). The unit vector $\hat{\mathbf{r}}_s$ points from the Earth's center to the Sun. The angle λ is the celestial longitude.

and a^{-2} is very nearly the average of r^{-2} (to fourth order in powers of e). The value of λ_0 determines the time of year of the closest approach of the Earth to the Sun (perihelion). If we arbitrarily choose that $\lambda = 0$ at winter solstice ($\approx$December 22 in the present epoch), then at present $\lambda_0 \approx 0$, as perihelion occurs at present within a few weeks of (Northern Hemisphere) winter solstice. However, λ_0 increases linearly through 2π in a time of 22 000 years, leading to a passing of perihelion through the seasons with a period of 22 000 years.

In what follows, we can take λ_0 to be fixed and consider the time dependence of λ. If the Earth's orbit were circular, λ would be linear with time:

$$\lambda_{\text{circ}} - \lambda_0 = 2\pi t. \tag{6.88}$$

Since the Earth's orbit is really elliptical, we must use the conservation of angular momentum to compute $\lambda(t)$. This is expressed as $r^2\, d\lambda/dt = $ constant. After some manipulation, it can be shown that, to the first order in powers of e,

$$\lambda(t) - \lambda_0 = 2\pi t - 2e \sin 2\pi(t - t_0). \tag{6.89}$$

6.9.2 Relation Between Declination and Obliquity

The *obliquity* or tilt of the Earth's orbit is the angle between $\hat{\mathbf{k}}$ and $\hat{\mathbf{r}}_{ecl}$ or $\hat{\mathbf{k}} \cdot \hat{\mathbf{r}}_{ecl} = \cos\delta_0$. At present, this angle is about $23.47°$. The declination on a given day is the angle between $\hat{\mathbf{k}}$ and $\hat{\mathbf{r}}_s$, that is, $\cos\delta = \hat{\mathbf{k}} \cdot \hat{\mathbf{r}}_s$. With some straightforward geometry, we can show that

$$\sin\delta = -\sin\delta_0 \cos(\lambda - \lambda_0). \tag{6.90}$$

6.9.3 Expansion of $S(\mu, t)$

Consider now the expansion of $S(\mu, t)$ into a series of $P_n(\mu)$ and $\sin 2\pi nt$, $\cos 2\pi nt$, the latter being a Fourier series as the function is strictly periodic in t (units of years here).

$$S(\mu, t) = \sum_{k=0}^{K} \sum_{n=0}^{N} (a_{nk} \cos 2\pi kt + b_{nk} \sin 2\pi kt) P_n(\mu). \tag{6.91}$$

Table 6.1 Fourier–Legendre coefficients for the present distribution of incident solar radiation $S(\mu, t) = \sum (a_{nk} \cos 2\pi kt + b_{nk} \sin 2\pi kt) P_n(\mu)$.

n	k	a_{nk}	b_{nk}
0	0	1.0001	0.0000
	1	0.0327	0.0067
	2	0.0006	0.0003
1	0	0.0000	0.0000
	1	−0.7974	−0.0054
	2	−0.0261	−0.0052
2	0	−0.4760	0.0000
	1	−0.0180	−0.0026
	2	0.1486	−0.0022
4	0	−0.0444	0.0000
	1	−0.0029	0.0000
	2	0.0909	−0.0012

a) The coefficients are zero for odd values of n greater than one. $t = 0$ corresponds to the Northern Hemisphere winter solstice.

In principle, the coefficients a_{nk}, b_{nk} can be computed numerically from (6.84) and (6.83); however, the main seasonal driving term can be computed analytically:

$$b_{11} = 0 + O(e), \tag{6.92}$$
$$a_{11} = -2 \sin \delta_0 + O(e). \tag{6.93}$$

Furthermore, it can be shown that

$$a_{nk}, b_{nk} = 0 \quad \text{for } n \geq 3, n \text{ odd}. \tag{6.94}$$

Table 6.1 presents the first few coefficients for the present orbital parameters. Truncating the series at $N = 2$ and $K = 2$ is an excellent approximation for many purposes. Note that the a_{01} is due to the eccentric orbit. In fact, $a_{01} \approx 2e$, where the present eccentricity is about 0.016. More numerical values of orbital changes can be found in North and Coakley (1979).

7

Nonlinear Phenomena in EBMs

Most of the energy balance models (EBMs) we have considered so far have been essentially linear. This allowed us the luxury of using the well-known methods of classical theoretical physics that have been developed over the last few centuries. Linear systems have many great advantages including the decomposition of fields into modes that in many cases are orthogonal, but even systems with non-orthogonal modes can be handled. Linear models can also be analyzed in very elegant forms when statistical noise drives the system, as we will see in Chapter 9. Linear systems described by partial differential equations in space and time often have symmetries that can be exploited as well. Large classes of these linear systems can be classified and their properties can be studied in great detail and generality. In particular, if the homogeneous problem is translationally invariant in time (e.g., no time-dependent coefficients), we can use Fourier methods. If in space, the geometry is simple such as translationally invariant along the line, or in the infinite plane, or rotationally on a spherical surface, we can make use of many established results. In many of these high-symmetry cases, our forebears have developed an encyclopedic archive of special functions that can be used. If no such symmetries are evident in the problem, we still can use numerical methods to find the modes, their shapes, time constants, and so on. For example, the type of equation most commonly encountered in this book is the damped diffusion equation with nonhomogeneous driving terms. A prototype is

$$C\frac{\partial T}{\partial t} = \nabla \cdot D \nabla T - BT + \text{drivers}. \tag{7.1}$$

Similar equations occur in electricity and magnetism, heat transfer, particle and molecular diffusion, as well as in many other fields. In this equation, C/B is a linear-decay timescale, $\sqrt{D/B}$ is a length scale, and so on. When the drivers are set to zero, the initial anomaly field decays to zero, spreading and smoothing as it decays. Typically, large spatial scales have longer decay times than small spatial scales. If the coefficients are time independent, the dynamical decay modes can be found and usually they form a basis set into which the solution field and the drivers can be expanded. Modes found in the drivers will excite response modes in the solution field. In particular, frequency components appearing in the drivers will appear as energy (in our case, variance) in the responses at the very same frequencies.

Another class of problems encountered in the climate system involves a second time derivative on the left-hand side of the last equation. In this case, the response modes are wavelike. If $B = 0$, the waves are not damped; if $B > 0$, the waves are damped. There can

Energy Balance Climate Models, First Edition. Gerald R. North and Kwang-Yul Kim.
© 2017 Wiley-VCH Verlag GmbH & Co. KGaA. Published 2017 by Wiley-VCH Verlag GmbH & Co. KGaA.

also be both a second and a first time derivative (an example is waves on a string, set in an air-resistant medium), the first derivative being a damper as well as the *B* term. Other than an occasional mention, we will not deal with wave equations in this book, although there are some excellent sources for that fascinating subject (e.g., Pedlosky, 2003) .

There are a few nonlinear problems that are amenable to analytical solutions in the EBM world. One obvious example is the σT^4 form in blackbody radiation rules. This one turns out to be a mild nonlinearity not of much interest here, but it can be handled easily in the context of the potential function discussed in Chapter 2. This nonlinearity does not produce new solutions or alter the stability characteristics of the solutions we have already found. Another nonlinear issue arises when we consider the ice–albedo feedback mechanism. This particular mechanism leads to the same multiple-solution structure studied in Chapter 2, but there are a couple of new twists that we would like to elaborate on here. Another nonlinear problem of interest to EBM modelers (especially those working in vertical atmospheric profiles and other planets) is that of the runaway greenhouse which we considered in Chapter 4.

When more than one steady-state solution can exist for the same values of the parameters, we are always interested in the stability of such steady states. We will solve this problem for the one dimensional zonally symmetric planet in this chapter.

7.1 Formulation of the Nonlinear Feedback Model

Consider a north–south symmetric planet with no zonal features. We let the coalbedo *a* depend on μ (cosine of the polar angle) along with a discontinuity at the ice-cap edge, μ_s (see Figure 7.1):

$$a(\mu, \mu_s) = (0.68 - 0.241 P_2(\mu)) \times \begin{cases} 1.00, & |\mu| < |\mu_s|; \\ 0.75, & |\mu| = |\mu_s|; \\ 0.50, & |\mu| > |\mu_s|. \end{cases} \quad (7.2)$$

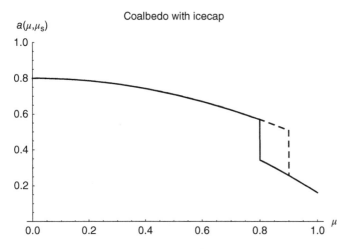

Figure 7.1 Shapes of the coalbedo as a function of sine of latitude, μ, in a solid line for the ice-cap edge at $\mu_s = 0.80$ and in a dashed line at $\mu_s = 0.90$.

The smooth μ dependence multiplying the step function mimics a zenith angle dependence and comes from satellite data (Graves *et al.*, 1993). The steady-state temperature field, $T(\mu)$, is governed by the heat conduction equation

$$-D\frac{d}{d\mu}(1-\mu^2)\frac{dT}{d\mu} + A + BT = QS_{\text{m.a.}}(\mu)a(\mu,\mu_s), \tag{7.3}$$

where D is a thermal diffusion coefficient; other symbols are as before. We require application of the boundary conditions:

$$\sqrt{1-\mu^2}\frac{dT}{d\mu}\bigg|_{\mu\to 0,1} = 0. \tag{7.4}$$

The equatorial boundary condition ensures symmetry, the polar condition states that no heat flux enters the poles. The necessity for the latter is, of course, just a consequence of our using the polar coordinate system. Basically, it forces only the regular solution of the energy balance equation at the pole. The function $S_{\text{m.a.}}(\mu)$ is the mean annual distribution of sunlight at the top of the atmosphere which is given approximately by

$$S_{\text{m.a.}}(\mu) \approx 1 - 0.477 P_2(\mu). \tag{7.5}$$

It is normalized, so that

$$\int_0^1 S(\mu)d\mu = 1. \tag{7.6}$$

Legendre polynomials are the eigenfunctions of the diffusion operator,

$$-\frac{d}{d\mu}(1-\mu^2)\frac{d}{d\mu}P_n(\mu) = n(n+1)P_n(\mu), \tag{7.7}$$

with eigenvalues $\lambda_n = n(n+1), n = 0, 2, \ldots$ Note that only evenly indexed modes are retained because of the north–south symmetry. Since they form a complete basis set, we can use them to express the temperature field

$$T(\mu) = \sum_{n=0,2,\ldots}^{\infty} T_n P_n(\mu), \tag{7.8}$$

with

$$T_n = (2n+1)\int_0^1 T(\mu)P_n(\mu)d\mu, \quad n \text{ even} \tag{7.9}$$

as

$$\int_0^1 P_n(\mu)P_m(\mu)d\mu = (2n+1)\delta_{nm}, \quad n \text{ even.} \tag{7.10}$$

Also define

$$H_n(\mu_s) \equiv (2n+1)\int_0^1 P_n(\mu)S(\mu)a(\mu,\mu_s)d\mu. \tag{7.11}$$

After inserting (7.8) into (7.3), multiplying through by $P_m(\mu)$ and integrating from 0 to 1, we find

$$[m(m+1)D + B]T_m = QH_m(\mu_s) - A\delta_{m,0}. \tag{7.12}$$

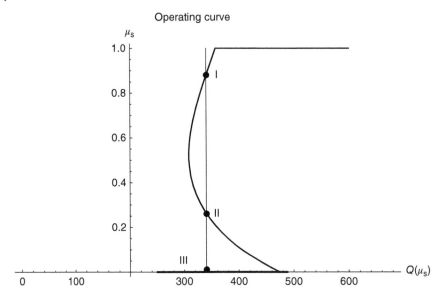

Figure 7.2 The solid curve is the sine of the latitude of the ice-cap edge, μ_s as a function of the solar irradiance $Q(\mu_s)$. The vertical thin line indicates the present value of the total solar irradiance ($\div 4$), 340 W m^{-2}. Note that there are three roots for this operating curve, I near the present climate, $\mu_s = 0.88$, II at $\mu_s = 0.26$, and III for an ice-covered planet at $\mu_s = 0$.

We may now solve for T_m and reconstruct $T(\mu)$ by use of (7.8).

$$T(\mu) = \sum_{n=0,2,\ldots}^{\infty} \frac{QH_n(\mu_s)P_n(\mu) - A\delta_{n,0}}{[n(n+1)D + B]}. \tag{7.13}$$

The problem is not yet solved, however, as we have not found the value of μ_s. This is done by enforcing the Budyko condition that the temperature at the ice-cap edge is $-10\,°C$. In other words,

$$T(\mu_s) = T_s = -10\,°C. \tag{7.14}$$

By evaluating (7.13) at $\mu = \mu_s$, we get an identity which constitutes a relation between Q and μ_s. We can plot $Q(\mu_s)$ as in Figure 7.2.

$$Q(\mu_s) = \frac{A + BT_s}{\sum_n \frac{H_n(\mu_s)P_n(\mu_s)}{n(n+1)D/B+1}}. \tag{7.15}$$

In the calculation that went into Figure 7.2, we included only the first two terms in the sum in the denominator of (7.15). Values of the parameters for the calculations were $D = 0.67$ (from Chapter 5), $a_f = 0.68 - 0.241P_2(\mu)$, $a_i = a_f/2$, $A = 208$ W m^{-2} (from Graves et al., 1993), $B = 2.00$ W m^{-2} K^{-1}.

7.2 Stürm–Liouville Modes

The differential equations we have encountered so far are amenable to solution by expansion into space–time modes. Usually, the time modes are simply the Fourier

Series amplitudes (harmonics). These amplitudes are often spatial modes whose graphs have more zero crossings the higher their index. There is a general class of these modes is called the *Stürm-Liouville system*. Consider the eigenvalue equation:

$$-\frac{d}{dx}\left(p(x)\frac{d\psi_n(x)}{dx}\right) + q(x)\psi_n(x) = \lambda_n w(x)\psi_n(x); \quad p(x), w(x) > 0, \quad (7.16)$$

where $q(x)$ is a continuous function on the interval. The interval of the domain can be taken to be symmetric: $-a \leq x \leq a$ with boundary conditions:

$$\alpha^{(+)}\psi_n(a) + \beta^{(+)}\psi'_n(a) = 0, \quad (7.17)$$
$$\alpha^{(-)}\psi_n(-a) + \beta^{(-)}\psi'_n(-a) = 0, \quad (7.18)$$

where $\psi'_n(\cdot)$ indicates the derivative evaluated at the argument in parentheses. The eigenvalues λ_n are bounded from below. To show this, multiply (7.16) by $\psi_n(x)$ and integrate by parts. The first term of the LHS becomes

$$-\int_{-a}^{a} \psi_n(x) \frac{d}{dx}\left(p(x)\frac{d\psi_n(x)}{dx}\right) dx$$
$$= \psi_n(x) p(x) \frac{d\psi_n(x)}{dx}\bigg|_{-a}^{a} + \int_{-a}^{a} p(x)\psi'_n(x)^2 \, dx. \quad (7.19)$$

The first term after the equal sign vanishes because of the boundary conditions and the second term is positive definite. We may write

$$\lambda_n \int_{-a}^{a} w(x)\psi_n^2(x) dx = \int_{-a}^{a} q(x)\psi_n^2(x) dx + \int_{-a}^{a} p(x)\psi_n'^2(x) dx \geq 0 \quad (7.20)$$

and $w(x), p(x) > 0$, and we assume $q(x) \geq 0$ (as $q(x)$ is the infrared radiative damping function in the EBMs, usually taken to be the positive constant coefficient B). All three integrals are positive, thus we have $\lambda_n \geq 0$. If the function $q(x)$ is negative over part of the interval, the λ_n will still be bounded from below, but not necessarily by a positive bound.[1]

7.2.1 Orthogonality of SL Modes

To prove orthogonality, begin with (7.16) and multiply through by $\psi_m(x)$. Rewrite the result with m interchanged with n. Now integrate both equations by parts from $-a$ to a. The LHS of the two are identical. This means we can subtract them to obtain

$$(\lambda_n - \lambda_m) \int_{-a}^{a} w(x)\psi_n(x)\psi_m(x) dx = N_{\text{norm},n}^2 \delta_{nm}, \quad (7.21)$$

where the coefficient $N_{\text{norm},n}^2$ is in case the functions are not normalized. They can be normalized by dividing $\psi_n(x)$ by $N_{\text{norm},n}$. Once the modes $\psi_n(x)$ are normalized, they form an *orthonormal set*. For the Legendre polynomials, $w(x) = 1, p(x) = (1 - x^2)$. Note that the Legendre polynomials are not normalized with respect to integration from -1 to 1, but rather their scale is set by the condition $P_n(\pm 1) = (\pm 1)^n$ for historical reasons.

---

1 The function $q(x)$ plays the role of the potential function in the steady-state Schrödinger equation. In that case, it could be negative over part of the interval, for example, the square well. In quantum mechanics, the energy levels are all bounded from below, otherwise we might fall into a bottomless pit.

Other important special functions such as the Hermite polynomials and Bessel functions utilize a weight function, $w(x) \neq 1$. We will encounter such a situation in Chapter 9 where the weight function $w(x)$ is related to the land–sea distribution on the planet.

We will soon encounter distinct modes that have the same eigenvalue. In that case, we say there is an ℓ-fold degeneracy, where ℓ is the number of distinct modes with that same eigenvalue. Two or more modes with the same eigenvalue have the property (easily shown by insertion in the eigen-equation) that $\alpha\psi_n^{(\lambda)}(x) + \beta\psi_m^{(\lambda)}(x)$ is also an eigenfunction with eigenvalue $(\alpha + \beta)\lambda$ where λ is the eigenvalue common to the two eigenfunctions. Knowing this, we can form two linear combinations (choices of α and β) that are mutually orthogonal and properly normalized. In this case, we can choose $\alpha = 1/\sqrt{2}$ and $\beta = \pm 1/\sqrt{2}$. Or

$$\phi_1 = \frac{1}{\sqrt{2}}(\psi_n + \psi_m), \tag{7.22}$$

$$\phi_2 = \frac{1}{\sqrt{2}}(\psi_n - \psi_m), \tag{7.23}$$

where ϕ_1 and ϕ_2 are the new orthonormal modes that have been orthonormalized.

7.3 Linear Stability Analysis

Once we have found that there are multiple solutions for the climate for a given value of the solar constant, we are obliged to examine the stability[2] of the solutions.[3] In this chapter, we first examine the linear stability of the solutions and find a slope-stability theorem analogous but not identical to that in Chapter 2. Next we construct and examine the potential function[al] for the one-dimensional problem. Consider now the EBM defined by

$$\frac{\partial I(\mu, t)}{\partial t} - \frac{\partial}{\partial \mu}\left[D(\mu)(1-\mu^2)\frac{\partial I(\mu, t)}{\partial \mu}\right] + I(\mu, t) = QS(\mu)a(\mu, \mu_s), \tag{7.24}$$

where we have set $C = 1$ to simplify the notation. Also, we have used $I(\mu, t) = A + BT(\mu, t)$ as our dependent variable, which simplifies the algebra in what follows. Note that this change means that the diffusion coefficient is different from the one where T is an dependent variable, that is, $D(\mu) = D_T/B$, where the subscript T indicates the case where T is the independent variable. This difference will have no effect on our stability analysis, as is the case where we have set $C = 1$. We have the ice-line constraint from Budyko,

$$I(\mu_s) = I_s > 0, \tag{7.25}$$

---

[2] There are several studies of stability of the solutions. For example, Budyko (1969) argued that the negative slopes should indicate a violation of the second law of thermodynamics. Others include Ghil (1976) and a more general study by Cahalan and North (1979). The proof given here follows that in North et al. (1981).
[3] The proof of the *slope-stability theorem* is rather involved such that some readers may wish to skip directly to the punch line (7.47) and Figure 7.3.

where $I_s = A + B \cdot (-10\,°C)$ and subject to the usual north–south symmetric boundary condition

$$D(\mu)\sqrt{1-\mu^2}\frac{dI}{d\mu}\bigg|_{\mu=0,1} = 0. \tag{7.26}$$

We are including a μ-dependence in $D(\mu) > 0$ to make the results as general as possible. As in Chapters 5 and 6 and the previous section of this chapter, we expand $I(\mu, t)$ into the eigenfunctions of the total convergence-of-flux operator of (7.24). It is an example of a Stürm–Liouville System, whose properties were outlined in the previous section.

$$\left\{-\frac{d}{d\mu}D(\mu)(1-\mu^2)\frac{d}{d\mu} + B\right\}f_n(\mu) = \ell_n f_n(\mu), \tag{7.27}$$

where the $f_n(\mu)$ are the Stürm–Liouville orthonormal eigenfunctions of degree n, and the eigenvalues ℓ_n are real and positive. We define

$$h_n(\mu_s) \equiv \int_0^1 S(\mu)a(\mu, \mu_s)f_n(\mu)d\mu. \tag{7.28}$$

The linear stability analysis begins by assuming the temperature field is perturbed by a small amount $\delta I(\mu, t)$:

$$I(\mu, t) = I^{eq}(\mu) + \delta I(\mu, t), \tag{7.29}$$

and the ice cap is similarly perturbed.

$$\mu_s(t) = \mu_0 + \delta\mu_s(t). \tag{7.30}$$

The equilibrium values satisfy

$$\ell_n I_n^{eq} = Q h_n(\mu_0), \tag{7.31}$$

$$\sum_n I_n^{eq} f_n(\mu_0) = I_s. \tag{7.32}$$

Inserting the small departure definitions and retaining only the linear terms, we have

$$\frac{d}{dt}\delta I_n + \ell_n \delta I_n = Q h'_n(\mu_0)\delta\mu_s, \tag{7.33}$$

where $h'_n(\mu_0)$ indicates the derivative of $h_n(\mu_s)$ evaluated at $\mu_s = \mu_0$ where μ_0 is the value of μ_s at equilibrium. Next, we use the relation

$$I_s = \sum_n I_n(\mu_s) f_n(\mu = \mu_s),$$

with the constraint $\delta I_s = 0$ to find

$$\delta I_s = \sum_n f_n \delta I_n + \left(\sum_n I_n^{eq} f'_n\right)\delta\mu_s = 0, \tag{7.34}$$

where we have now suppressed the argument of f_n to mean its value at μ_0 and f'_n, indicating the derivative of $f_n(\mu)$ evaluated at μ_0.

The last expression becomes

$$\delta\mu_s = -\left(\sum_k f_k \delta I_k\right)\left(\sum_m I_m^{eq} f'_m\right)^{-1}. \tag{7.35}$$

In the last formula and in what follows, we suppress the argument μ_0. We may substitute the last equation into (7.33) and after making the small-departure assumption that

$$\delta I_n(t) = \delta I_n \, e^{-\lambda t}, \tag{7.36}$$

as in Chapter 2, we find the eigenvalue problem

$$\sum_m M_{nm}\delta I_m = \lambda \delta I_n, \tag{7.37}$$

where

$$M_{nm} = \ell_n \delta_{nm} + \gamma f_n f_m, \tag{7.38}$$

with

$$\gamma = Q\Delta a S(\mu_0)\left(\sum_k I_k^{eq} f_k'\right)^{-1}. \tag{7.39}$$

To obtain the last equation, we assumed that $a(\mu, \mu_s)$ is a step function with discontinuity $\Delta a > 0$ at $\mu = \mu_s$,

$$h_n' = (\Delta a)S(\mu_0)f_n. \tag{7.40}$$

The stability of the system is determined by the sign of the eigenvalues λ. Because M_{mn} is real and symmetric, all eigenvalues are real and bounded from below.[4] If the lowest eigenvalue is negative, the system is unstable; $\delta T(t)$ grows exponentially in time. By casting the eigenvalue problem into a different form, we can determine the sign of the lowest root. From this point in the proof, all arguments are suppressed with the understanding that the function in question is to be evaluated at $\mu = \mu_0$.

To determine this sign, we rearrange (7.37) and use (7.38) to obtain

$$(\ell_n - \lambda)\delta I_n = -\gamma f_n \sum_m f_m \delta I_m. \tag{7.41}$$

Dividing this expression by $(\ell_n - \lambda)$, multiplying by f_n, and summing over n we obtain

$$1 = -\gamma \sum_n \left(\frac{f_n^2}{\ell_n - \lambda}\right). \tag{7.42}$$

This relation is a transcendental equation that is satisfied for certain discrete values of λ, the stability eigenvalues. By further rearrangement, we arrive at the sign of the lowest eigenvalue. Using the definition of γ, we find

$$\sum_n \left(\frac{f_n^2}{\ell_n - \lambda} + \frac{I_n^{eq} f_n'}{Q\Delta a S(\mu_0)}\right) = 0, \tag{7.43}$$

which, with the equilibrium condition, becomes

$$\sum_n \left(\frac{\Delta a S(\mu_0) f_n^2}{\ell_n - \lambda} + \frac{h_n f_n'}{\ell_n}\right) = 0. \tag{7.44}$$

---

4 This result is a well-known property of real symmetric matrices.

Since

$$\frac{d}{d\mu_0}I_s = 0 = \frac{d}{d\mu_0}\left[Q\sum_n \frac{h_n f_n}{\ell_n}\right], \quad (7.45)$$

this leads to

$$\sum_n \frac{h_n f_n'}{\ell_n} = -\frac{1}{Q^2}\frac{dQ}{d\mu_0}I_s - \sum_n \frac{h_n' f_n}{\ell_n}. \quad (7.46)$$

Finally, substituting (7.46) into (7.44) we obtain

$$\frac{dQ}{d\mu_0} = \lambda \frac{Q^2}{I_s} \sum_n \frac{\Delta a S(\mu_0) f_n^2}{\ell_n(\ell_n - \lambda)}. \quad (7.47)$$

Figure 7.3 shows a plot of the LHS and the function on the RHS as functions of the stability parameter λ. Intersections of the curve and the horizontal line indicate roots of the equation above and are therefore the eigenvalues corresponding to the stability problem. If all roots are positive, the climate corresponding to μ_0 is stable. However, if even one root is negative, the climate at that point on the operating curve will be unstable. We see from the graph that as long as the slope of the operating curve (μ_s vs Q) is positive, the climate will be stable. But if the dashed line falls below the horizontal axis (meaning there is a negative slope of the operating curve), the left-most root in the figure will become negative, indicating instability. This is referred to as the *slope-stability theorem*. Note that it differs from the case treated in Chapter 2 in which the operating curve T versus Q was used to calculate the slope. In the one-dimensional model, the slope must be calculated from the μ_s versus Q curve.

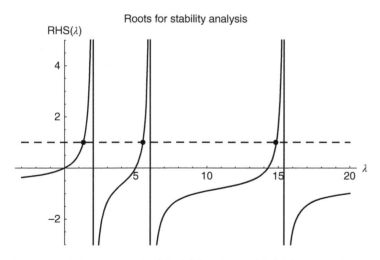

Figure 7.3 Schematic graph of the left-hand side of the stability equation versus the right right-hand side, where $(Q^2/I_s)\Delta aS(\mu_0)f_n^2$ are taken to be positive quantities for all n. The ℓ_n are taken to be the eigenvalues for a constant diffusion coefficient model ($n(n + 1)D + B$). Intersections of the horizontal line (nominally $dQ/d\mu_0$) with the continuous curves represent roots of the system and therefore eigenvalues $\lambda_m, m = 0, 2, 4, \ldots$. The horizontal line will lie above the abscissa so long as the slope of the operating curve is positive. All these positive roots imply stability. When the slope of the operating curve is negative (horizontal dashed line below the λ-axis), there will be one root λ_{dd} which is negative, leading to instability.

Another proof of the slope-stability theorem somewhat simpler in complexity than that presented here is given by Shen and North (1999). The most general proof is by Cahalan and North (1979). Drazin and Griffel (1977) examined the solutions in the case where the hemispheres might not be constrained to be symmetric. Since the length scale is short compared to the circumference of the planet, an ice cap in one hemisphere cannot feel the effect of an ice cap in the other. Indeed, they found this to be the case by careful numerical solution of the problem (using the "shooting method" in which one fits the polar condition in one hemisphere and adjusts the ice cap in the other by trial and error until a match is made). Our proof of the slope-stability theorem would not hold in such a situation, as for some solutions the ice-cap size in one hemisphere is not the same size as that in the other.

One wonders what the stability conditions would be if there is nontrivial geography as in the next chapter. This problem has not been addressed to our knowledge except incidentally during numerical solutions in paleoclimatology. We do not even have a proof as yet for the case of a perturbation of the ice-cap edge that has wave number above zero for the homogeneous planet case. Such proofs await the creative efforts of others.

7.4 Finite Perturbation Analysis and Potential Function

In analogy to the potential function analysis of Chapter 2, we present here its generalization to the case of one-dimensional climate models (North *et al.*, 1979). In this case, we must deal with a *functional* instead of a function, as the potential will depend on the function $T(\mu)$ at each point in the interval $0 < \mu < 1$. We can best describe what is meant by a functional by directly proceeding with our example. Consider the integral

$$F[T] = \int_0^1 d\mu \left[\frac{1}{2} D(1 - \mu^2) T_\mu^2 + R(T) - QS(\mu) G(T, \mu) \right], \qquad (7.48)$$

where for the sake of compactness we have introduced subscripts for partial derivatives $T_\mu = dT/d\mu$,

$$R(T) = \int^T I(T') dT', \qquad (7.49)$$

$$G(T, \mu) = \int^T a(\mu, T') dT'. \qquad (7.50)$$

Here we have chosen to write $a(\mu, \mu_s)$ as a function of μ and T, for example,

$$a(\mu, T) = (a_0 + a_2 P_2(\mu)) \left(\theta(T(\mu) - T_s) + \frac{1}{2} \theta(T_s - T(\mu)) \right), \qquad (7.51)$$

where $\theta(y)$ is the unit step function,

$$\theta(y) = \begin{cases} 0, & \text{if } y < 0, \\ 1, & \text{if } y \geq 0. \end{cases} \qquad (7.52)$$

Note that as

$$\int_{-\epsilon}^{\epsilon} \delta(y' - y) \, dy' = \theta(y); \quad \text{then } \frac{d\theta(y)}{dy} = \delta(y). \qquad (7.53)$$

Consider the functional for an arbitrary small variation $\delta T(\mu)$; that is, $T(\mu)$ is replaced by $T(\mu) + \delta T(\mu)$ to form $F[T] + \delta F[T]$. After subtracting $F[T]$, we have

$$\delta F[T] = \int_0^1 d\mu \, [D(1-\mu^2)T_\mu(\delta T(\mu))_\mu + R'(T)\delta T(\mu) + QS(\mu)G'(T)\delta T(\mu)], \tag{7.54}$$

where we used only the first-order terms in Taylor expansions of R and G in δT. Now noting that

$$\delta(T_\mu) = (\delta T)_\mu, \tag{7.55}$$

we integrate the first term above by parts (endpoint contributions vanish) to obtain

$$\delta F[T] = \int_0^1 d\mu \, [-(D(1-\mu^2)T_\mu)_\mu + R'(T) + QS(\mu)G'(T)]\delta T(\mu), \tag{7.56}$$

where the prime denotes differentiation with respect to T. If the functional $F[T]$ is to be stationary for an arbitrary but small variation $\delta T(\mu)$, it must vanish. Since the functional form of $\delta T(\mu)$ is arbitrary, the quantity in brackets in the integrand above must vanish. The latter is just the expression for the energy balance equation. In other words, our functional is a quantity which is a local extremum when the energy balance equation is satisfied.

It is easier to visualize the functional $F[T]$ in spectral form, $T = \sum T_n f_n(\mu)$. In the spectral form, we may think of F as an ordinary function of the variables $T_0, T_2, \ldots$; an infinite number of variables. An extremum of $F(T_0, T_2, \ldots)$ may be expressed as $\partial F/\partial T_n = 0$ for all n.

Substitution of the spectral form into the definition of the functional $F[T]$ leads to

$$F(T_0, T_2, \ldots) = \sum_n \frac{1}{2}\ell_n T_n^2 - M(T_0, T_2, \ldots), \tag{7.57}$$

where

$$M(T_0, T_2, \ldots) = Q\int_0^1 S(\mu)(T - T_s)[a_f\theta(T - T_s) + a_i\theta(T_s - T)]d\mu. \tag{7.58}$$

For simplicity, we have taken a_f and a_i, the values of coalbedo over ice-free and ice-covered surfaces, to be constants. It is understood that $T = \sum T_n f_n(\mu)$ is to be substituted for T in the last expression.

The condition that the $\partial F/\partial T_n$ vanish simultaneously leads to

$$\ell_n T_n = Q\int_0^1 S(\mu)f_n(\mu)[a_f\theta(T(\mu) - T_s) + a_i\theta(T_s - T(\mu))]d\mu,$$
$$= Qh_n(\mu_s), \tag{7.59}$$

which takes us back to the slope-stability theorem.

7.4.1 Neighborhood of an Extremum

Consider a particular extremum (steady state) centered at the point

$$T_0^{(0)}, T_2^{(0)}, \ldots.$$

Nearby (in function space), we may write $T(\mu) = T^{(0)} + \phi(\mu)$, or the deviation may be written in terms of its spectral components $\phi_0, \phi_2, \ldots$. Expanding $F[T]$ about the local extremum, we obtain

$$F(T_0, T_2, \ldots) = F_0 + \sum_n \left(\frac{\partial F}{\partial T_n}\right)_0 \phi_n + \frac{1}{2}\sum_{n,m}\left(\frac{\partial^2 F}{\partial T_n \partial T_m}\right)_0 \phi_n \phi_m + \cdots, \quad (7.60)$$

where the subscript zero denotes evaluation at the extremum. The terms linear in ϕ_n vanish because $\partial F/\partial T_n$ vanishes at the extremum. Up to the terms considered, F is locally a quadratic in ϕ_n. The matrix elements,

$$N_{nm} = \left(\frac{\partial^2 F}{\partial T_n \partial T_m}\right)_0 \quad (7.61)$$

are the structure constants for the quadratic geometric surface, $F(T_0, T_2, \ldots)$. If all eigenvalues of N_{nm} are positive, the surface is concave upward; if one or more of the eigenvalues are negative, the surface is locally a saddle point. We proceed to show that these eigenvalues are the stability eigenvalues studied earlier.

First note that if the temperature field is allowed to be a function of time, then by following the approach for the zero-dimensional models, we have

$$\dot{T}_n = -\left(\frac{\partial F}{\partial T_n}\right); \quad (7.62)$$

that is, the time derivative is given by the gradient in this multidimensional space. For infinitesimal departures from steady state, we set $T_n(t) = T_n^{eq} + \phi_n\, e^{-\lambda t}$ above, expand about $\phi_n = 0$, and obtain

$$-\lambda \phi_n = -\sum_m \left(\frac{\partial^2 F}{\partial T_n \partial T_m}\right)_0 \phi_m, \quad (7.63)$$

$$-\lambda \phi_n = -\sum_m N_{nm} \phi_m. \quad (7.64)$$

The latter equation concludes the proof that the local geometrical structure constants of $F(T_0, T_2, \ldots)$ yield the stability eigenvalues for that particular steady state. The last equation is the analog of the simple zero-dimensional equation (2.90) in Chapter 2.

Finally, as a conclusion to this section, consider the time behavior of the value of $F(T_0, T_2, \ldots)$ when the point $(T_0, T_2, \ldots)$ is governed by the time-dependent energy balance equation

$$\frac{dF}{dt} = \sum_n \frac{\partial F}{\partial T_n} \dot{T}_n, \quad (7.65)$$

$$\frac{dF}{dt} = -\sum_n (\dot{T}_n)^2, \quad (7.66)$$

where we inserted the equation of motion. This latter result is the multidimensional analog of (2.91) in Chapter 2. It has a corresponding interpretation: initial departures of the state $(T_0, T_2, \ldots)$ from a local extremum of F lead to a trajectory of the system point such that F decreases. The point will continue down the gradient of F until a local extremum is found. Clearly, saddle points are unstable—if perturbed infinitesimally, the system point can leak out into some neighboring basin.

7.4.2 Relation to Gibbs Energy or Entropy

Solutions of the energy balance equation satisfy a minimum principle reminiscent of Gibbs energy in thermodynamics (when pressure and temperature are held fixed). If we take the negative for $F[T]$, it might be interpreted as the maximum of entropy production. Then the extremum condition would be somewhat like Prigogine's theory of nonequilibrium thermodynamic states (Prigogine, 1968). Golitsyn and Mokhov (1978) showed that the linear climate models satisfy Prigogine's conditions. The idea is tantalizing, but has so far not been satisfactorily implemented in general terms.

7.4.3 Attractor Basins—Numerical Example

It is possible to work out in some detail an example illustrating the concept of the potential surface for a two-mode model. In this case, the functional F becomes the truncated version of (7.57). The function M is

$$M(T_0, T_2) = Q \int_0^{\mu_s} S(\mu')(T_0 + T_2 P_2 - T_s) a_f \, d\mu'$$
$$+ Q \int_{\mu_s}^1 S(\mu')(T_0 + T_2 P_2 - T_s) a_i \, d\mu', \tag{7.67}$$

where, in the two-mode model, the ice edge can be expressed in terms of T_0, T_2:

$$\mu_s = \frac{1}{\sqrt{3}} \left(1 + \frac{2(T_s - T_0)}{T_2}\right)^{\frac{1}{2}}. \tag{7.68}$$

Together with the other terms in (7.57), we have enough information to plot the surface corresponding to $F(T_0, T_2)$ as a contour diagram in the T_0, T_2 plane. We choose the parameters such that there is a cusp: $Q = 340\,\text{W m}^{-2}$, $A = 220\,\text{W m}^{-2}$, $B = 1.90\,\text{W m}^{-2}\,\text{K}^{-1}$, $a_f = 0.68$, $a_i = 0.38$, $D/B = 0.30$. The operating curve for this system is shown in Figure 7.4a. This choice of parameters leads to five solutions labeled a, b, c, d, and e. The potential surface is mapped in (b) with the steady-state points indicated. It can be clearly seen that d is a saddle point corresponding to an unstable solution. Figure 7.4c and d shows successively higher resolution focusing on the right-hand basin. The higher resolution shows that there are two distinct minima: a and c with a saddle point b. The ice-free solution a corresponds to a very shallow minimum and a very small but finite agitation would push it over the hill (b) into the deeper (more-stable) basin c.

The relative minima in a purely dissipative system such as those we are studying here are isolated points called *attractors*. It is possible to map out the attractor basins for this class of problems as illustrated by the two-mode approximation with the basins shown in Figure 7.4.

7.5 Small Ice Cap Instability

Consider the operating curve in Figure 7.5. This two-mode operating curve has five equilibrium solutions for $Q = 340\,\text{W m}^{-2}$. Because of the slope-stability theorem, only the ones with positive or zero slope are stable. The cusp at the top of the graph is magnified in the right-hand panel as the solid line. To illustrate the sensitivity of the cusp, we can

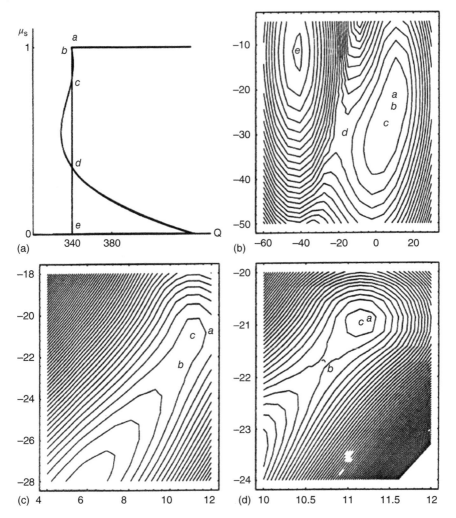

Figure 7.4 Potential function for a two-mode, one-dimensional nonlinear climate model. (a) The operating curve indicating five solutions (labeled a, b, c, d, e) for $Q = 340$ W m^2. (b) A coarse resolution plot of the potential function in the T_0, T_2 plane. (c) and (d) Successively higher resolution around the right-hand basin which turns out to have two local minima. The relative minima are stable climate states, the saddle points correspond to unstable states. Further discussion in the text.

vary the coalbedo of ice cover. According to Figure 7.6 (take the solid curve parameters), the operating curve has a negative slope for values of $\mu_s > 0.89$, which corresponds to a latitude of about 63°. The negative slope of the operating curve implies that polar ice caps whose radius is smaller than about 25–37° on a great circle will be unstable. This point is worth examining more closely as it may have implications for paleoclimate and for future climates. For example, the Antarctic ice cap is of about this size and the seasonal sea ice in the North Pole area is also of roughly this size. Is it on the verge of unstable collapse if the planet is heated slightly by the greenhouse effect? Moreover, did it form suddenly? What parameters control the size of the least stable ice cap?

Figure 7.5 An operating curve with five equilibrium solutions for $Q = 340\,\text{W}\,\text{m}^{-2}$. In this illustration, the $D/B = 0.30$, $A = 210\,\text{W}\,\text{m}^{-2}$; $B = 1.90\,\text{W}\,\text{m}^{-2}\,\text{K}^{-1}$, and the coalbedo is taken to be flat as function of latitude with $a_f = 0.68$ and $a_i = 0.38$.

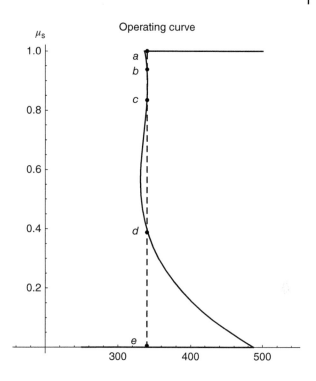

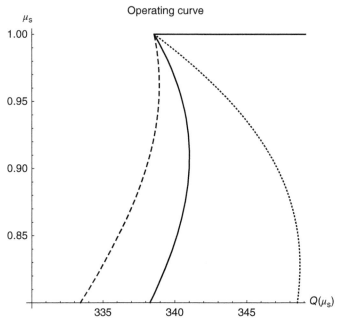

Figure 7.6 Illustration of the sensitivity of the cusp's shape to small changes in the coalbedo of ice: The solid curve is for the same values as in Figure 7.5 ($a_f = 0.68$, $a_i = 0.38$). The long-dashed (leftmost) curve is for dark ice ($a_i = 0.42$), while the rightmost curve is for bright ice ($a_i = 0.30$).

7.5.1 Perturbation of an Exact Ice-Free Solution

Let us imagine a situation where we are on the ice-free branch of the solution and slowly lower the solar constant until we are at $Q = 340$ W m^{-2}. Our solution will constitute a point on the upper branch of Figure 7.6. The model is linear in the (infinitesimal) neighborhood of this solution as there is no ice cap. Using the parameters of Figures 7.5 and 7.6, we find that the temperature is given by

$$T_a(\mu) = 11.2 - 20.7 P_2(\mu). \tag{7.69}$$

At every point, the temperature is above $-10\,°$C (Figure 7.7). At the pole, the temperature is $-9.5\,°$C, which is close to the critical value. Consider next the effect of adding a heat source centered at the pole whose density is $q(\mu)$ (W m^{-2}). Let the departure from the ice-free solution be $T'(\mu)$. It satisfies

$$-D\frac{d}{d\mu}\left[(1-\mu^2)\frac{dT'}{d\mu}\right] + BT' = q(\mu). \tag{7.70}$$

This equation can be solved by use of Legendre polynomial series on the sphere, but it is more instructive to look at the solution on the plane tangent to the pole where solutions in closed form can be obtained. Let r be the plane polar coordinate corresponding to the distance from the pole. And let $H(r)$ be the response function to the thermal perturbation $q(r)$. Then $H(r)$ satisfies

$$\frac{1}{r}\frac{d}{dr}\left[r\frac{d}{dr}H(r)\right] - \frac{H(r)}{\lambda_{dd}^2} = -\frac{q(r)}{D}, \tag{7.71}$$

where the length scale is defined as $\lambda_{dd} = \sqrt{D/B}$.

For a small patch of ice (diameter $\ll \lambda_{dd}$), the solution[5] is

$$H(r) \approx \frac{1}{D}K_0\left(\frac{r}{\lambda_{dd}}\right)\int_0^{r_0} r'q(r')dr', \quad r > r_0, \tag{7.72}$$

where r_0 is the radius of the ice patch and $K_0(z)$ is the zero-order modified Bessel function which is pictured as the solid curve of Figure 7.8. This particular Bessel function looks like a decaying exponential and in fact has the asymptotic form $K_0(z) \sim (\pi/2z)^{1/2}e^{-z}$ for large z. The important point is that the influence of the small patch extends from its edge a distance the order of λ_{dd}. It is hardly surprising that stable ice caps smaller than about λ_{dd} in radius did not exist in model solutions.

Imagine the no-icecap solution to be in place; this state's temperature ($-9.5\,°$C) is above the critical value ($-10\,°$C). Now "by hand," we add a small ice patch at the pole. The patch as a source of cooling has an influence a distance $\sim \lambda_{dd}$ away from the pole. If the albedo of the ice is high enough, the patch can pull the temperature down below critical over a distance comparable to λ_{dd}. This means that two solutions can exist, one with no ice and another with a patch no smaller than approximately λ_{dd}. This is the underlying meaning of the small ice cap instability. More details can be found in North (1984).

---

5 The exact solution can be found by the Green's function technique. It is given in North (1984).

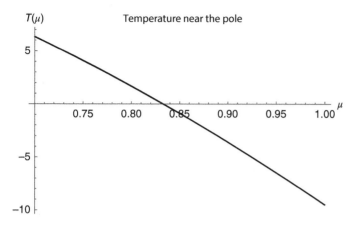

Figure 7.7 Plot of the temperature near the pole as function of μ, the cosine of the polar angle in the case where the temperature at the pole is just above the critical value of $-10\,°C$. The value of the curve $T_a(\mu) = 11.2 - 20.7 P_2(\mu)$; at the pole, this yields $T_a(1) = -9.5\,°C$. What happens if a small patch of ice is placed at the pole?

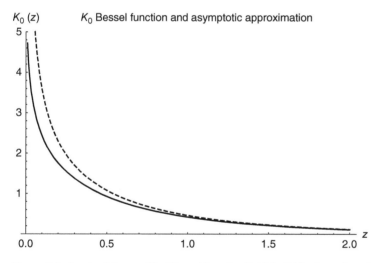

Figure 7.8 Graph of the modified Bessel function $K_0(z)$ (solid line) as a function of z and an asymptotic approximation, $\sqrt{(\pi/2z)}e^{-z}$ (dashed line). The $K_0(z)$ Bessel function is the response to a point source in plane polar coordinates. The figure shows that it is nearly proportional to ze^{-z} for large z. It partially answers the question posed in the previous figure.

7.5.2 Frequency Dependence of the Length Scale

Next consider the solution to a small scale source in the heating problem, when the strength of heating has a sinusoidal time dependence: $q(r,t) = q_0(r)e^{2\pi i f_0 t}$. Then we should use the time-dependent version of the energy balance equation:

$$C\frac{\partial H(r,t)}{\partial t} - D\frac{1}{r}\frac{d}{dr}\left[r\frac{d}{dr}H(r,t)\right] + BH(r,t) = +q(r,t). \tag{7.73}$$

By the assumption that $H(r,t)$ has the same sinusoidal dependence as $q(r,t)$, we find the same result except that $1/\lambda_{dd}^2$ is replaced by

$$\frac{1}{\lambda_{dd}'^2} = \frac{1}{\lambda_{dd}^2}(1 + 2\pi i f_0 \tau_0), \tag{7.74}$$

or its magnitude is

$$\left|\frac{\lambda_{dd}'}{\lambda_{dd}}\right| \sim \frac{1}{\sqrt{1 + (2\pi f_0 \tau_0)^2}}. \tag{7.75}$$

To see the r dependence of the solution, we plot the magnitude of $K_0(r/\lambda_{dd}')$ in Figure 7.9. We see that the characteristic length scale is strongly dependent on the frequency of the forcing as the forcing frequency is raised to $1/\tau_0$. Over this interval of frequency, the length scale shrinks by a factor of 2. It continues to shrink toward zero as the frequency is increased. The magnitude of the K_0 Bessel function decreases similarly to an exponential as a function of distance from the origin, but the real and imaginary parts of the function oscillate as dampened pulses diffuse away from the origin.

An application of the shortened length scale for oscillating point sources is in the seasonal cycle. In this case, over land τ_0 is about $1/12$ year and $f = 1$ year^{-1}; hence with respect to a land surface, the seasonal cycle is very low frequency and the low frequency limit is a reasonable approximation. On the other hand, over ocean, τ_0 is about 5 years and $f = 1$ year^{-1}, hence, we are far from the low frequency limit over oceans. The consequence is that in seasonal cycle applications the reach or fetch of a seasonally oscillating source is only a fraction of its low frequency or static value. During the seasonal cycle, the oceans and land surfaces have different lags behind the heating. The above considerations mean that a neighboring lagging ocean surface has an influence

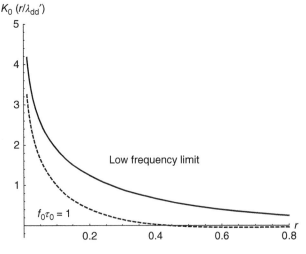

Figure 7.9 *Solid line*: The magnitude of the complex modified Bessel function $K_0(r/\lambda_{dd}')$ as a function of r, where $\lambda_{dd}' = \lambda_{dd}/\sqrt{1 + 2\pi f_0 \tau_0}$ for $\lambda_{dd} = 1.0$ and $f_0 = 0$ (the low frequency limit). The dashed line indicates how the length scale is shortened for higher frequency forcing. In this latter case, $f_0 \tau_0 = 1$. The length scale tends to zero as the frequency increases further.

far inland into a continent. But the influence of the land temperatures on neighboring ocean surfaces extends only a fraction of that distance off the shore. A larger study of the frequency dependence of the length scale can be found in North *et al.* (2011).

One caveat about the small ice cap instability is its sensitivity to the parameters chosen. Even tiny changes in the parameters of the model can cause it to appear or not in the operating curve. One might conclude that it is an artifact of such a simple model. However, it might be indicative of phenomena that do actually occur in the climate system. The fact that it is so sensitive to the parameter values suggests that it might be there at some times in history and not at others depending on the environment at the time.

7.6 Snow Caps and the Seasonal Cycle

The latitude of the steady-state snow line varies through the seasons following roughly the freezing line. By steady state of course, we mean the ensemble average over many seasonal cycles as before. We postpone the issue of the snow-line realizations in the presence of climate fluctuations. The seasonal progression of the snow line and its modification under changed external forcing conditions has important consequences as we will see. A very important issue is whether snow lingers through the summer at a given location, say the pole. If this does happen, there is the possibility of growing an ice sheet if there is persistent snow over land.

First consider an all-ocean planet. In this situation, the seasonal cycle is small and the pole will remain below freezing all year round. The Arctic Ocean may be an analog to this case and it is interesting to note that there is perennial sea ice at the North Pole and that this may be suggestive of a small ice cap poised for abrupt removal under the influence of a warm forcing such as the greenhouse effect or low frequency changes in oceanic conditions. Removal of perennial sea ice at the North Pole would have negligible direct thermal effects on the planet but it might cause profound changes in the circulation of the Arctic Ocean and consequently the formation of deep water in the North Atlantic.

Next consider an all-land planet. It is easy to show that, in this case, with the present obliquity (tilt of the Earth's spin axis with respect to the ecliptic plane) the summers are very hot at the poles (exceeding 30 °C). Hence, for the all-land planet there will be no persistent snow at the poles and the planet is much too hot for a small ice cap summer solution.

7.7 Mengel's Land-Cap Model

An interesting intermediate case is that of a symmetrical cap of land at the pole surrounded by open ocean (Mengel *et al.*, 1988). The analog of this case is the continent of Antarctica whose size is about the size of an annual cycle length scale in the simple climate models.

As an introduction to problems with partial land and sea geographies, consider a planet with a disklike island centered at the pole. The rest of the planet is ocean covered. If the solar irradiance is in a certain range, the temperature will fall below freezing in winter in which case the albedo will become higher in the snow-covered areas.

This is a model that still has only one horizontal dimension, the latitude. The way of implementing the land–sea contrast is to introduce an effective heat capacity $C(\mu)$ that is small over land and large over ocean. This means the coefficient of the $\partial T/\partial t$ term has a strong discontinuity at the land cap's edge. Solving this boundary value problem analytically is not likely to be possible; hence, one must resort to numerical methods. Other problems include how to handle the heat capacity when an area of ocean surface is ice covered. In this initial study, this latter effect has been ignored.

Mengel *et al.* (1988) used an expansion of the temperature field into Legendre modes. Because of the discontinuity in albedo and heat capacity, many modes (29) had to be retained in the expansion, but experiments with half that many did not reveal different results. Details of the methodology can be found in the paper.

Figures 7.10 and 7.11 show results of numerical calculations for such a zonally symmetric planet. The figure shows that for warm values of the solar constant, the snow line retreats rapidly to the pole in spring and reappears in fall, moving well out into the ocean areas in winter. In other words, we have a snow-free summer over the land cap. However, as the solar constant is lowered, an abrupt transition occurs to a climate configuration with persistent snow throughout the summer. This transition is a form of the small ice cap instability but is more complicated than the simple static ice cap studied earlier in this chapter. It is clearly affected by the seasonality and the geography. The smallness of the land cap makes it possible for the oceanic influence to stretch all the way to the pole, keeping summers marginally cool enough to support ice in summer if the snow albedo is present. On the basis of this model finding (and it persists in general circulation model (GCM) simulations), one might expect that the transition to full glaciation on Antarctica might have been rather sudden as the parameters passed slowly over the critical point. Of course, in reality, it was not the changing solar

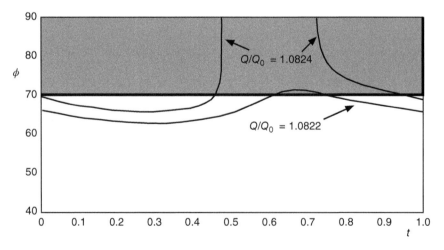

Figure 7.10 Illustration of the extreme sensitivity of the seasonal cycle of the snow line to solar brightness when a bifurcation (tipping point is near). Shown is a plot of the seasonal variation of the snow line for a land-cap planet (analogous to Antarctica) for two neighboring values of normalized solar constant. The shaded area indicates land area. The ordinate is the latitude ϕ in degrees and the abscissa is time t in years. The origin is at NH winter solstice. (Mengel *et al.* (1988). Reproduced with permission of Springer.)

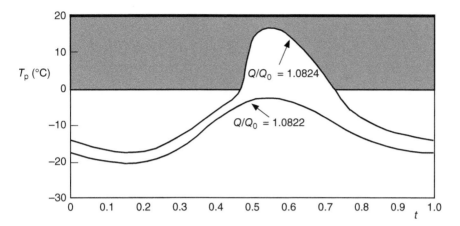

Figure 7.11 Similar to Figure 7.10, except that the variable plotted is the temperature at the pole. Shown is a plot of the seasonal variation of the polar temperature as a function of time of the year. The shaded area indicates land area. It shows the temperature region above freezing. The origin is at winter solstice. The values of total solar irradiance for this pair of curves coincides with those of Figure 7.10. (Mengel *et al.* (1988). Reproduced with permission of Springer.)

constant that induced the Antarctic transition but some other parameter (continental drift) acting in a similar way. We will return to this paleoclimate question in Chapter 8.

An example of counterintuitive behavior can be found in Figure 7.12. Here the model was initialized on one side of the bifurcation but the parametric conditions were on the other side. The relaxation to the seasonal cycle appropriate to the parametric conditions did not occur with the linear timescale of a few years, but actually took tens of years. This is because the solution and parametric conditions were very close to a bifurcation where nonlinear effects are dominant. This is a problem often ignored in climate simulation experiments. For example, we may well have already passed the bifurcation for sea ice-free summers in the Arctic, but the planet is still adjusting to this condition. Similar long-delayed adjustments may be underway in Greenland and Antarctica.

The work of Mengel *et al.* (1988) (see Figure 7.12) was extended by Lin and North (1990) to several other geographies (see also, Huang and Bowman, 1992). The techniques used by these authors were also different and that might be interesting for some readers. Lin used the Fourier harmonic technique to solve for the steady seasonal cycle. This turned out to be more efficient. The nonlinearity was taken into account by iteration. First, Lin demonstrated that she could repeat Mengel's work for a disklike land mass centered at the pole. She investigated how the bifurcation occurs as it is expressed in an operating curve, shown here in Figure 7.13. Using the same method, Lin investigated the case of a planet with a polar ocean but several cases of zonally symmetric bands of land configurations. Figure 7.14 shows an example of a band of land configuration where the land surface lies between 50° and 75° latitude. In this case, as indicated in Figure 7.14, there are two bifurcations, one with an ice-free Arctic Ocean as well as snow-free land for warmest values of solar irradiance. As the solar irradiance is lowered to 0.9316 of the present value, the former ice-free polar ocean exhibits summer ice cover, and the land surface is nearly snow free for most of the summer. Finally, at solar irradiance 0.9255, the entire polar area is ice covered year round.

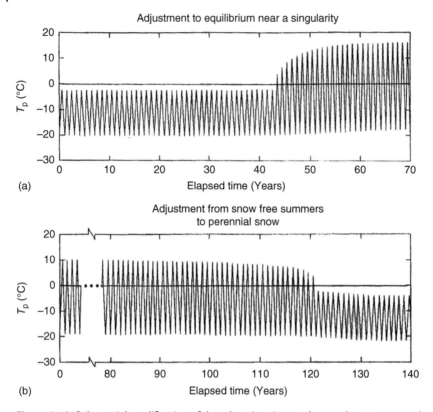

Figure 7.12 Substantial modification of the relaxation time to the steady-state seasonal cycle by the nearness of a bifurcation (tipping point). Shown is a plot of the time series for two realizations of the model for initial conditions near a singularity or bifurcation in the solution. (a) Initially the model is set to be in the icy pole case, but suddenly the solar constant is switched to the ice-free pole case. The time constant for the transition in this case is of the order of 43 years, but it is sensitive to the initial condition. (b) The opposite case where the solution is initially for the ice-free pole, but the total solar irradiance is suddenly switched to the icy pole value. This time the adjustment takes about 120 years. (Mengel *et al.* (1988). Reproduced with permission of Springer.)

The Mengel-like experiment has also been examined in a GCM (the Genesis Model of the early 1990s) by Crowley and Baum (1993). They used a geography from the Carboniferous period which was similar to the continent at the pole configuration in the Mengel *et al.* (1988) work (Figure 7.15).

7.8 Chapter Summary

The chapter began with a discussion of the ease of studying linear systems compared to the difficulty of dealing with nonlinear systems. Hence, it is wise to work linear problems when possible. In many cases, when nonlinearity is small, one can get by with such an approximation. On the other hand, the great power garnered from experience based on linear methods fails spectacularly when the nonlinear effects are large enough

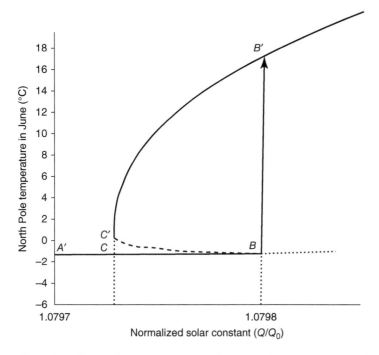

Figure 7.13 Graph of the operating curve for the north polar temperature (time = 0 at winter solstice). Operating curve for the polar temperature at summer solstice for the Mengel's land-cap model (analogous to Antarctica). (Lin and North (1990). Reproduced with permission of Springer.)

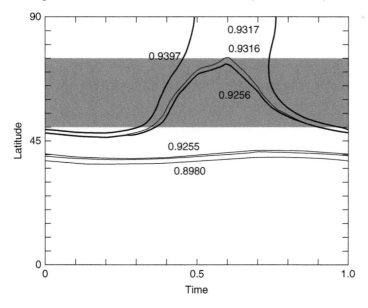

Figure 7.14 The complicated bifurcation structure of a seasonal cycle of a "band of land" model. Similar to previous figures except for a different land/sea configuration. Snow-line edge versus time of year (in tenths). The shaded latitudes are land, unshaded are ocean. Winter solstice is at $t = 1$. Note that there are two bifurcations for this geography. The numbers beside the snow-line curves indicate the total solar irradiance in ratio to its present value for the particular case. (Lin and North (1990). Reproduced with permission of Springer.)

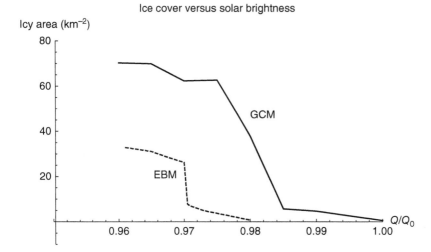

Figure 7.15 Bifurcation in the operating curve for the Genesis GCM with a Carboniferous land/sea distribution that is roughly comparable to a land mass centered at the South Pole. Both EBM and GCM show summer snow lines that have a discontinuous break at a certain value of the solar constant (shown here as Q/Q_0, the ratio to its value in the Carboniferous period. (Crowley and Kim (1994). Reproduced with permission of AAAS.)

to cause bifurcations in steady-state solutions especially when parameter conditions bring the solution very near to a bifurcation or other form of singularity.

The first problem taken up in the chapter was the nonlinear ice-cap problem where one finds three and sometimes five solutions depending sensitively on parameter values. An interesting problem consists in finding the stability of steady-state solutions. It turns out that with ice-cap models, the stability of a solution depends on the sign of the slope of the operating curve at the solution point. This is the point that depicts the polar ice-cap edge's latitude versus the solar irradiance. If the ice cap increases with increasing solar irradiance, the solution is unstable, a result anticipated heuristically by Budyko. The nonlinear ice cap's stability properties can also be studied by means of the potential function introduced in Chapter 2, only this time it must be multidimensional (usually, infinitely so).

The idea of a length scale suggests an explanation for the cusp-like behavior of the operating curve near small ice-cap sizes. It turns out that because a tiny patch of high-albedo material such as snow can induce cool surface air over large regions. If the albedo change is large enough even for a small patch, the surrounding region up to a radius of the length scale can become below freezing. Hence, ice caps can form.

The chapter closes with a seasonal model with a mix of land and sea surface. It requires a new addition to EBCMs, the position-dependent effective heat capacity. Position-dependent heat capacity is to be explored in more detail in Chapter 8. This addition keeps the model linear but it was introduced here to solve an interesting seasonal model, the Mengel model, which includes a strong nonlinear albedo feedback, owing to the high albedo of seasonal snow cover. The first model explored is one with a cap of land centered at the pole analogous to Antarctica. The model solutions are for a steadily repeating seasonal cycle that includes a snow line hooked to the freezing point. The model solutions indicate a strong bifurcation at a particular value of total solar irradiance. The seasonal cycle warmer than a critical value of the control parameter leads to snow-free summers over the polar land cap. Values lower than the critical value lead to snow cover large enough to cover the polar continent.

A second model by Lin and North (1990) expands the work to include more geographical configurations including zonally symmetric bands of land. Again interesting bifurcations are found with even more richness of behavior. This effect brings to mind a theory for how glaciers or ice sheets might initiate. When the control parameter is below critical, the summers are perpetually snow covered, allowing the accumulation of ice as the temperature never goes above freezing. Another effect discovered in these models is an explicit demonstration that relaxation times near one of these bifurcations can exceed decades, departing by an order of magnitude from the linear timescales of EBCMs. There were a number of experiments on the two-dimensional model (Chapter 8) with realistic geography with nonlinear snow/ice albedo feedback Hyde et al. (1990), which will be discussed in Chapter 12.

Notes for Further Reading

There are many books on nonlinear systems, but a particularly good one is that by Drazin (1992).

Exercises

7.1 On the basis of (7.11), find the closed form for $H_n(\mu_s)$ as a function of μ_s for $n = 0$ and $n = 2$.

7.2 On the basis of your answer in Exercise 7.1, determine the insolation for which the ice-cap edge is at $\mu = \mu_s$. Use $A = 208$ W m^{-2}, $B = 2.00$ W m^{-2} (°C)$^{-1}$, and $D = 0.67$.

7.3 Consider a time-dependent, one-dimensional energy balance model with unit heat capacity C in the form

$$\frac{\partial I}{\partial t} - \frac{\partial}{\partial \mu}\left(D(\mu)(1-\mu^2)\frac{\partial I}{\partial \mu}\right) + I = QS(\mu)a(\mu,\mu_s),$$

where $I = A + BT(\mu, t)$ is the outgoing radiation flux density, which is a function of latitude and time, $D(\mu)$ is the latitude-dependent diffusivity and $a(\mu, \mu_s)$ is the coalbedo depending on both latitude and the latitude of the ice-cap edge. Assume further that

$$\left[-\frac{\partial}{\partial \mu}\left(D(\mu)(1-\mu^2)\frac{\partial}{\partial \mu}\right) + 1\right]\phi_n(\mu) = \lambda_n \phi_n(\mu),$$

namely, there exist orthonormal eigenfunctions $\phi_n(\mu)$ and eigenvalues λ_n satisfying

$$\int_0^1 \phi_n(\mu)\phi_m(\mu)dx = \delta_{nm}, \quad \phi_0(\mu) = 1.$$

Then, the outgoing radiation field can be expanded in terms of eigenfunctions $\phi_n(\mu)$, that is,

$$I(\mu, t) = \sum_n I_n \phi_n(\mu).$$

Solve the energy balance model to determine the radiation (and therefore the temperature) as a function of μ, in closed form. Find the equilibrium solution for a given ice-cap-edge location $\mu_s = \mu_0$.

7.4 In Exercise 7.3, imagine that the equilibrium infrared radiation to space, $I(\mu, t)$, is perturbed by a small amount $\delta I(\mu, t)$; that is

$$I(\mu, t) = I^{eq}(\mu, t) + \delta I(\mu, t).$$

As a result of this perturbation, the latitude of the ice-cap edge changes slightly by an amount $\delta \mu_s$, that is,

$$\mu_s = \mu_0 + \delta \mu_s.$$

(a) Show that the perturbed temperature, to a first-order approximation, satisfies

$$\frac{d}{dt}\delta I_n(t) + \lambda_n \delta I_n(t) = Q h'_n(\mu_0)\delta \mu_s(t).$$

(b) Show that the perturbation of the ice-cap edge is determined to be

$$\delta \mu_s(t) = -\left(\sum_k \phi_k(\mu_0)\delta I_k(t)\right)\left(\sum_m I_m^{eq}\phi'_m(\mu_0)\right)^{-1}.$$

7.5 In Exercise 7.4, a linearized equation for a small temperature departure, $\delta I(\mu, t)$, for a one-dimensional nonlinear model is shown to satisfy

$$\frac{d}{dt}\delta I_n + \lambda_n \delta I_n = Q h'_n(\mu_0)\delta \mu_s(t),$$

where

$$h_m(\mu_s) = \int_0^1 S(\mu)a(\mu, \mu_s)\phi_m(\mu)\, d\mu$$

and the coalbedo is defined as a step function with a jump $\Delta a > 0$

$$a(\mu, \mu_s) = a_0 + a_2 P_n(\mu) - (\Delta a) H(\mu - \mu_s),$$

where

$$H(x) = \begin{cases} 0, & \text{for } x < 0, \\ 1, & \text{for } x \geq 0. \end{cases} \tag{7.76}$$

Assume a solution of the form

$$\delta I_n(t) = \delta I_n(0) e^{-\lambda t},$$

show that

$$\sum_n \frac{\Delta a \, S(\mu_0) \phi_n^2(\mu_0)}{(\lambda_n - \lambda)} + \frac{((h_n(\mu_0) - (A/Q))\delta_{n0}) \phi_n'(\mu_0)}{\lambda_n} = 0.$$

7.6 Show that

(a)

$$\frac{d}{d\mu_0} I_s = 0 = \frac{d}{d\mu_0} \left[Q \sum_n \frac{h_n(\mu_0) \phi_n(\mu_0)}{\lambda_n} \right].$$

(b)

$$\sum_n \frac{h_n(\mu_0) \phi_n'(\mu_0)}{\lambda_n} = -\frac{1}{Q^2} \frac{dQ}{d\mu_0} I_s - \sum_n \frac{h_n'(\mu_0) \phi_n(\mu_0)}{\lambda_n}.$$

(c)

$$\frac{dQ}{d\mu_0} = \lambda \frac{CQ^2}{I_s} \sum_n \frac{\Delta a \, S(\mu_0) \phi_n^2(\mu_0)}{\lambda_n (\lambda_n - \lambda)}.$$

7.7 Plot the right-hand side of Exercise 7.6(c), and find the solutions as a function of the value and sign of the left-hand side ($dQ/d\mu_0$). Interpret the result in terms of the stability of the solution $\delta I_n(t) = \delta I_n(0) e^{-\lambda t}$.

8

Two Horizontal Dimensions and Seasonality

In this chapter, we consider climate models with a latitude and longitude dependence. This suggests that we find an appropriate basis set for expansions on the sphere. The basis set most often used in this kind of application is the set of spherical harmonics, which are derived in Section 8.3.

8.1 Beach Ball Seasonal Cycle

The Beach Ball Model[1] (North and Coakley, 1979), referred to as BBM, is an intermediate model that falls between one- and two-dimensional energy balance models (EBMs). In this section, we sketch a summary of results of the BBM as a kind of bridge to the full two-dimensional models that will follow. This chapter suggests that extending the EBCM to two horizontal dimensions might be a worthwhile effort. We call it the BBM because it takes the land and ocean borders to be along meridians. The zonally averaged seasonal cycles of the two hemispheres are quite different because the NH has about 40% land and the SH has only 20% land. The plan is to do one model for the NH, fixing the free parameters, then turn to the SH using the same parameters, but different geography, as a test.

When modeling the NH, we take the whole planet to be 40% covered by a single continent with borders pole-to-pole along meridians. The NH model consists of modeling both ocean- and land-seasonal cycles with a longitudinal heat transport term, proportional to the difference between the land temperature at that latitude and time of the year and the corresponding ocean temperature. These terms couple the equations for the land temperature (at that latitude and time of year) to that of the ocean. To compare with data for the NH model, we reflect the NH observed fields across the equator, but lagging the time by 6 months in the opposite hemisphere. In Chapter 6, we found simple expressions for the seasonally dependent insolation function. We also know that Legendre modes 0 and 2 fit the mean annual climate (for either NH or SH, Chapter 5). The main difference between the two hemispheres then comes in the response amplitude and phase lag in the Legendre mode 1, $T_{11} \sin(2\pi t - \phi_{11}) P_n(\mu)$, where t is time, with values 0 and 1 at winter solstice, $\mu = \sin(\text{latitude})$, and δ_0 is the present obliquity (tilt of the rotation axis with respect to the orbital plane). The orbit is assumed to be circular in this exercise.

---

1 This name was coined and given to GRN by the late Prof. Lev S. Gandin, then at the Main Geophysical Observatory, Leningrad, USSR.

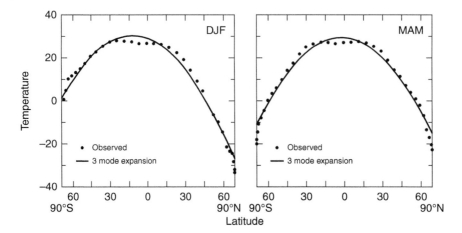

Figure 8.1 Observed zonally averaged surface temperatures of the symmetrized (dots) and the representation of the surface temperatures obtained with the 00, 11, and 20 modes (solid curves). The mode labels are: first digit is the Legendre index, the second digit is the harmonic (0 = mean annual, 1 = annual harmonic). (North and Coakley (1979). © American Meteorological Society. Used with permission.)

As motivation, consider the representation of the NH winter and spring zonal averages (these correspond also to the summer and fall in the other hemisphere) as depicted in Figure 8.1. Note the quality of fit of the simple mode structure with only the first three modes (index 0, 1, 2) in the expansion into Legendre polynomials. Given the expansion coefficients taken from the data, one can adjust the coupling coefficient between land areas and sea areas to make an excellent fit as shown by the solid curves in Figure 8.1.[2] This tells us that it should be possible to construct a three-mode seasonal model that will fit this data by adjusting the amplitude of the seasonal mode response (Legendre mode 1) to have the appropriate amplitude and phase lag. Note that in this scheme Legendre modes 0 and 2 are not excited by the sinusoidal time-dependent seasonal cycle of solar heating because those two modes do not have a time dependence.

The model consists of two dependent variables $T_L(\mu)$ and $T_W(\mu)$, the land and ocean surface temperatures averaged around the appropriate segments of their latitude belts. The two energy balance equations are distinguished by the use of C_L and C_W the heat capacities over land and ocean. The governing equation for land is

$$C_L \frac{\partial T}{\partial t} - D_0 \frac{\partial}{\partial \mu}(1-\mu^2)\frac{\partial T}{\partial \mu} + \frac{\nu}{f_L}(T_L - T_W) + A + BT_L = QS(\mu,t)a_L(\mu,t), \quad (8.1)$$

where the term $\frac{\nu}{f_L}(T_L - T_W)$ indicates that the heat is transferred from ocean to land proportionally to the difference in those temperatures at the same latitude, but averaged only over the land fraction (hence, the denominator f_L). The ocean equation is similar. The temperature as seen in Figure 8.1 is

$$T(\mu,t) = f_L T_L(\mu,t) + f_W T_W(\mu,t). \quad (8.2)$$

---

2 This introduces a new adjustable parameter to our formulation. We now have three, the diffusion coefficient, the heat capacity over land, and now the exchange coefficient across the seams of the beachball. We will be able to drop this parameter when we go to the full two-dimensional geography later in this chapter.

The three mode amplitudes are governed by the system of equations:

$$A + BT_0 = QH_0, \tag{8.3}$$

$$C_{L,W} \frac{dT_1^{L,W}}{dt} + (2D_0 + B)T_1^{L,W} + \frac{\nu}{f_{L,W}}(T_1^{L,W} - T_1^{W,L}) = QH_1 \tag{8.4}$$

and

$$(6D_0 + B)T_2 = QH_2. \tag{8.5}$$

By adjusting the value of the land–sea coupling ν/B to 0.226 and the heat capacity over land $C_L = 0.16$ year, one brings the amplitude of T_1^L to the observed value of 15.5 K. Since the phase lag is about a week too long, one can further reduce C_L to such a value that both the amplitude and phase of T_L match the data. The ocean amplitude is only about 3 K for a value of $C_W/B \approx 4$ years and the phase lag is 3 months. Note that for an infinitely deep ocean the phase lag would be a quarter of a cycle and that happens to be 3 months. This says that for our problem of forcing at the 1 year period, this mixed-layer model responds with a phase lag as though the mixed layer were infinitely deep. The seasonal cycle amplitude of the land and/or zonal average temperature is insensitive to the size of C_W.

Further experiments with a symmetrized SH show that the same values of these parameters serve to fit the SH seasonal cycle as well. The BBM is interesting, but suffers from too many choices about the phenomenological (adjustable) parameters. Its success in the examples of North and Coakley (1979) are suggestive that a fully two-dimensional model might just work in simulating the seasonal cycle. Before we undertake this task, we must introduce basis functions that will be useful on the two-dimensional plane and then the spherical surface. These steps form the building blocks and generalize from the Legendre polynomials already considered in previous chapters.

8.2 Eigenfunctions in the Bounded Plane

Before tackling the sphere it is useful to solve a simpler eigenvalue problem with less forbidding geometry and with constant heat diffusion coefficient. Imagine our climate to be the temperature distribution on a flat square whose corners are at $(0,0)$, $(0,1)$, $(1,1)$ and $(1,0)$ in the (x,y) plane. The appropriate operator for the heat transport term (divergence of heat flux) in an EBM is

$$-\nabla^2 = -\left\{\frac{\partial^2}{\partial x^2} + \frac{\partial^2}{\partial y^2}\right\}. \tag{8.6}$$

We seek the eigenfunctions defined by

$$-\left\{\frac{\partial^2}{\partial x^2} + \frac{\partial^2}{\partial y^2}\right\}\phi_i(x,y) = \lambda_i \phi_i(x,y), \tag{8.7}$$

subject to boundary conditions that we will take to be zero on the boundaries:

$$\phi_i(x,0) = 0, \tag{8.8}$$

$$\phi_i(x, 1) = 0, \tag{8.9}$$
$$\phi_i(0, y) = 0, \tag{8.10}$$
$$\phi_i(1, y) = 0, \tag{8.11}$$

where $\phi_i(x, y)$ is an eigenfunction with index i and λ_i is the corresponding eigenvalue.

We proceed by using the *method of separation of variables*. This consists of making the assumption that the solution for a particular eigenfunction is factorable into a part that is a function only of x and a part that is a function only of y:

$$\phi_i(x, y) = X(x)Y(y). \tag{8.12}$$

Substituting into (8.7) and dividing by XY we have

$$\frac{X''(x)}{X(x)} + \frac{Y''(y)}{Y(y)} = -\lambda_i. \tag{8.13}$$

The first term is a function only of x and the second is a function only of y. The only way these functions can be independent of each other and have this dependence is for each to be equal to a constant. First consider the x-dependent term:

$$\frac{X''(x)}{X(x)} = -c^2, \tag{8.14}$$

where we have taken the separation constant to be $-c^2$, anticipating that it will need to be negative. The solution to this class of ODEs is

$$X(x) = E \cos(cx) + F \sin(cx), \tag{8.15}$$

where E and F are arbitrary constants. To fit the boundary conditions, we must have $E = 0$. In addition, we must force c to be such that $X(1) = 0$. This can be accomplished by setting $c = n\pi$ where $n = 1, 2, 3, \cdots$. We then have

$$X_n(x) = F_n \sin(n\pi x), \tag{8.16}$$

where now the index n is used to distinguish the fact that we have a separate function for each (positive) integer value of n. We can choose the value of F_n to normalize the functions:

$$\int_0^1 X_n(x)^2 \, dx = 1, \quad \rightarrow F_n = \sqrt{2}. \tag{8.17}$$

Consider next the $Y(y)$ equation:

$$Y''(y) = -\gamma^2 Y(y), \tag{8.18}$$

where

$$\gamma^2 = \lambda_i - n^2\pi^2. \tag{8.19}$$

By applying the same procedure on the boundaries, we find

$$\gamma = m\pi, \quad m = 1, 2, 3, \cdots. \tag{8.20}$$

This implies

$$\lambda_{n,m} = (n^2 + m^2)\pi^2, \quad m, n = 1, 2, 3, \cdots, \tag{8.21}$$

where we have replaced the index i by the double index n, m to indicate that actually two integers are involved in specifying the eigenvalue. After normalizing both eigenfunctions, we have

$$\phi_{n,m}(x,y) = 2\sin(n\pi x)\sin(m\pi y). \tag{8.22}$$

Note that these eigenfunctions individually satisfy the boundary conditions and they are orthonormal.

$$\int_0^1 \int_0^1 \phi_{n,m}(x,y)\phi_{n',m'}(x,y)\, dx\, dy = \delta_{nn'}\delta_{mm'}. \tag{8.23}$$

In addition, the eigenfunctions satisfy a condition known as the *completeness relation*:

$$\sum_{m=1}^{\infty}\sum_{n=1}^{\infty} \phi_{m,n}(x,y)\phi_{m,n}(x',y') = \delta(x-x')\delta(y-y'). \tag{8.24}$$

We note that a two-dimensional problem on a finite domain requires two discrete indices for its specification. It is interesting that there is a *degeneracy* in the eigenvalues, in that two eigenfunctions $\phi_{n,m}$ and $\phi_{m,n}$ have the same eigenvalue: $(n^2 + m^2)\pi^2$. This degeneracy is traceable to the invariance of the problem under finite rotation about the domain center by 90°. Such rotational symmetries always lead to degeneracy of eigenvalues (more than one eigenfunction belonging to the same eigenvalue).

The last two equations imply that an arbitrary reasonably well-behaved (it can have simple discontinuities) function $g(x,y)$ can be expanded into the eigenfunctions:

$$g(x,y) = \sum_{m=1}^{\infty}\sum_{n=1}^{\infty} g_{n,m}\phi_{n,m}(x,y). \tag{8.25}$$

Knowing the eigenfunctions allows us to solve many boundary value problems on the same domain having the same boundary conditions. For example, the EBM-like problem

$$-\nabla^2 T(x,y) + \eta^2 T(x,y) = g(x,y) \tag{8.26}$$

(η is a constant with dimension (length)$^{-1}$) on the square with the same boundary conditions as above (the edges are in contact with a reservoir held at 0 °C) can be solved by expanding $T(x,y)$ and the heat source $g(x,y)$ into the eigenfunctions, then multiplying through by $\phi_{n,m}$ and integrating over the square. We are left with an equation for each n and m:

$$\left[(n^2+m^2)\pi^2 + \eta^2\right] T_{n,m} = g_{n,m} \tag{8.27}$$

or

$$T_{n,m} = \frac{g_{n,m}}{(n^2+m^2)\pi^2 + \eta^2} \tag{8.28}$$

and finally,

$$T(x,y) = \sum_{n=1}^{\infty}\sum_{m=1}^{\infty} \left(\frac{g_{n,m}}{(n^2+m^2)\pi^2 + \eta^2}\right) 2\sin(n\pi x)\sin(m\pi y), \tag{8.29}$$

which is the complete solution to the problem. Figure 8.2 shows the first four modal shapes for the square plate.

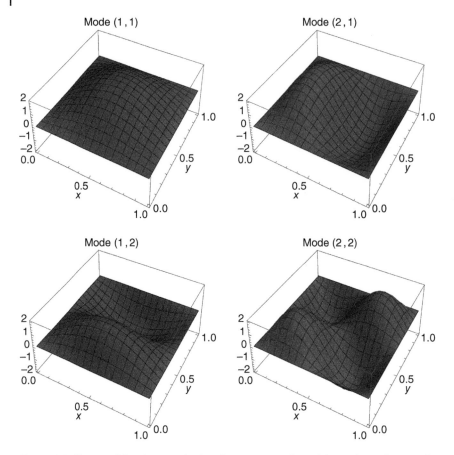

Figure 8.2 Shapes of the eigenmodes (n, m) on a square plate with zero boundary conditions at the edges. (© Amer. Meteorol. Soc., with permission.)

8.3 Eigenfunctions on the Sphere

8.3.1 Laplacian Operator on the Sphere

The divergence of heat flux for a linear heat conductor with constant thermal conductivity is the Laplacian operator ∇^2. In spherical coordinates, this is

$$\nabla^2 = \frac{1}{R_e^2}\left(\frac{\partial}{\partial \mu}(1-\mu)\frac{\partial}{\partial \mu} + \frac{1}{1-\mu^2}\frac{\partial^2}{\partial \varphi^2}\right). \tag{8.30}$$

Consider now the eigenfunctions of the two-dimensional Laplacian operator

$$-R_e^2 \nabla^2 \phi(\mu, \varphi) = \lambda \phi(\mu, \varphi). \tag{8.31}$$

The functions ϕ will have to be indexed as we shall soon see. As before, each eigenfunction ϕ will have an eigenvalue λ that will also have an index label. Our strategy will be to assume that $\phi(\mu, \varphi)$ can be factored into

$$\phi(\mu, \varphi) = \Phi(\varphi)P(\mu). \tag{8.32}$$

We are again using the method of separation of variables. Inserting this form into the defining equation,

$$\Phi[(1-\mu^2)P']' + \frac{1}{1-\mu^2}\Phi''P = -\lambda\Phi P. \tag{8.33}$$

Dividing through by $P\Phi$ leads to

$$(1-\mu^2)\left(\frac{[(1-\mu^2)P']'}{P} + \lambda\right) = -\frac{\Phi''}{\Phi}. \tag{8.34}$$

8.3.2 Longitude Functions

Note[3] that the left-hand side is a function only of μ and the right-hand side is a function only of φ. The only way this can be so is if each side is a constant, which we set arbitrarily equal to m^2. First consider the right-hand side:

$$\frac{d^2\Phi}{d\varphi^2} + m^2\Phi = 0. \tag{8.35}$$

We can now find the solution for $\Phi(\varphi)$

$$\Phi_m(\varphi) = A_m\, e^{im\varphi} + B_m\, e^{-im\varphi}, \tag{8.36}$$

where A_m and B_m are arbitrary constants. Note that in order for $\Phi_m(\varphi)$ to be single valued, m *must* be an integer.

8.3.3 Latitude Functions

We turn to the longitude-dependent equation that now becomes

$$\frac{d}{d\mu}\left((1-\mu^2)\frac{dP}{d\mu}\right) + \left(\lambda - \frac{m^2}{1-\mu^2}\right)P = 0. \tag{8.37}$$

It is clear that $P(\mu)$ needs an index that indicates it is for the particular integer m. There is yet to come an index associated with the discretization of λ. Anticipating the result, we set $\lambda = n(n+1)$.

$$\frac{d}{d\mu}\left((1-\mu^2)\frac{dP_n^m}{d\mu}\right) + \left(n(n+1) - \frac{m^2}{1-\mu^2}\right)P_n^m = 0. \tag{8.38}$$

When appropriately normalized, these $P_n^m(\mu)$ are known as the *associated Legendre Functions*.

In finding an explicit form for the associated Legendre functions, the first step is to substitute

$$P_n^m(\mu) = (1-\mu^2)^{m/2} u(\mu), \quad m > 0. \tag{8.39}$$

After substituting, we arrive finally at the differential equation for $u(\mu)$:

$$(1-\mu^2)u'' - 2\mu(m+1)u' + (n(n+1) - m(m+1))u = 0. \tag{8.40}$$

---

[3] Readers uninterested in the mathematical derivations may wish to skip directly to Section 8.4 where the functions on the sphere are described.

We proceed as in the last chapter with the assumption that $u(\mu)$ can be written as a power series:

$$u(\mu) = \sum_{l=0}^{\infty} a_l \mu^l. \tag{8.41}$$

Similar to what we did earlier, after some manipulation, we arrive at the recursion formula for the coefficients a_k:

$$a_{k+2} = \frac{[k(k-1) + 2k(m+1) - n(n+1) + m(m+1)]}{(k+2)(k+1)} a_k. \tag{8.42}$$

Hence, if a_0 and a_1 are provided, the rest of the coefficients are determined. As before, we encounter the divergence problem unless the numerator above cuts off at some level so that the rest of the terms vanish and $u(\mu)$ becomes a finite-degree polynomial. We find that this will happen if $k = n - m$. In other words, n must be an integer larger than or equal to m. We break up the functions into odd- or even-order polynomials, depending upon whether $n - m$ is odd or even. When it is even, we choose $a_1 = 0$ and proceed to have an even polynomial. When $n - m$ is odd, we set $a_0 = 0$ and proceed to have an odd-order polynomial. We now have a prescription for finding the $P_n^m(\mu)$. Hence, we may write

$$\phi(\mu, \varphi) = (A_{nm} e^{im\varphi} + B_{nm} e^{-im\varphi}) P_n^m(\mu); \tag{8.43}$$
$$n = 0, 1, 2, \cdots; \quad m = 0, 1, 2, \cdots; \quad n \geq m.$$

By convention, we simplify the notation further by letting m run from $-\infty$ to $+\infty$, then we can also define the $P_n^{-m}(\mu) \propto P_n^m(\mu)$, and we must then restrict $n \geq |m|$. This allows us to write

$$\phi(\mu, \varphi) = A_{nm} e^{im\varphi} P_n^m(\mu); \quad n \geq |m|. \tag{8.44}$$

Note that the normalization A_{n0} can be chosen such that

$$P_n^0(\mu) = P_n(\mu). \tag{8.45}$$

It is possible to show by direct substitution into the defining differential equation the important rule

$$P_n^m(\mu) = (-1)^m (1 - \mu^2)^{m/2} \frac{d^m}{d\mu^m} P_n(\mu). \tag{8.46}$$

If we employ Rodrigues' formula for the $P_n(\mu)$,

$$P_n(\mu) = \frac{1}{2^n n!} \frac{d^n}{d\mu^n} (\mu^2 - 1)^n, \tag{8.47}$$

we can show

$$P_n^m(\mu) = \frac{(-1)^m}{2^n n!} (1 - \mu^2)^{m/2} \frac{d^{n+m}}{d\mu^{n+m}} (\mu^2 - 1)^n, \tag{8.48}$$

which leads to evaluation of the aforementioned proportionality coefficient:

$$P_n^{-m}(\mu) = (-1)^m \frac{(n-m)!}{(n+m)!} P_n^m(\mu). \tag{8.49}$$

It is possible to derive an orthogonality relation for the $P_n^m(\mu)$ when the m values agree:

$$\int_{-1}^{1} P_{n'}^m(\mu) P_n^m(\mu) \, d\mu = \frac{2}{2n+1} \frac{(n+m)!}{(n-m)!} \delta_{n'n}. \tag{8.50}$$

8.4 Spherical Harmonics

We define the functions[4]

$$Y_{nm}(\hat{\mathbf{r}}) = \sqrt{\frac{2n+1}{4\pi} \frac{(n-m)!}{(n+m)!}} P_n^m(\mu) e^{im\varphi}, \tag{8.51}$$

where we have used the shorthand notation $\hat{\mathbf{r}}$ to stand for the point on the sphere corresponding to (ϑ, φ). From this definition, we can establish

$$Y_{nm}(\hat{\mathbf{r}}) = (-1)^m Y_{nm}^*(\hat{\mathbf{r}}). \tag{8.52}$$

8.4.1 Orthogonality

Normalization and orthogonality conditions are then

$$\int\int_{4\pi} Y_{n'm'}^*(\hat{\mathbf{r}}) Y_{nm}(\hat{\mathbf{r}}) \, d^2\Omega_{\hat{\mathbf{r}}} = \delta_{nn'} \delta_{mm'}, \tag{8.53}$$

where we have used the further shorthand notation $d^2\Omega_{\hat{\mathbf{r}}} = d(\cos\vartheta)\, d\varphi$, the element of solid angle on the Earth's surface. Another important relation is the *completeness* relation:

$$\sum_{n=0}^{\infty} \sum_{m=-n}^{n} Y_{nm}^*(\hat{\mathbf{r}}') Y_{nm}(\hat{\mathbf{r}}) = \delta(\varphi - \varphi') \delta(\cos\vartheta - \cos\vartheta'). \tag{8.54}$$

The first few spherical harmonics are

$$Y_{00} = \frac{1}{\sqrt{4\pi}},$$

$$Y_{11} = -\sqrt{\frac{3}{8\pi}} \sin\vartheta \, e^{i\varphi},$$

$$Y_{10} = \sqrt{\frac{3}{4\pi}} \cos\vartheta,$$

$$Y_{1-1} = \sqrt{\frac{3}{8\pi}} \sin\vartheta \, e^{-i\varphi},$$

---

[4] For spherical harmonics, we use the notation of Jackson (1962), Kelly (2006), and MATHEMATICA. Some reference books (e.g., Arfken and Weber, (2005); Byron and Fuller, 1992) use the same formula defining the Y_{nm} but with a factor $(-1)^m$. Yet others, especially in atmospheric sciences, tend to use the sine and cosine forms instead of $e^{im\varphi}$.

$$Y_{22} = \frac{1}{4}\sqrt{\frac{15}{2\pi}}\sin^2\vartheta\, e^{2i\varphi},$$

$$Y_{21} = -\sqrt{\frac{15}{8\pi}}\sin\vartheta\,\cos\vartheta\, e^{i\varphi},$$

$$Y_{20} = -\frac{1}{2}\sqrt{\frac{5}{4\pi}}(3\cos^2\vartheta - 1),$$

$$Y_{2-1} = \sqrt{\frac{15}{8\pi}}\sin\vartheta\,\cos\vartheta\, e^{-i\varphi},$$

$$Y_{2-2} = \frac{1}{4}\sqrt{\frac{15}{2\pi}}\sin^2\vartheta\, e^{-2i\varphi}.$$

Figure 8.3 shows contour maps of the real and imaginary parts of $Y_{1m}(\mu, \varphi), m = -1, 0, 1$.

8.4.2 Truncation

The truncation level of an eigenfunction expansion usually can be associated with a level of smoothing. In the case of a spherical harmonic expansion on the sphere, we have two indices m and n representing longitudinal and latitudinal levels in the series. There are good reasons to truncate the series at some value of n, the *degree* of the spherical harmonic at the end of the series. This is known as *triangular truncation*. It is indicated schematically in Figure 8.4a. Also shown in Figure 8.4b is a truncation known as *rhomboidal truncation*. This latter truncation was used in some early numerical models of the atmosphere. Nearly all spherical harmonic representations are truncated in the triangular form today.

Figure 8.5 gives a further idea of the way spatial scales are included in a T3 truncation. Note in the figure the different phasing between the imaginary part in the top row of components. The vertical column entries at $m = 0$ are just the Legendre polynomials.

8.5 Solution of the EBM with Constant Coefficients

We are now in a position to solve the energy balance equation on the sphere when the coefficients are not functions of position. We consider the spherical harmonic series expansion

$$T(\hat{\mathbf{r}}, t) = \sum_{n=0}^{\infty}\sum_{m=-n}^{n} T_{nm}(t) Y_{nm}(\hat{\mathbf{r}}). \tag{8.55}$$

The energy balance equation reads

$$C\frac{\partial T(\hat{\mathbf{r}}, t)}{\partial t} - D\nabla^2 T(\hat{\mathbf{r}}, t) + A + BT(\hat{\mathbf{r}}, t) = QS(\hat{\mathbf{r}}, t)a(\hat{\mathbf{r}}, t), \tag{8.56}$$

where now we have allowed for the possibility of S and a to depend on $\hat{\mathbf{r}}$ and t. Inserting the Laplace series expansion, we have

$$C\frac{dT_{nm}(t)}{dt} + [n(n+1)D + B]T_{nm}(t) = QH_{nm}(t) - \sqrt{4\pi}A\delta_{n0}, \tag{8.57}$$

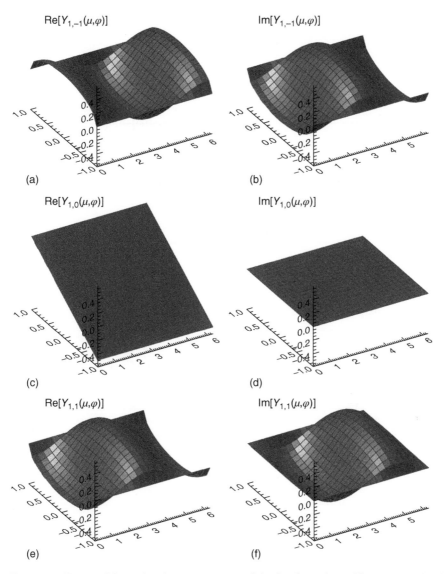

Figure 8.3 Shapes of the real and imaginary parts of the first few spherical harmonics. (a) Re($Y_{(1,-1)}$; (b) Im($Y_{(1,-1)}$); (c) Re($Y_{(1,0)}$); (d) Im($Y_{(1,0)}$); (e) Re($Y_{(1,1)}$); (f) Im($Y_{(1,1)}$).

where

$$H_{nm}(t) = \int\int_{4\pi} Y_{nm}^*(\hat{\mathbf{r}})S(\hat{\mathbf{r}},t)a(\hat{\mathbf{r}},t)\ \mathrm{d}^2\Omega_{\hat{\mathbf{r}}}. \tag{8.58}$$

The special case for which the time dependence of S and a is suppressed leads to the equilibrium solution,

$$T_{nm}^{eq} = \frac{QH_{nm} - \sqrt{4\pi}A\delta_{n0}}{n(n+1)D + B}. \tag{8.59}$$

8 Two Horizontal Dimensions and Seasonality

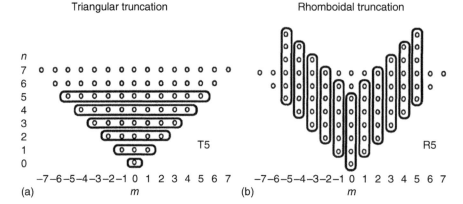

Figure 8.4 Truncating a spherical harmonic expansion. Each panel shows an array of 0s with the vertical column labeled by the degree n and the horizontal representing the longitudinal index m. Each 0 in the diagram represents a term retained in the truncated series. (a) Illustration of how the triangular truncation works in an example labeled T5. (b) alternative rhomboidal truncation, denoted by R5.

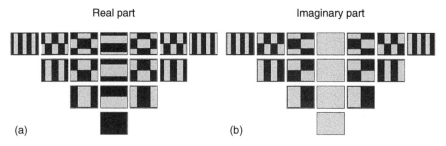

Figure 8.5 Triangular (T3) diagram showing the real part of spherical harmonic patterns in (a) and the imaginary part in (b) up to T3. The dark shading indicates positive values.

If we perturb the solution from equilibrium, we find the interesting result for uncoupled exponential decay modes with time constants given by

$$\tau_{nm} = \frac{\tau_0}{n(n+1)(D/B)+1} = \tau_n. \tag{8.60}$$

In other words, the decay time for mode (n, m) depends only on the degree n and not on the longitudinal wave number m.

8.6 Introducing Geography

In this section, we briefly introduce the geographical input that we will use extensively later. Consider the possibility that the heat capacity density C depends upon position on the globe, $C(\hat{\mathbf{r}})$. It is reasonable that $C(\hat{\mathbf{r}})$ depends strongly on whether the surface type is continental or oceanic. For an all-land planet we have seen that in many applications, the time constant can be taken to be about 30 days, as that is roughly equivalent to taking about half of the Earth's atmospheric mass into account. On the other hand, over ocean, heat mixes very quickly (few days) down to a depth of about 80 m. This means that the amount of mass involved is about 60 times as much as in the atmosphere alone. One way to account for geography in these models is to let $C(\hat{\mathbf{r}})$ take on these drastically different magnitudes depending upon the local surface type. Let us take $C/B = 30$ days over land surfaces and $C/B = 5$ years over oceanic surfaces. This leads to a strongly position-dependent step function at the continental borders all around the spherical surface. A handy way to incorporate this in our formulation is to expand in a Laplace series

$$C(\hat{\mathbf{r}}) = \sum_{n=0}^{N} \sum_{m=-n}^{n} C_{nm} Y_{nm}(\hat{\mathbf{r}}), \qquad (8.61)$$

where N is the *triangular truncation degree*. By not carrying the sum to ∞ and truncating it at a finite level, we have effectively smoothed out some of the sharp edges in the function – the truncation acts like a smoothing filter, excluding features smaller than a certain size inversely proportional to N (recall that n is the number of zeros from pole to pole in $P_n(\mu)$). Figure 8.6 shows contour maps of $C(\hat{\mathbf{r}})$ at two different degrees of truncation. There is some arbitrariness in the way one truncates a Laplace series. In atmospheric science, there have appeared two truncation conventions: the rhomboidal and the triangular. The triangular is easiest as it is the one we have used in the previous sections.

$$T(\hat{\mathbf{r}}) \approx \sum_{n=0}^{N} \sum_{m=-n}^{n} T_{n,m} Y_{nm}(\hat{\mathbf{r}}) \qquad (8.62)$$

is known as *triangular truncation* at degree N or simply TN. Most atmosphere/ocean climate model simulations cited in the recent Intergovernmental Panel on Climate Change (IPCC) reports require at least a resolution of T84. Typical weather forecast models at the time of this writing are at T200+. The advantage of triangular truncation is that this truncation preserves some of the rotational symmetry properties associated

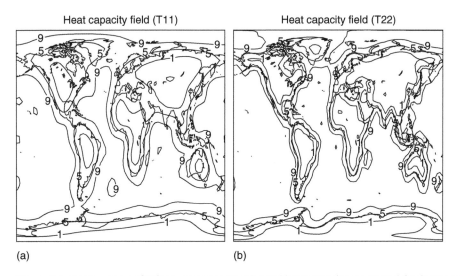

Figure 8.6 Contour map of $C(\hat{\mathbf{r}})$ truncated at (a) spherical harmonic degree 11 and (b) degree 22.

with the spherical harmonics. For example, under a rotation of the spherical coordinate system, the members of a given m multiplet transform into each other. This is a useful property as will be seen in a later chapter on fluctuations on the uniform sphere.

An alternative way of truncating the series that has been utilized in many early general circulation models is the rhomboidal truncation. In this case, we reverse the order of the summation:

$$T(\hat{\mathbf{r}}) \approx \sum_{m=-M}^{M} \sum_{n=|m|}^{|m|+M} T_{n,m} Y_{nm}(\hat{\mathbf{r}}). \tag{8.63}$$

We call this *rhomboidal truncation at level M* or simply *RM*. The two methods of summation are compared in the diagram in Figure 8.4. The advantage of rhomboidal truncation has been mostly computational. The length of the columns in the figure are clearly equal in this scheme and this has been useful in efficiently vectorizing some computer algorithms in the numerical solutions.

8.7 Global Sinusoidal Forcing

Consider the two-dimensional EBM forced by a sinusoidal forcing function in time by modifying the outgoing radiation constant term:

$$A \to A - A_f\, e^{i2\pi ft}. \tag{8.64}$$

This is similar to a global forcing by such an agent as greenhouse gas forcing, but we will allow it to be sinusoidal in time. Similarly to an engineer probing a black box, we drive the electrodes of the box with a sinusoidal forcing and study the response amplitude and phase lag. We can insert this forcing into our governing equation and as the temperature responds at the same frequency as the forcing,

$$T(\hat{\mathbf{r}}, t) - T^{(\mathrm{eq})}(\hat{\mathbf{r}}) = T_f(\hat{\mathbf{r}}) e^{i2\pi ft}, \tag{8.65}$$

with $T_f(\hat{\mathbf{r}})$ a complex function of position only. After canceling the common exponential factor, we have

$$(2\pi i f C(\hat{\mathbf{r}}) - D\nabla^2 + B) T_f(\hat{\mathbf{r}}) = A_f. \tag{8.66}$$

This means that each frequency component of the temperature field is uncoupled from every other one. On the other hand, we now find that because of the presence of the spatial dependence of $C(\hat{\mathbf{r}})$, the situation is much more complicated than in the rotationally invariant cases studied in previous sections. Inserting the Laplace series for $T_f(\hat{\mathbf{r}})$ and $C(\hat{\mathbf{r}})$, we obtain

$$\sum_{n}\sum_{m}\left(2\pi i f \sum_{n'}\sum_{m'} C_{n'm'} Y_{n'm'}(\hat{\mathbf{r}}) + n(n+1)D + B\right) T^{(f)}_{nm} Y_{nm}(\hat{\mathbf{r}}) = A_f. \tag{8.67}$$

Now, on multiplying through by $Y^*_{n''m''}(\hat{\mathbf{r}})$ and integrating over the sphere, we encounter some new objects:

$$\Gamma(n, m; n', m'; n'', m'') \equiv \int\!\!\int_{4\pi} Y_{nm}(\hat{\mathbf{r}}) Y_{n'm'}(\hat{\mathbf{r}}) Y^*_{n''m''}(\hat{\mathbf{r}})\, \mathrm{d}^2 \Omega_{\hat{\mathbf{r}}}, \tag{8.68}$$

the so-called *spherical harmonic coupling coefficients*. The result is

$$\sum_{n'm'} \mathcal{M}_{m,n;n',m'} T^{(f)}_{n'm'} = \sqrt{\frac{1}{4\pi}} A_f \delta_{n0} \delta_{m0}, \qquad (8.69)$$

where the coupling matrix $\mathcal{M}$ is given by

$$\mathcal{M}_{m,n;n',m'} = (B + n(n+1)D)\delta_{nn'}\delta_{mm'} + \sum_{m''n''} 2\pi i f C_{n''m''} \Gamma(n,m;n'm';n'',m''). \qquad (8.70)$$

In principle, the matrix $\mathcal{M}$ has an inverse so that, formally,

$$T^{(f)}_{nm} = \sum_{n'm'} (\mathcal{M}^{-1})_{n,m;n',m'} \sqrt{\frac{1}{4\pi}} A_f \delta_{n0} \delta_{m0}, \qquad (8.71)$$

or, in more compact form,

$$T^{(f)}_{nm} = (\mathcal{M}^{-1})_{nm;00} \sqrt{\frac{1}{4\pi}} A_f. \qquad (8.72)$$

This is the formal solution to the problem. The matrix $\mathcal{M}$ has to be computed by first obtaining $C(\hat{\mathbf{r}})$ from a map of the land–sea geography, then it must be expanded into the C_{nm} and finally this must be combined with values of the $\Gamma(n,m;n',m';n'',m'')$ (from readily available tables or computer algorithms). The matrix $\mathcal{M}$ also depends on the driving frequency f. Since the complex element i appears explicitly, the matrix $\mathcal{M}$ is a complex matrix that has to be inverted on a computer. Once the geography is set, the list C_{nm} are fixed once and for all. The $\Gamma(n,m;n',m';n'',m'')$ are only computed once and stored. However, the linear weighting of these in the computation of $\mathcal{M}$ will depend on forcing frequency f, requiring that $\mathcal{M}$ be reinverted for each experiment f.

Once the complex amplitude $T^{(f)}_{nm}$ has been computed, we must recompose the Laplace series to obtain the complex function $T^f(\hat{\mathbf{r}})$. The magnitude or modulus of this function tells us the amplitude of the response at the point $\hat{\mathbf{r}}$ on the Earth. The phase of the complex function at $\hat{\mathbf{r}}$ tells us the phase lag in radians behind the forcing phase.

It is instructive to examine the behavior of the amplitude of the response as a function of position on the Earth at different forcing frequencies. Figure 8.7 shows response amplitude maps for four different frequencies (indicated by their periods). The amplitude of the forcing was chosen such that it corresponds roughly to a doubling of CO_2 ($A_f = 5.3$ W m^{-2}). The equilibrium response ($f \to 0$) to this forcing should be about 2°. Figure 8.7a indicates that the amplitude is large over the large continental interiors because the thermal inertia ($C(\hat{\mathbf{r}})$) is smaller in those locations. As the period of the forcing is increased (Figure 8.7b–d), the amplitude of the response becomes more uniform over the globe. The relevant parameter is the relaxation time of the particular surface compared to the period of the forcing. Observe that the length scale in the response field increases with longer periods of forcing.

8.8 Two-Dimensional Linear Seasonal Model

The two-dimensional model can be formulated as in the previous chapter, only here we allow for the spatial dependence of thermal diffusion, $D(\hat{\mathbf{r}})$. We make Laplace–Fourier

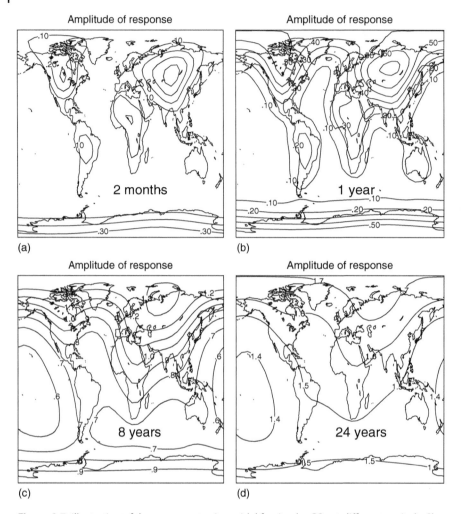

Figure 8.7 Illustration of the response to sinusoidal forcing by CO_2 at different periods. Shown are the contour maps of amplitude of response for a sinusoidal CO_2 doubling. All four cases are shown for easy comparison: (a) Period = 2 months; (b) Period = 1 year; (c) Period = 8 years; (d) Period = 24 years. As the period of the forcing increases, the spatial pattern of the response increases in scale.

expansions of the quantities

$$C(\hat{\mathbf{r}}) = \sum_{nm} C_{nm} Y_{nm}(\hat{\mathbf{r}}), \tag{8.73}$$

$$D(\hat{\mathbf{r}}) = \sum_{nm} D_{nm} Y_{nm}(\hat{\mathbf{r}}), \tag{8.74}$$

$$a(\hat{\mathbf{r}}) = \sum_{nm} a_{nm} Y_{nm}(\hat{\mathbf{r}}), \tag{8.75}$$

$$S(\hat{\mathbf{r}}, t) = \sum_{k} \sum_{nm} S_{nm}^{(k)} Y_{nm}(\hat{\mathbf{r}}) e^{i 2\pi k t}, \tag{8.76}$$

$$T(\hat{\mathbf{r}}, t) = \sum_{k} \sum_{nm} T_{nm}^{(k)} Y_{nm}(\hat{\mathbf{r}}) e^{i 2\pi k t}, \tag{8.77}$$

for insertion into the EBM

$$C(\hat{\mathbf{r}})\frac{\partial}{\partial t}T(\hat{\mathbf{r}},t) - \nabla \cdot (D(\hat{\mathbf{r}})\nabla T(\hat{\mathbf{r}},t)) + A + BT(\hat{\mathbf{r}},t) = QS(\hat{\mathbf{r}},t)a(\hat{\mathbf{r}}). \tag{8.78}$$

This model has been solved (North *et al.*, 1983; Hyde *et al.*, 1989) and the solutions recovered by the very same technique as in the last chapter except that the model must be solved separately for each temporal harmonic k. Once the mode amplitudes $T_{nm}^{(k)}$, $n \leq 11; |k| \leq 2$ have been found by inverting the response matrix for each k, the solution field can be recomposed from the aforementioned.

8.8.1 Adjustment of Free Parameters

The model now has several new parameters whose values must be determined by fitting to the present seasonal cycle. The diffusion parameter has the form (Mengel *et al.*, 1988)

$$D(\mu) = D_0(1 + D_2\mu^2 + D_4\mu^4). \tag{8.79}$$

Different values of the parameters D_0, D_2, and D_4 have been used in different publications over the years. The function $D(\mu)$ is plotted in Figure 8.8 with the solid line representing the thermal conductivity used by Mengel *et al.* (1988) and Kim and North (1992) and the short dashed curve is form used by Graves *et al.* (1993). The long-dashed line is the constant value used in Chapter 5 in fitting the one dimensional model. The coefficients D_0, D_2, and D_4 had to be adjusted to give a reasonable fit to the present seasonal cycle. One conjecture is that the diffusion parameter has to be larger in the tropics to account for the larger vapor pressure of water there and the increased efficiency of the Hadley cell (Lindzen and Farrell, 1977).

Another problem with the simulations is that the Arctic Ocean is mostly covered with sea ice. This has the heat capacity of neither land nor sea. Hence, we must introduce another step in the step function $C(\hat{\mathbf{r}})$ to account for the fact that sea ice has puddles, open cracks (leads), and other features that are not easily modeled in the linear form we have taken. Nevertheless, we attempt to take these into account in the present context by simply assigning those areas with perennial sea ice to have a value of $C(\hat{\mathbf{r}}_{\text{sea ice}}) = C_i = C_w/6.5$. This last value came from adjusting its magnitude until the annual harmonic of the surface temperature field came into agreement with the data.

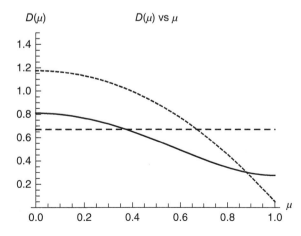

Figure 8.8 Illustration of the choices of the latitude dependence of diffusion coefficient in different models, with the diffusion parameter $D(\mu)$ as a function of sin(*latitude*). The long-dashed curve is the choice used in the one-dimensional model of Chapter 5, the solid curve is from Mengel *et al.* (1988) and Kim and North (1992), and the short-dashed curve is from Graves *et al.* (1993), who took $D_4 = 0$, by adjusting C_L to a smaller value. An alternative not considered was to give C_L a latitude dependence and keeping $D(\mu)$ a constant.

8.9 Present Seasonal Cycle Comparison

The next three subsections contain discussions on how the model simulates the annual and semiannual harmonics of the surface temperature field. In these simulations, the seasonal cycle forcing includes the important semiannual harmonic contribution to the second Legendre polynomial mode $(S_{02} + S_{22} \cos 4\pi t)P_2(\mu)$. This term makes an important contribution especially near the equator over which the Sun passes twice a year. It also contributes significantly in the polar regions where there is perpetual darkness in winter and perpetual daylight in summer. The convergence of the Fourier–Laplace series is slow in the polar regions because of the discontinuous derivative at the latitude of perpetual day (or night). Including this additional term improves the behavior in those regions.

8.9.1 Annual Cycle

The amplitude of the annual cycle agrees well with the smoothed data as seen in Figure 8.9. The agreement is especially noteworthy over the two Northern Hemisphere continents. It is interesting that the data indicate a hint of the Himalayan plateau but the model has no such feature because no allowance is made in this model for topography. The agreement is also good in and around Antarctica. The tiny 1 K amplitude curves in the tropics show some conformity with the observations. The overall coincidence of the contours with the continental boundaries is indicative of the very strong influence of the land–sea contrast in heat capacity. The phase lag is more sensitive to local geography and it is not mimicked as well as seen in Figure 8.10. First, the phase lag in the tropics should be ignored as the annual harmonic is very small in the model and in the data. The phase lag of about or less than 30 days in the interior of each Northern Hemisphere continent is very good except for some details. In North America, the 30 days contour is obviously influenced by the Rocky Mountain chain. There is similar disagreement that is easily explained over the mountains in South America (both due to shorter lags over high terrain). A similar high plateau error occurs over the Himalayas. Agreement over the ocean is moderate, being more extreme in the model regarding the quarter cycle (90 days) lag. We conclude that the annual cycle is modeled rather well except for some features that are not expected to be faithfully represented in such a comparison.

8.9.2 Semiannual Cycle

The first thing to notice about the semiannual cycle as shown in Figure 8.11 is how small it is. This is indicative of the rapid convergence of the Fourier series representing the seasonal surface temperature field. Over ocean surfaces, it is smaller than 1 K except in polar regions, where, as noted above, it feels the second harmonic forcing from the polar day/night term. This same day/night forcing of the amplitude over polar regions, especially Antarctica, leads to large responses over both model and data.

8.10 Chapter Summary

The expansion of EBCMs from one to two dimensions on the spherical surface was first motivated by a discussion of the BBM, which is a planet with continents whose borders are pole to pole along meridians. This led to a model that is symmetric across the

Figure 8.9 Plots illustrating the agreement of the modeled versus observed first harmonic of the seasonal cycle. Shown are contour plots of the amplitude of the surface temperature field of the annual harmonic. (a) From observations as determined from smoothed data by truncating the spherical harmonic expansion at T11. (b) The same as in (a) except for the two-dimensional EBCM. Contours are in intervals of 5 K except for the 1 K curve. (Kim and North (1992). Reproduced with permission of Wiley.)

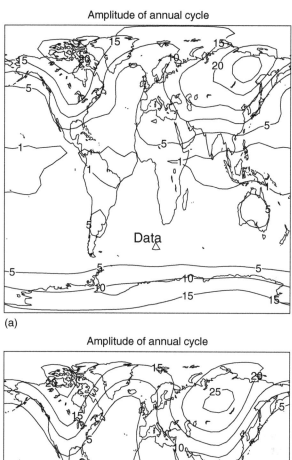

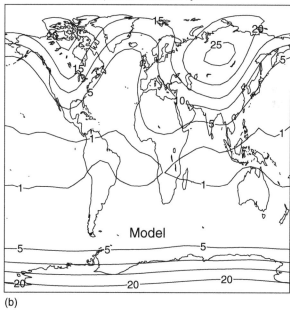

equator. An EBCM was concocted that mimicked a symmetrized NH. The effective heat capacities of land and sea along with a parameter that coupled ocean and land temperatures at the same latitudes could be adjusted to form a seasonal cycle of the model that agreed with the seasonal cycle of observations. The fact that the data could be well represented by three Legendre polynomial modes ($n = 0, 1, 2$), where the seasonal cycle is

8 Two Horizontal Dimensions and Seasonality

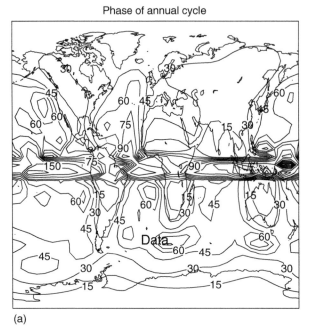

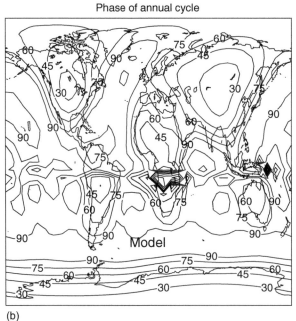

Figure 8.10 Plots featuring the agreement of the modeled versus observed first harmonic phase lag of the seasonal cycle. Shown are contour plots of the phase lag (days) of the annual harmonic. (a) From observations as determined from smoothed data by truncating the spherical harmonic expansion at T11. (b) The same as (a) except for the two-dimensional EBCM. Note that over large land masses, the phase lag is about 1 month while over large ocean expanses the lag is 2–3 months. (Kim and North (1992). Reproduced with permission of Wiley.)

carried only by mode index unity, and that mode was sinusoidal in time suggested that this simple approach was on the right track. Unfortunately, the BBM required too many adjustable parameters for it to be useful.

The next step was clearly to try allowing the heat capacity to be a function of position on the sphere: $C(\hat{r})$. Before building the model, we introduced two-dimensional

Figure 8.11 Contour plot of the surface temperature field of the semiannual harmonic. (a) From observations as determined from smoothed data by truncating the spherical harmonic expansion at T11. (b) The same as (a) except for the two-dimensional EBCM. The semiannual harmonic is important in the tropics and in the polar regions. In the polar regions, this is the first correction toward resolving the polar day and night features. This graphic also helps to indicate the rapid convergence of the Fourier series in time. (Kim and North (1992). Reproduced with permission of Wiley.)

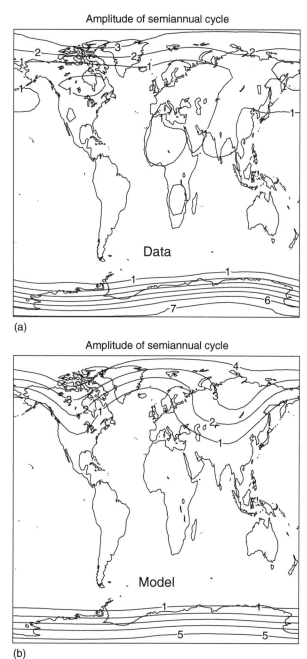

eigenfunctions of the Laplace operator ($-\nabla^2$), first on a plane square then onto the sphere, where we encounter the spherical harmonics. Spherical harmonics turn out to be the modal shapes of the EBCM if the heat capacity function $C(\hat{\mathbf{r}})$ is not a function of position. When it is a function of position, we must resort to numerical methods. This has been carried out by many studies over the last three decades by a number of different numerical approaches (North et al., 1983; Hyde et al., 1989; Bowman and Huang, 1991; Kim and North, 1991; Stevens and North, 1996; North and Wu, 2001; Zhuang et al., 2014).

The solutions to the two-dimensional model, as shown, were obtained by analyzing the model and data into mean annual, annual harmonic, and semi-nnual harmonic as a function of position. The Fourier series of these harmonics converges very rapidly so that only these few harmonics are necessary to obtain a very good representation of both model and data. The shape of the annual cycle fits the data extremely well with large amplitude over land masses and smaller ones over the ocean surfaces. The phase lag between insulation and response of the annual harmonic is only about 1 month over the interiors of the large continents and roughly three months over the ocean surfaces. While some details differ, the gross features of the phase lag are captured. The semiannual harmonic amplitude is also in very good agreement with the data with very small amplitude (< 2 K) over most of the globe but larger as one approaches either pole.

The general success of the two-dimensional EBCM suggests that it might be good enough for some applications in paleoclimatology and others such as designing observational networks or in detection of climate signals. Given the success of the two-dimensional EBCM in simulating the seasonal cycle, we push the two-dimensional EBCM to include random (weather noise) forcing.

Notes for Further Reading

The book by Washington and Parkinson (2005) discusses spherical harmonics in the context of three-dimensional climate models, as well as numerical methods for general circulation models. Most books on mathematical physics such as Arfken and Weber (2005) discuss linear boundary value problems and the use of spherical harmonics and other special functions.

Exercises

8.1 (a) Show that the eigenvalue problem

$$-\frac{\partial^2}{\partial x^2}\phi_n(x) = \lambda_n \phi_n(x), \quad x \in [0,1]$$

allows a solution in the form

$$\phi_n(x) = (\phi_n^c(x), \phi_n^s(x)) = (\sqrt{2}\cos \pi n x, \sqrt{2}\sin \pi n x).$$

Determine the corresponding eigenvalue.

(b) Determine the solution of

$$\frac{\partial^2 \phi(x)}{\partial x^2} = -q(x), \quad x \in [0,1]$$

with boundary conditions

$$\phi(0) = \phi(1) = 0.$$

8.2 Show that the solution of

$$\left(\frac{\partial^2}{\partial x^2} + \frac{\partial^2}{\partial y^2}\right)\psi(x,y) = -q(x,y)$$

with boundary conditions

$$\psi(x,0) = \psi(x,1) = 0, \quad \psi(0,y) = \psi(1,y) = 0$$

is given by

$$\psi(x,y) = \sum_{nm} \frac{2H_{nm}}{(\pi n)^2 + (\pi m)^2} \sin 2\pi nx \sin 2\pi my,$$

where

$$H_m = \int_0^1 \int_0^1 q(x,y) dx\, dy.$$

8.3 Prove by deduction that

$$\sum_m \sum_n \psi_{nm}(x,y)\psi_{mn}(x',y') = \delta(x-x')\delta(y-y').$$

8.4 The Poisson equation in spherical coordinates is give by

$$\frac{1}{\cos\theta}\left(\frac{\partial}{\partial \phi}\left(\frac{1}{\cos\theta}\frac{\partial}{\partial \phi}\right) + \frac{\partial}{\partial \theta}\left(\cos\theta\frac{\partial}{\partial \theta}\right)\right)\psi(\phi,\theta) = -q(\phi,\theta),$$

where ϕ is longitude and θ is latitude. Boundary conditions are given by

$$\frac{\partial}{\partial \theta}(\psi(\phi,\theta)) = 0 \quad \text{at } \theta = \pm\pi/2.$$

Obtain a zonally symmetric solution of the Poisson equation.

8.5 Consider the Laplace equation in spherical coordinates:

$$\frac{1}{\cos^2\theta}\frac{\partial^2\psi}{\partial \phi^2} + \frac{1}{\cos\theta}\frac{\partial}{\partial \theta}\left(\cos\theta\frac{\partial\psi}{\partial \phi}\right) + \frac{\partial}{\partial r}\left(r^2\frac{\partial\psi}{\partial r}\right) = 0,$$

where ϕ is longitude, θ is co-latitude, and r is radius.
(a) Assuming a separable solution, $\psi(\phi,\theta,r) = \Phi(\phi)\Theta(\theta)R(r)$, rewrite the Laplace equation above in terms of the separable solution.
(b) Let us assume a radial solution in the form $R(r) = r^n$, rewrite the governing equation in Part (a).
(c) Set up the equation for a zonally symmetric solution. Solve the resulting problem.
(d) Without the assumption of zonal symmetry, a separable solution of the equation in Part (b) can be written as $\psi(\phi,\theta) = \Phi(\phi)\Theta(\theta)$. Use the result in Part (c), set up the governing equation for each component of the separable solution. Obtain the longitudinal component of the solution.

(e) Let us assert that the solution for the latitudinal component of the equation is given by

$$\Theta(\mu) = (1-\mu^2)^{m/2}\frac{d^m P_n(\mu)}{d\mu^m} = P_n^m(\mu),$$

where $P_n^m(\mu)$ ($0 \le m \le n$) is called the *associated Legendre function* of order n and rank m. Prove that the associated Legendre function satisfies latitudinal component of the Laplace equation.

8.6 Legendre functions are generated by using

$$P_n(\mu) = \frac{1}{2^n n!}\frac{d^n}{d\mu^n}(\mu^2-1)^n,$$

which is known as the *Rodrigues' formula*. Using the formula, generate the Legendre polynomials up to order 4 ($n = 0, 1, \cdots, 4$).

8.7 For this problem, use the Rodrigues' formula in Exercise 8.6 and the definition of associated Legendre functions in Exercise 8.5.
(a) Develop the Rodrigues' formula for associated Legendre functions.
(b) Show that the associated Legendre functions with a negative rank defined by

$$P_n^{-m}(\mu) = \frac{1}{2^n n!}(1-\mu^2)^{-m/2}\frac{d^{n-m}}{d\mu^{n-m}}(\mu^2-1)^n,$$

satisfies

$$P_n^{-m}(\mu) = (-1)^m \frac{(n-m)!}{(n+m)!} P_n^m(\mu).$$

(c) Using the Rodrigues' formula in Parts (a) and (b), derive the associated Legendre functions for up to order $n = 3$.

8.8 Spherical harmonic basis functions are the solutions of Laplace equations in spherical coordinates. As discussed in Exercises 8.4 and 8.5, a most general solution is given by

$$\Psi(\phi, \theta, r) = \Phi(\phi)\Theta(\theta)R(r) = A\, e^{im\phi} P_n^m(\theta) r^n = A Y_n^m(\phi, \theta) r^n,$$

where

$$Y_n^m(\phi, \theta) = \sqrt{\frac{4\pi}{2n+1}\frac{(n-m)!}{(n+m)!}} e^{im\phi} P_n^m(\theta)$$

is called the *spherical harmonics* of order n and rank m. Any horizontal spatial patterns (at the same elevation r) can be decomposed in terms of spherical harmonic basis functions.
(a) Show that

$$\nabla^2 Y_n^m(\phi, \theta) = -n(n+1) Y_n^m(\phi, \theta)$$

(b) Consider an equilibrium solution of a two-dimensional EBM for a perpetual January 1 simulation with a constant albedo:

$$-\nabla \cdot (\nabla D T(\hat{\mathbf{r}})) + A + B T(\hat{\mathbf{r}}) = Q\, a\, s(\hat{\mathbf{r}}),$$

where $\hat{\mathbf{r}}$ is a unit vector pointing from the center of the planet to a point (ϕ, θ) on the surface, the insolation distribution function $s(\hat{\mathbf{r}})$ is at its perpetual configuration and A, B, D, and a are simply constants. Set up an EBM equation by using spherical harmonics and obtain a closed-form solution.

9

Perturbation by Noise

In this chapter, we take up the problem of the climate system's energy balance disturbed by noise. The noise term is taken to imitate weather and other small space–time scale perturbations of the energy balance. First, we consider the basic case of the model with constant coefficients on the sphere. This starting point is in line with our approach throughout the book of step-by-step understanding of the convergence toward a more comprehensive and realistic climate model. The new element in our process is an attempt to capture the fluctuations of the system about the ensemble mean. The idea was introduced in Chapter 2 for the global average model. In Section 6.5, we introduced the idea of fluctuations in the system due to random winds (departures from normal mean circulations) whose timescales are shorter than the relaxation timescale of a column of air. There are other physical elements besides the winds, which fluctuate with such short timescales, for example, areal extent and height of cumulus clouds, passage of mid-latitude weather systems, water vapor concentration in three dimensions, and aerosol particle concentrations. Some of these can be lumped on the driver's side of the EBM governing equation, while the effects of horizontal heat transport fluctuations are located in the advection term (divergence of heat flux). If we make the assumption that the heat advection term can be decomposed,

$$\mathbf{v}(\hat{\mathbf{r}}, t) \cdot \nabla T(\hat{\mathbf{r}}, t) \approx -\nabla \cdot D\nabla \langle T(\hat{\mathbf{r}}, t) \rangle - F_{\text{noise}}(\hat{\mathbf{r}}, t), \qquad (9.1)$$

where the angle brackets mean ensemble averaging (or expectation value in the probabilistic sense), and the energy balance climate model (EBCM) surface temperature in all the previous chapters is now $\langle T(\hat{\mathbf{r}}, t) \rangle$, and $F_{\text{noise}}(\hat{\mathbf{r}}, t)$ is a random field representing timescales much shorter than the relaxation time, τ_0, and spatial scales less than those of a spherical harmonic of degree 11. These latter would include all the fluctuations we might associate with "weather."

In most of the treatment in this chapter, we deal with anomalies, that is, the departures from the ensemble mean at a given point on the sphere, $\hat{\mathbf{r}}$.

$$T'(\hat{\mathbf{r}}, t) = T(\hat{\mathbf{r}}, t) - \langle T(\hat{\mathbf{r}}, t) \rangle. \qquad (9.2)$$

We drop the prime hereafter and refer simply to $T(\hat{\mathbf{r}}, t)$ as the anomaly.

Our next task (as usual) is to solve some models with ideal geography and then proceed to more complicated ones. The models will have the governing equation we are accustomed to, except for the noise driver, $F(\hat{\mathbf{r}}, t)$ on the RHS. We will be seeking statistical quantities from the solutions to problems such as for a given weather noise forcing, what is the response in terms of the distribution of variance in the response modes.

Energy Balance Climate Models, First Edition. Gerald R. North and Kwang-Yul Kim.
© 2017 Wiley-VCH Verlag GmbH & Co. KGaA. Published 2017 by Wiley-VCH Verlag GmbH & Co. KGaA.

9.1 Time-Independent Case for a Uniform Planet

In this section we first consider an imaginary Earth that is spatially uniform with respect to the time-independent problem dealt with earlier but forced by spatially dependent noise:

$$-D\nabla^2 T(\hat{\mathbf{r}}) + BT(\hat{\mathbf{r}}) = F_{\text{noise}}(\hat{\mathbf{r}}), \tag{9.3}$$

where $F(\hat{\mathbf{r}})$ is spatially white noise and that means it satisfies

$$\langle F_{\text{noise}}(\hat{\mathbf{r}})F_{\text{noise}}(\hat{\mathbf{r}}')\rangle = \sigma_F^2 \delta(\hat{\mathbf{r}} - \hat{\mathbf{r}}'), \tag{9.4}$$

where $\delta(\hat{\mathbf{r}} - \hat{\mathbf{r}}') = \delta(\cos\theta - \cos\theta')\delta(\phi - \phi')$. This kind of random field evaluated at one point is *uncorrelated* with the field evaluated at another point even for very tiny separation distances. In our problem, it means the variability is in the form of eddies in space that are small compared to the natural length scales in the problem; basically, the natural length scales are much smaller than $\lambda_{dd} = \sqrt{D/B}$, expressed in units of the Earth's radius.

We say the forcing is *spatially white noise*. We can expand the white noise random forcing field into a Laplace series (i.e., into spherical harmonic components as in Chapter 8). Note that from here we drop the subscript "noise" to keep the notation simpler.

$$F(\hat{\mathbf{r}}) = \sum_{n=0}^{\infty}\sum_{m=-n}^{n} F_{nm} Y_{nm}(\hat{\mathbf{r}}), \tag{9.5}$$

where the Laplace series components F_{nm} are complex random numbers which we will take to be normally distributed. Using the orthogonality of the spherical harmonics, we obtain the inverse to be

$$F_{nm} = \iint_{4\pi} Y_{nm}^*(\hat{\mathbf{r}}) F(\hat{\mathbf{r}}) d^2\Omega. \tag{9.6}$$

Consider the covariance between these components with different indices. Using the properties, we find that

$$\langle F_{nm} F_{n'm'}^* \rangle = \iint_{4\pi}\iint_{4\pi} \langle F(\hat{\mathbf{r}})F(\hat{\mathbf{r}}')\rangle Y_{nm}(\hat{\mathbf{r}})Y_{n'm'}^*(\hat{\mathbf{r}}')d^2\Omega\, d^2\Omega'. \tag{9.7}$$

Next substitute the expression for the quantity in angular brackets in (9.4) and the resulting expression reduces to

$$\langle F_{nm} F_{n'm'}^* \rangle = \sigma_F^2 \delta_{nn'}\delta_{mm'}. \tag{9.8}$$

This last equation tells us a lot about the covariances between the Laplace components of spatially white noise. The covariance vanishes unless $n = n'$ and $m = m'$. There are no cross-covariances! Each component F_{nm} is statistically independent of every other component. In addition, the variance associated with each component F_{nm}, σ_F^2, is the same for every component indexed. For each spherical harmonic degree, n, there are $2n + 1$, m components, each of which has the same variance.

White noise is a special case of the more general condition of statistically rotationally invariant random fields on the sphere. When ensemble averages of the mean and some second moments of a random field are rotationally invariant on the sphere, it is possible to decompose the variance of the random field into a spectrum of variances analogous to the treatment of a stationary time series where the symmetry or invariance is along the timeline. We will find that in some cases this can be utilized on the uniform-sphere models to follow.

Returning to the white noise spatial process we will use to perturb the energy balance, it is helpful to think of *realizations* of the white noise field $F(\hat{\mathbf{r}})$. We can generate a realization of the field by first going to a complex Gaussian random number generator and pulling out one random number whose variance is σ_F^2 (square of the sum of the real and imaginary parts of F) for F_{00}, then repeating this to draw statistically independent values for $F_{-1,1}, F_{0,1}, F_{1,1}$ each with exactly the same variance σ_F^2, and so on. We then take these complex random numbers and enter them in the formula for $F(\hat{\mathbf{r}})$ given by (9.5). Now turn to the solution for the temperature components. We can expand

$$T(\hat{\mathbf{r}}) = \sum_{n=0}^{\infty} \sum_{m=-n}^{n} T_{nm} Y_{nm}(\hat{\mathbf{r}}) \tag{9.9}$$

and its inverse,

$$T_{nm} = \iint_{4\pi} T(\hat{\mathbf{r}}) Y_{nm}^*(\hat{\mathbf{r}}) \mathrm{d}^2 \Omega. \tag{9.10}$$

By substituting and using the fact that the $Y_{nm}(\hat{\mathbf{r}})$ are the eigenfunctions of $-\nabla^2$ with eigenvalue $n(n+1)$ and then using the orthogonality of the $Y_{nm}(\hat{\mathbf{r}})$, we have

$$T_{nm} = \left(\frac{1}{n(n+1)\lambda_{\mathrm{dd}}^2 + 1} \right) \frac{F_{nm}}{B}, \tag{9.11}$$

with $\lambda_{\mathrm{dd}} = \sqrt{D/B}$, where we have taken the Earth's radius to be unity. The T_{nm} are complex random numbers and the randomness comes from the factor F_{nm}. The factor within the large parentheses weights the (n, m) components of T_{nm} according to the n dependence in the denominator. Note that the proportionality factor does not contain any m dependence.

The covariance between different (n, m) components can be readily calculated:

$$\langle T_{nm}^* T_{n'm'} \rangle = \sigma_T^2(n) \delta_{nn'} \delta_{mm'} \geq 0, \tag{9.12}$$

with

$$\sigma_T^2(n) = \left(\frac{1}{(n(n+1)\lambda_{\mathrm{dd}}^2 + 1)^2} \right) \frac{\sigma_F^2}{B^2}. \tag{9.13}$$

Note that there is no m dependence. There are two important parameters: σ_F^2/B^2 and λ_{dd}^2. The first governs the overall variance of T_{nm}, the second determines how the variance from the white noise forcing is apportioned by the proportionality factor in (9.11), or more physically the inverse of the operator $(-D\nabla^2 + B)$ (see 9.3) which operates on $F(\hat{\mathbf{r}})$. The factor in the large parentheses can be thought of as a "filter" that modifies the input variance, σ_F^2/B^2. This filter allows low modes (larger scales) to pass from the stimulus to the corresponding modes of the response but reduces the power (or variance) passed to the higher index modes. This filter defines the dynamical character of the damped diffusion operator. The damped diffusion filter smooths out the highly erratic (high-mode-index) white noise input.

The covariance of the temperature field is given by

$$\langle T(\hat{\mathbf{r}})T(\hat{\mathbf{r}}')\rangle = \sum_{n,m}\sum_{n',m'} \delta_{nm}\delta_{n'm'}\sigma_T(n)\sigma_T(n')Y_{nm}(\hat{\mathbf{r}})Y^*_{n'm'}(\hat{\mathbf{r}}'). \tag{9.14}$$

After making use of the Kronecker deltas, we have

$$\langle T(\hat{\mathbf{r}})T(\hat{\mathbf{r}}')\rangle = \sum_{n} \sigma_T(n)^2 \sum_{m=-n}^{n} Y_{nm}(\hat{\mathbf{r}})Y^*_{nm}(\hat{\mathbf{r}}'). \tag{9.15}$$

Now we can employ a wonderful theorem called the *addition theorem for spherical harmonics*.[1] The theorem states

$$\sum_{m=-n}^{n} Y_{nm}(\hat{\mathbf{r}})Y^*_{nm}(\hat{\mathbf{r}}') = \frac{2n+1}{4\pi} P_n(\hat{\mathbf{r}}\cdot\hat{\mathbf{r}}'). \tag{9.16}$$

Inserting this result, we find that

$$\langle T(\hat{\mathbf{r}})T(\hat{\mathbf{r}}')\rangle = \sum_{n=0}^{\infty} \frac{(2n+1)}{4\pi} \sigma_T(n)^2 P_n(\hat{\mathbf{r}}\cdot\hat{\mathbf{r}}'). \tag{9.17}$$

This last formula tells us that the covariance of the temperature field with itself depends only on the opening angle from the Earth's center between the two points at the surface. (remember $R_e = 1$ here). The opening angle is simply $\cos^{-1}(\hat{\mathbf{r}}\cdot\hat{\mathbf{r}}')$. This is the condition for rotational invariance on the sphere. It is hardly a surprise for the case with constant coefficients, as the operator ∇^2 (being a scalar product $\nabla\cdot\nabla$) is also rotationally invariant along with all the other terms on the sphere. Figure 9.1a shows an example of a degree spectrum of the white noise variance as a function of degree n. The spectrum increases as a function of n because there are $(2n+1)$ modes for each value of n. Figure 9.1b shows the responding temperature field utilizing a value of $\lambda_{dd} = 0.30$. Note that the degree spectrum begins to turn over at $n = 1$ and the growing terms in the denominator begin to dominate and filter out the high wavenumber stimulus of the white spatial noise in the numerator.

We cannot ignore the opportunity to draw a parallel with empirical orthogonal functions (EOFs), which have the property that if a random field is expanded into these orthogonal functions, the expansion coefficients are statistically independent. We have just proved (with the help of the spherical harmonic addition theorem) that the $Y_{nm}(\hat{\mathbf{r}})$ are the EOFs of any random field whose statistics are rotationally invariant on the sphere. There is nothing empirical here, so a comment is called for. Karhunen and Loève studied the basis sets of random fields, not just generated from data but from theoretically generated continuous random fields. The basis sets that do it are now called the *Karhunen–Loève* functions. When we use EOFs to reduce the dimension of an empirical "random" field we call the basis set the *EOFs*. The EOFs are just the eigenvectors of the cross-covariance matrix. For the case of the sphere, the Karhunen–Loève functions are the eigenfunctions of the kernel $C_G(\hat{\mathbf{r}}\cdot\hat{\mathbf{r}}') = \langle G(\hat{\mathbf{r}})G(\hat{\mathbf{r}}')\rangle$:

$$\iint_{4\pi} C_G(\hat{\mathbf{r}}\cdot\hat{\mathbf{r}}')Y_{nm}(\hat{\mathbf{r}}')d\Omega' = \lambda_n Y_{nm}(\hat{\mathbf{r}}), \tag{9.18}$$

---

[1] This theorem can be found in any quantum mechanics or advanced electricity and magnetism book. An excellent source is Arfken and Weber (2005). The theorem is often compared to its trigonometric cousin: $\sin(\alpha+\beta) = \sin\alpha\cos\beta + \sin\beta\cos\alpha$.

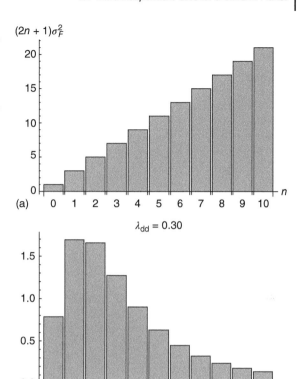

Figure 9.1 (a) Illustration of how spatial white noise that is homogeneous on the sphere is expressed in terms of the variance attributable to individual spherical harmonic components. In the homogeneous (statistically rotationally invariant, and spatially white noise) case, this means that each component has the same amount of variance analogous to the spectrum of a stationary white noise time series. Shown is the chart of the degree variance for spatially white noise on the sphere. The ordinate is $\sum_{m=0}^{2n+1} \langle |F_{nm}|^2 \rangle = \sigma_F^2 \sum_{m=0}^{2n+1} \delta_{nn'} \delta_{mm'} = (2n+1)\sigma_F^2$. (b) Spectral density of variance versus degree number n for a climate model of a uniform Earth with no time dependence and with length scale $\lambda_{dd} = 0.30$ (a value chosen for illustrative purposes).

where λ_n is the eigenvalue and is proportional to the variance of the particular EOF coefficient labeled n. This result can be proved by expanding $C_G(\hat{\mathbf{r}} \cdot \hat{\mathbf{r}}')$ into Legendre polynomials and using the addition theorem.[2] We have the remarkable coincidence (rotational invariance) in our climate model (with uniform properties) that the KL functions $Y_{nm}(\hat{\mathbf{r}})$ are also the dynamical normal modes—or in our damped diffusive problem, the decay modes. This happens in the present case because the rotational symmetry forces it. The variances associated with the modes of T also do not depend on the index m, which is a consequence of the rotational invariance as well.

We turn to the n dependence of $\sigma_T^2(n, m)$ (note: there is no m dependence):

$$\sigma_T^2(n) = \frac{\sigma_F^2/B^2}{(n(n+1)\lambda_{dd}^2 + 1)^2}. \tag{9.19}$$

The *degree variance* which includes the variance contributed by longitudinal modes is

$$S_n = \frac{2n+1}{4\pi} \frac{\sigma_F^2/B^2}{(n(n+1)\lambda_{dd}^2 + 1)^2}, \tag{9.20}$$

which is shown in Figure 9.1b for $\lambda_{dd} = 0.30$. For reference, Figure 9.1a shows the degree spectrum of the white noise driving force. If λ_{dd} were vanishing, the degree spectrum would expand into the white noise spectrum in Figure 9.1a. Instead, contributions from

---

2 A proof can be found in North and Cahalan (1981); the original proof was in the doctoral dissertation of Obukhov (1947).

the upper part of the variance index n are shut out by the damped-diffusive operator for $\lambda_{dd} > 0$. We call operators like this *low-pass filters*. In other words, only the low-index modes from the forcing are passed to the response spectrum of variances.

9.2 Time-Dependent Noise Forcing for a Uniform Planet

Now we admit the heat storage term $C\, dT/dt$ to the energy balance equation to permit time dependence in the problem. We continue with C = constant for a uniform planet. The noise agent forcing the system will also be white in time as well as in space, keeping the conditions for rotational invariance on the sphere as well as the conditions of stationarity for the time series representing the temperature field. The spatial white noise is quite different from that of planet Earth as, in reality, the conditions for weather noise would not be rotationally invariant on our planet. Weather noise intensity (variance) is seasonally dependent and is restricted to a seasonally cycling band in the middle latitudes of each hemisphere.

We prescribe the noise to be white in time as well as in space. This means the noise has a very short autocorrelation time (equal to its relaxation time, e.g., over land, $\tau_0 \sim 30$ days) compared to that of actual weather whose characteristic time is a few days. We might think of the noise here as the fluctuations of middle-latitude weather, which has an autocorrelation timescale of about 3 days. The mean of the noise function vanishes:

$$\langle F(\hat{\mathbf{r}}, t) \rangle = 0, \tag{9.21}$$

and its autocovariance function in space–time is given by

$$\langle F(\hat{\mathbf{r}}, t) F(\hat{\mathbf{r}}', t') \rangle = \sigma_F^2 \delta(t - t') \delta(\hat{\mathbf{r}} - \hat{\mathbf{r}}'). \tag{9.22}$$

We must allow the forcing and response (random) fields to be decomposed in frequency as well as spherical harmonic components. This will be possible as the model forcing as well as the solution will be a stationary and continuous time series. The time series span is $-\infty < t < \infty$. Since the time is continuous (as opposed to jumping in discrete steps), we are obliged to use the continuous Fourier transform.

$$F(\hat{\mathbf{r}}, t) = \sum_{n=0}^{\infty} \sum_{m=-n}^{n} Y_{nm}(\hat{\mathbf{r}}) \int_{-\infty}^{\infty} \tilde{F}_{nm}^f e^{-2\pi i f t}\, df, \tag{9.23}$$

$$T(\hat{\mathbf{r}}, t) = \sum_{n=0}^{\infty} \sum_{m=-n}^{n} Y_{nm}(\hat{\mathbf{r}}) \int_{-\infty}^{\infty} \tilde{T}_{nm}^f e^{-2\pi i f t}\, df, \tag{9.24}$$

where we have used the superscript f to denote the Fourier component corresponding to frequency f. We also employ a tilde, $\tilde{F}_{nm}^f$ to denote the Fourier-transformed variable as opposed to the Fourier mirror image, $F_{nm}(t)$. By substituting and using the fact that the $Y_{nm}(\hat{\mathbf{r}})$ are the eigenfunctions of ∇^2 and then using the orthogonality of the $Y_{nm}(\hat{\mathbf{r}})$, we have

$$\tilde{T}_{nm}^f = \frac{\tilde{F}_{nm}^f / C}{2\pi i f \tau_0 + n(n+1)\lambda_{dd}^2 + 1}, \tag{9.25}$$

and we have, in addition,

$$\langle (\tilde{F}_{nm}^f)^* \tilde{F}_{n'm'}^{f'} \rangle = \sigma_F^2 \delta(f - f') \delta_{nn'} \delta_{mm'}. \tag{9.26}$$

Because the time is continuous, we need the Dirac delta function, $\delta(f-f')$ in the autocovariance of white noise. Just as before, we can write the relations, but now including frequency:

$$\langle(\tilde{T}^f_{nm})^*\tilde{T}^{f'}_{n'm'}\rangle = \sigma^2_T(n,m,f)\delta(f-f')\delta_{nn'}\delta_{mm'}, \tag{9.27}$$

with

$$\sigma^2_T(n,f) = \frac{\sigma^2_F/B^2}{4\pi^2\tau_0^2 f^2 + (n(n+1)\lambda^2_{dd}+1)^2}$$

$$= \frac{\sigma^2_F\tau_n^2\tau_0^{-2}B^{-2}}{4\pi^2\tau_n^2 f^2 + 1} \geq 0, \tag{9.28}$$

where

$$\tau_n = \frac{\tau_0}{1+n(n+1)\lambda^2_{dd}}. \tag{9.29}$$

We can now compute the autocovariance of the temperature field evaluated at two separated points on the sphere $\hat{\mathbf{r}}$ and $\hat{\mathbf{r}}'$ and at two times t and t'. The separations are lag $\tau = |t-t'|$ and $\hat{\mathbf{r}}\cdot\hat{\mathbf{r}}'$, the latter being the great circle distance from the two points on the unit sphere. We start with

$$\langle T(\hat{\mathbf{r}},t)T(\hat{\mathbf{r}}',t')\rangle = \iint e^{2\pi i(tf-t'f')}\sum_{n,m}\sum_{n',m'}Y_{nm}(\hat{\mathbf{r}})Y^*_{n'm'}(\hat{\mathbf{r}}')\langle T^f_{nm}T^{f'}_{n'm'}\rangle \mathrm{d}f\,\mathrm{d}f'. \tag{9.30}$$

After use of the techniques above, including the addition theorem, we obtain for $\hat{\mathbf{r}} = \hat{\mathbf{r}}'$:

$$\mathrm{Covar}_T(\hat{\mathbf{r}},\tau) = \sum_{n=0}^\infty \frac{(2n+1)}{4\pi}\int_{-\infty}^\infty \sigma^2_T(n,f)e^{2\pi i\tau f}\,\mathrm{d}f. \tag{9.31}$$

Note that the result does not depend on $\hat{\mathbf{r}}$ (all points on this sphere are the same). We have one more integral to deal with (we can turn to tables or MATHEMATICA):

$$\int_{-\infty}^\infty \sigma^2_T(n,f)e^{2\pi i\tau f}\,\mathrm{d}f = \int_{-\infty}^\infty \frac{\sigma^2_F\tau_n^2\tau_0^{-2}B^{-2}}{4\pi^2\tau_n^2 f^2+1}e^{2\pi i\tau f}\,\mathrm{d}f = \frac{\sigma^2_F\tau_n}{2\tau_0^2 B^2}e^{-|\tau|/\tau_n}. \tag{9.32}$$

The result to be used in (9.31) is a sum of exponentially decaying terms. The solid curve in Figure 9.2 shows a log-plot of the sum $(2n+1)\tau_n\,e^{-|\tau|/\tau_n}$, using $\lambda_{dd}=0.64$, $\tau_0=1$ and retaining 10 terms (more terms do not affect the result). The solid curve is the sum, and the dashed curve is the leading term in the sum.

9.3 Green's Function on the Sphere: $f=0$

Consider the response of the temperature field to a steady heat source located at the point $\hat{\mathbf{r}}'$. We examine first the time-independent case

$$-D\nabla^2 G(\hat{\mathbf{r}};\hat{\mathbf{r}}') + BG(\hat{\mathbf{r}};\hat{\mathbf{r}}') = \delta(\hat{\mathbf{r}}-\hat{\mathbf{r}}'), \tag{9.33}$$

where $G(\hat{\mathbf{r}};\hat{\mathbf{r}}')$ denotes the thermal response field to the point source. The function $G(\hat{\mathbf{r}};\hat{\mathbf{r}}')$ is called *Green's function* for the field. It can be seen by symmetry that it has the

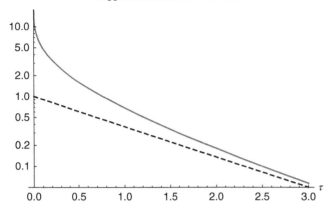

Figure 9.2 Graphic illustration of the lag covariance between one time and a lagged one both taken at the same point on the sphere and the dominant contribution from the largest (global) scale. The condition is for the uniform sphere with a space–time white noise driver. The solid curve shows a log-plot of the sum $(2n + 1)\tau_n\, e^{-|\tau|/\tau_n}$ using $\lambda_{dd} = 0.64$. The solid curve is the sum, and the dashed curve is the leading term in the sum. Note that at lag zero in both space and time, the variance diverges.

property that its value depends only on the great circle distance from the source $\hat{\mathbf{r}} \cdot \hat{\mathbf{r}}'$. It is especially of interest as the thermal field response to an arbitrary distribution of heat, say, $h(\hat{\mathbf{r}})$ leads to a thermal anomaly $T(\hat{\mathbf{r}})$ and that anomaly can be related to the Green's function in a relatively simple way. According to our definitions,

$$-D\nabla^2 T(\hat{\mathbf{r}}) + BT(\hat{\mathbf{r}}) = h(\hat{\mathbf{r}}). \tag{9.34}$$

Now expand all the functions into their Laplace series:

$$G(\hat{\mathbf{r}} \cdot \hat{\mathbf{r}}') = \sum_{nm} G_n Y_{nm}(\hat{\mathbf{r}}') Y^*_{nm}(\hat{\mathbf{r}}), \tag{9.35}$$

$$T(\hat{\mathbf{r}}) = \sum_{nm} T_{nm} Y_{nm}(\hat{\mathbf{r}}), \tag{9.36}$$

$$h(\hat{\mathbf{r}}) = \sum_{nm} h_{nm} Y_{nm}(\hat{\mathbf{r}}), \tag{9.37}$$

$$\delta(\hat{\mathbf{r}} - \hat{\mathbf{r}}') = \sum_{nm} Y^*_{nm}(\hat{\mathbf{r}}') Y_{nm}(\hat{\mathbf{r}}). \tag{9.38}$$

The expansion of $G(\hat{\mathbf{r}} \cdot \hat{\mathbf{r}}')$ comes about by first expanding into Legendre polynomials $P_n(\hat{\mathbf{r}} \cdot \hat{\mathbf{r}}')$. The others are conventional expansions that can be checked from definitions. By the usual insertion and projection of spherical harmonic components,

$$G_n = \frac{1/B}{n(n+1)\lambda_{dd}^2 + 1}, \tag{9.39}$$

$$T_{nm} = \frac{h_{nm}/B}{n(n+1)\lambda_{dd}^2 + 1}. \tag{9.40}$$

Now consider the thermal anomaly

$$\begin{aligned}
T(\hat{\mathbf{r}}) &= \sum_{nm} T_{nm} Y_{nm}(\hat{\mathbf{r}}) \\
&= \sum_{nm} \frac{h_{nm} Y_{nm}(\hat{\mathbf{r}})/B}{n(n+1)\lambda_{dd}^2 + 1} \\
&= \sum_{nm} h_{nm} G_n Y_{nm}(\hat{\mathbf{r}}) \\
&= \int G(\hat{\mathbf{r}} \cdot \hat{\mathbf{r}}') h(\hat{\mathbf{r}}') d\Omega'.
\end{aligned} \qquad (9.41)$$

The last, which is our desired result, can be obtained by inserting the Laplace series for the factors in the integrand.

9.4 Apportionment of Variance at a Point

Next consider the fluctuations of the surface temperature at a point. These fluctuations may be thought of as being composed of contributions from all space and timescales. For example, consider the uniform earth case (C and D are constant). The Fourier component corresponding to frequency f of the temperature at point $\hat{\mathbf{r}}$ is given by

$$T(\hat{\mathbf{r}},f) = \sum_{n,m} \tilde{T}^f_{n,m} Y_{n,m}(\hat{\mathbf{r}}). \qquad (9.42)$$

The variance at point $\hat{\mathbf{r}}$ and frequency f is given by

$$\begin{aligned}
\langle T^2(\hat{\mathbf{r}},f) \rangle &= \sum_{n',m'} \sum_{n,m} \langle (T^f_{n',m'})^* T^f_{n,m} \rangle Y^*_{n',m'}(\hat{\mathbf{r}}) Y_{n,m}(\hat{\mathbf{r}}) \\
&= \sum_n \sigma_T^2(n,f) \sum_m |Y_{n,m}(\hat{\mathbf{r}})|^2 \\
&= \sum_n \sigma_T^2(n,f) \frac{2n+1}{4\pi} P_n(1) \\
&= \sum_n \sigma_T^2(n,f) \frac{2n+1}{4\pi}.
\end{aligned} \qquad (9.43)$$

The total variance at degree n (including contributions for all m modes of the degree n level) is

$$S_n^f = \left(\frac{2n+1}{4\pi}\right) \frac{\sigma_F^2/B^2}{4\pi^2 \tau_0^2 f^2 + (n(n+1)\lambda_{dd}^2 + 1)^2}. \qquad (9.44)$$

Figure 9.3 shows the fraction of total variance contained in a given spherical harmonic of degree n for $f = 1/\tau_0$ and $f = 3/\tau_0$ as a function of n. As the frequency is increased from $f = 1/\tau_0$ to $f = 3/\tau_0$, the power moves to higher-degree indices. This figure is to be compared with Figure 9.1 where the cases for $f = 0$ and the case for the integral over all frequencies (equivalent to the distribution of variance at a point in time).

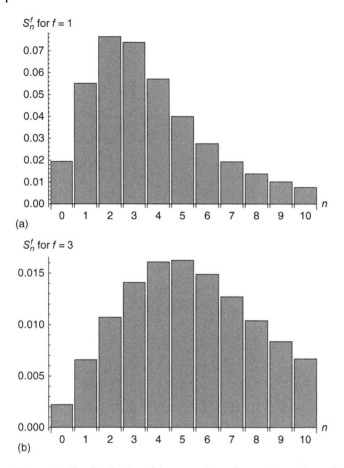

Figure 9.3 The distribution of degree variance for a point on the uniform sphere forced by space–time white noise. (a) The distribution of degree variance S_n^f for $f = 1/\tau_0$. (b) Same as (a) except for $f = 3/\tau_0$. In each case, $\tau_0 \equiv 1$. Note also that, in each case, the factor $(2n + 1)$ is the number of m modes for a given degree index n and each has equal variance for the spherically symmetric globe.

9.5 Stochastic Model with Realistic Geography

As in the last chapter, models with a realistic land–sea distribution cannot be solved analytically. One must turn to numerical methods to obtain solutions. The introduction of noise as a forcing agent is not too difficult. First of all, the EBCM is basically a linear system (if we ignore snow and other nonlinear feedbacks) with time-independent coefficients. This means that the model that simulates the seasonal cycle (Chapter 8) can be used by simply removing the seasonal driving term and inserting the noise field.

Leung and North (1991) compared a general circulation model (GCM) simulating climate on a bald, land-covered planet (referred to in Chapter 1 as *Terra Blanda*) run at equinox conditions with models of the all land models of the type studied earlier in this chapter. The two modeling schemes had similar spatial statistics. As an example, consider the relaxation time for Legendre modes as shown in Figure 9.4. The fit to the

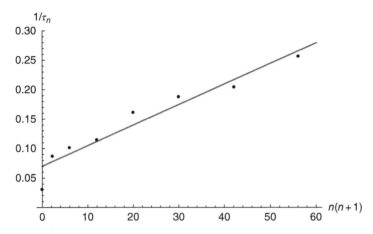

Figure 9.4 Illustration of the agreement for the characteristic times of individual modes between the EBM and a GCM (CCM0) both for a uniform planet. Shown is such a plot of $1/\tau_n$ as a function of $n(n + 1)$ for the all-land GCM (points) and the EBCM values (line). Only the global mode is significantly off the regression line. (Leung and North (1991). © American Meteorological Society. Used with permission.)

relaxation times in GCM and EBCM for a bald planet is remarkable. Note that the GCM runs were set at equinox conditions and the EBCM is set at mean annual values (in the linear ECBM, this does not matter!). Moreover, the GCM statistics are hardly rotationally invariant on the sphere. Because of the featureless geography, its solution statistics are longitudinally stationary, but not the latitudinal ones. The results of this study by Leung and North together with the seasonal modeling success of the full EBM suggests that we might have a chance at modeling the statistics of the response to white space–time noise forcing.

The linear 2-D EBCM of North *et al.* (1983), which used the spherical harmonic basis, and many later versions of it have been solved by various methods including finite differencing on the spherical surface (Wu and North, 2007), finite differencing employing multigrid relaxation (Bowman and Huang, 1991; Stevens and North, 1996). A novel means of solving the model with geography involves using the nonorthogonal decay modes as a basis set. The first study to attempt including noise with the current land–sea distribution, this was carried out by Kim and North (1991), who employed the spherical harmonic basis set (Wu and North, 2007).

We take figures here from the two-dimensional EBCM of Kim and North (1991). This model employed a simple mixed-layer ocean and the forcing noise was white in space and time. Note especially that the space–time white noise was uniformly distributed over the globe in variance and with zero mean. Some later model studies attempted to apportion the forcing noise to be only in the mid-latitudes, but here we show only the uniformly distributed case, as otherwise we would have to introduce more phenomenological parameters. The only adjustable parameter is then the variance of the noise field, σ_F^2. The variance of the temperature field is everywhere proportional to this variance. We adjust this variance to match the observed variance to that of the model's simulated temperature fluctuations. Note that in all cases in this chapter, the noise-forced EBCM has a mixed-layer (slab) ocean. Low-frequency variability would be somewhat different in a model with deeper oceanic elements.

240 | *9 Perturbation by Noise*

Figures 9.5–9.7 show a sequence of maps of the variance of the surface temperature field in the observations. The left panels indicate observations and the right panels are for EBCM simulations. Both fields were smoothed to the same level of T11. The data and observations were band-pass filtered to include the periods between the limits

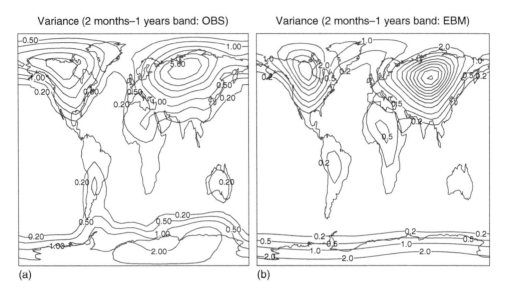

Figure 9.5 Variance of the surface temperature field in the observations (a) and the EBCM simulations (b). Both fields were smoothed to the same level of T11. The data and observations in Figure 9.5 were band-pass filtered to include the periods between 2 months and 1 year. The contours are spaced at 0.5 °C intervals. (Kim *et al.* (1996). © American Meteorological Society. Used with permission.)

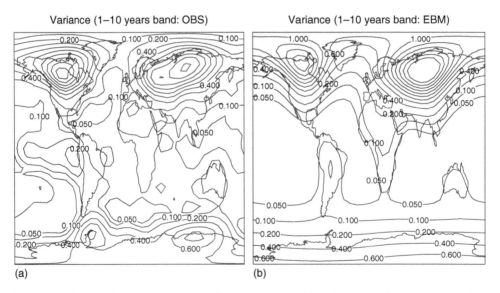

Figure 9.6 Same as the last figure, only the frequency band width is from periods of 1 to 10 years. The contours are every 0.02 °C except for values below 0.100 °C (mainly over oceans). (Kim *et al.* (1996). © American Meteorological Society. Used with permission.)

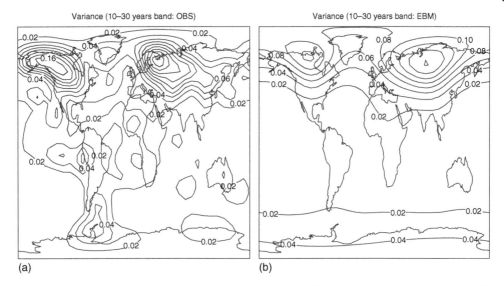

Figure 9.7 Same as the last figure only the frequency band width is from periods of 10 to 30 years. The contours are every 0.02 °C (values below 0.020 °C are mainly over oceans). (Kim *et al.* (1996). © American Meteorological Society. Used with permission.)

indicated on the maps. Procedures for the filtering are found in the paper by Kim and North (1991). It is interesting that once the model's free parameters are adjusted (tuned) to match the variance of the observations in the middle of Asia, the maps come into reasonable agreement elsewhere, for example 2 °C in Antarctica and 0.2 °C in South America (Figure 9.5). As in the seasonal model, the response to high-frequency forcing is dominated by the land–sea configuration. Note also the influence of the Himalayan plateau on the observations, but not on the simulated field, which of course has no topography. Topography is likely to be at work also in northwestern North America. The agreement in this procedure emphasizes the dominant nature of the geographical imprint of the land–sea distribution of the surface temperature.

Figure 9.6 is the same as the last figure except that the frequencies lie between periods of 1 and 10 years. In this figure, we also note the strong effects of the Himalayas and also a hint of the ENSO pattern in the data, but not, of course, in the mixed-layer ocean model. Figure 9.7 showing fluctuations having periods between 10 and 30 years shows some strong indications in the data of sea ice fluctuations around the northern edges of the continents. There is also a hint of ENSO in the data, but most ENSO is at higher frequencies than this band permits. Note the overall washing out of the features of the EBCM simulation in these low-frequency components. A deeper ocean than the mixed layer shown here would show more continental–oceanic contrast.

Figures 9.8 and 9.9 show the spatial correlation between six fixed points and their surrounding areas (mid-Asia, Africa, South Pole, mid-Atlantic, equatorial Pacific, West coast-America). In each case, the correlation decays from the fixed point out to where it falls off to 1/e indicated by the heavy line. Features to note are that at high frequencies (Figure 9.8), the length scales are long over land and short over ocean. Since the ocean has a timescale of a few years, we are in the high-frequency regime of the oceans in this band. On the other hand, we are in the low-frequency band for land areas and the

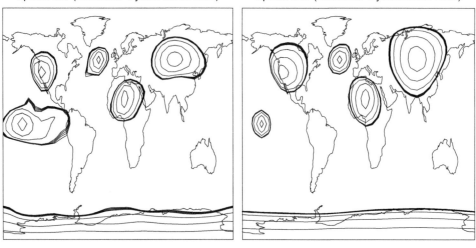

Figure 9.8 Correlation between the surface temperature fluctuations at high frequencies at six fixed sites with neighboring points. The six fixed sites are mid-Asia, Africa, South Pole, mid-Atlantic, equatorial Pacific, West coast America. (Kim *et al.* (1996). © American Meteorological Society. Used with permission.)

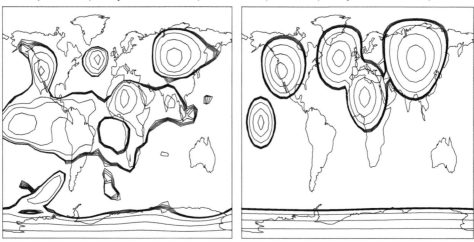

Figure 9.9 Correlation between the surface temperature fluctuations at a broad band of frequencies (between periods corresponding to 1 and 10 years) at six fixed sites with neighboring points. The six fixed sites are mid-Asia, Africa, South Pole, mid-Atlantic, equatorial Pacific, West coast America. (Kim *et al.* (1996). © American Meteorological Society. Used with permission.)

autocorrelation distances are larger. This is shown dramatically for the fixed point at San Francisco where the correlation out to sea is short and inland it is long. The effect is prominent in both data and model simulation. Note that ENSO is prominent in the data, but missing completely as expected in the EBCM simulation.

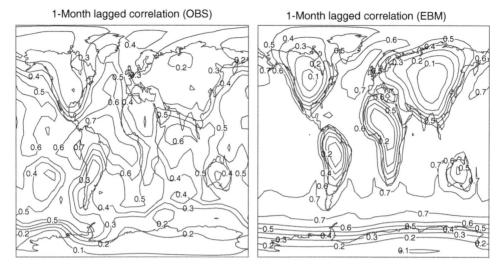

Figure 9.10 The one-month lagged correlation between the surface temperature at a point on the map and the same lagged by one month. (Kim et al. (1996). © American Meteorological Society. Used with permission.)

The situation in Figure 9.9 is quite different. The correlation lengths over the North Atlantic are too large in the EBCM simulation because of the short timescale of the mixed-layer model. ENSO dominates all of the tropics in the data, but is absent in the EBCM. Our last comparison is for the one-month lagged correlation map between the observations (Figure 9.10a) and the EBCM (Figure 9.10b). In this comparison, we do find some serious discrepancies although the lags over continental interiors is pretty good. The transition from land to oceans is quite abrupt in the EBCM upto the level of about 0.6, while in the data it is smoother to roughly this same value. Also shown in the paper by Kim et al. (1996) are comparisons of the same statistics as in Figures 9.6–9.10. (Figures modified from Kim et al. (1996). (© Amer. Meteorol. Soc., with permission.))

9.6 Thermal Decay Modes with Geography

In this section,[3] we consider an alternative modal decomposition of the time-dependent problem with real land–sea geography. The model is linear. We begin with the equation for departures of the local surface temperature $T(\hat{\mathbf{r}}, t)$ from steady state:

$$C(\hat{\mathbf{r}})\frac{\partial T}{\partial t} - \nabla \cdot (D(\hat{\mathbf{r}})\nabla T) + BT = F(\hat{\mathbf{r}}, t). \tag{9.45}$$

Next, insert the exponential time dependence $T(\hat{\mathbf{r}}, t) = \psi(\hat{\mathbf{r}})e^{-\lambda t}$. This leads to the eigenvalue problem

$$(-\nabla \cdot D(\hat{\mathbf{r}})\nabla + B)\psi(\hat{\mathbf{r}}, t) = \lambda C(\hat{\mathbf{r}})\psi(\hat{\mathbf{r}}), \tag{9.46}$$

---

3 This section follows Wu and North (2007).

with the boundary conditions that the solution be finite and without divergence at the poles. This is called a *generalized Stürm–Liouville system* because of the space-dependent factor $C(\hat{\mathbf{r}})$ on the RHS. Under certain conditions, this system yields a set of eigenfunctions $\psi_n(\hat{\mathbf{r}})$ corresponding to eigenvalues λ_n, $n = 0, 1, 2, \ldots$ (see Horn and Johnson, 1985). The main conditions are that the domain be finite (the spherical surface, in this case) and that the operator on the LHS of the last equation be Hermitian. An operator $\mathcal{O}[\hat{\mathbf{r}}]$ in this context is one in which

$$\int_{4\pi} \phi(\hat{\mathbf{r}})\mathcal{O}[\hat{\mathbf{r}}]\eta(\hat{\mathbf{r}})\mathrm{d}^2\Omega = \int_{4\pi} \eta(\hat{\mathbf{r}})\mathcal{O}[\hat{\mathbf{r}}]\phi(\hat{\mathbf{r}})\mathrm{d}^2\Omega, \tag{9.47}$$

which can be demonstrated by using integration by parts or equivalently the two-dimensional divergence theorem on the spherical surface. Note that the inverse of the eigenvalues are just the relaxation times for the modes, $\psi_n(\hat{\mathbf{r}})$.

The modes $\psi_n(\hat{\mathbf{r}})$ are *not* orthogonal. Nevertheless, we can form series representations because of the following relation (which can be derived from the above expressions):

$$\int_{4\pi} \psi_m(\hat{\mathbf{r}})C(\hat{\mathbf{r}})\psi_n(\hat{\mathbf{r}})\,\mathrm{d}^2\Omega = \delta_{m,n} = (\psi_m, C\psi_n), \tag{9.48}$$

where the *inner product notation* $(\psi_m, C\psi_n)$ is introduced as a notational simplification. These steps lead us to

$$T(\hat{\mathbf{r}}, t) = \sum_n a_n(t)\psi_n(\hat{\mathbf{r}}), \tag{9.49}$$

with

$$a_n(t) = \int_{4\pi} \psi_n(\hat{\mathbf{r}})C(\hat{\mathbf{r}})T(\hat{\mathbf{r}}, t)\mathrm{d}^2\Omega = (\psi_n, CT). \tag{9.50}$$

We can now insert (9.57) into the governing equation to find

$$\dot{a}_n + \lambda_n a_n = \int_{4\pi} \psi_n(\hat{\mathbf{r}})F(\hat{\mathbf{r}}, t)\mathrm{d}^2\Omega = (\psi_n, F). \tag{9.51}$$

Note the absence of $C(\hat{\mathbf{r}})$ in the integral on the RHS compared to the previous equation. Here we can see explicitly that if $F(\hat{\mathbf{r}}, t)$ is set to zero the amplitude of mode n decays exponentially with time constant λ_n^{-1}.

Returning to the eigenvalue relation, we can use values from the map of $C(\hat{\mathbf{r}})$ in (9.48). The heat capacity map is represented by 64 (longitude) × 31 (sin(latitude)) plus 2 polar grid points. Once the heat capacity matrix is formed, one can use standard methods on (9.48) to recover the eigenvalues, λ_n, and the eigenvectors, $\psi_n(\hat{\mathbf{r}})$. Hereafter, we will refer to the $\psi_n(\hat{\mathbf{r}})$ as the *thermal decay modes* (TDMs). We will now examine a few of the results. First, consider the spectrum of relaxation times, λ_n^{-1}, as shown in Figure 9.11. The dotted curve shows the values that would be obtained from observational data projected onto the modal shapes. The essential difference between these modes and the EOF modes (Kim and North, 1992) is that they are spatial physical modes and their shapes do not depend on frequency as the statistical (EOF) modes do. Also, they are not strictly orthogonal as are the EOFs. ($C(\hat{\mathbf{r}})$ is the weighting function).

Consider the log–log spectra of relaxation times shown in Figure 9.11. The number of spectral components is 1986, equal to the number of grid points. The spectrum is ordered from the longest time at $n = 1$ to the shortest at $n = 1986$. The dashed curve

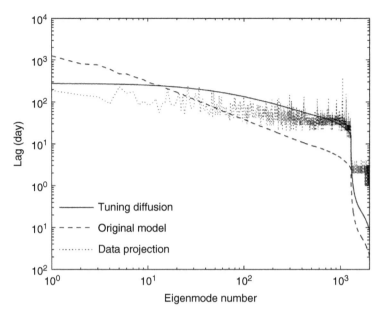

Figure 9.11 Log–log spectra of timescales as a function of mode index n. The dashed line is the spectrum of theoretical decay timescales calculated by using the parameters in Hyde *et al.* (1990). "Original Model" indicates using the parameter values in the EBCM model used in Hyde *et al.* (1990). The dotted line is the spectrum of the estimated autocorrelation timescales calculated by data projection onto the TDMs. The solid line is the spectrum of theoretical decay timescales from the tuned model (model parameters in this chapter) with much shorter length scales over the oceans. (Wu and North (2007). Reproduced with permission of Taylor and Francis.))

shows a smoothed estimate calculated from a model based on the parameterization of Hyde *et al.* (1990). It is interesting that there is a sharp fall in the relaxation times at $n = 1282$. This represents the transition from the family of oceanic modes with long timescales to land-dominated modes for $n \geq 1282$. The dotted curve in Figure 9.11 is an estimate of the relaxation time spectrum data. This is done by projecting the observed data onto the model-calculated eigenmode and calculating the autocorrelation times from the resulting time series for each decay mode. Note that the "observed" spectrum is much flatter than the model-generated spectrum (dashed line). One can adjust the parameter values in $D(\hat{\mathbf{r}})$ as well as B to bring the spectra more into line. Numerical experimentation shows that the shapes of the decay modes are not significantly affected by a fairly wide range of choices for these parameters. Given the problem with matching the model to data with respect to one-month lags in Figure 9.10, it is not surprising that we might need to make adjustments. More details are contained in Wu and North (2007).

When we examine the oceanic modes ($n \leq 1281$), we find that virtually all of the non-negligible amplitudes are over ocean as shown in Figure 9.12. The lowest modes shown there have time constants between 0.763 and 0.735 year. Since length scales are short over this mixed-layer ocean, these response modes as driven by white noise are virtually white in space as well (no correlation between one point and another on this sparse grid). Next, turn to the land modes, $n \geq 1282$. These modes group themselves into families according to continental clusters. For example, Figure 9.13 shows four

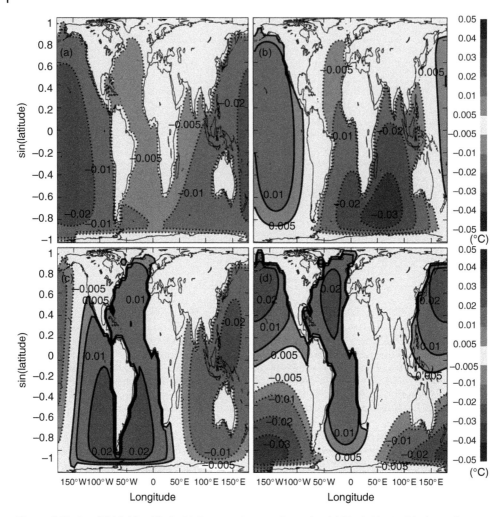

Figure 9.12 Four TDMs identified with large-scale oceanic modes: (a) Mode No. 1 with decay time $\tau = 0.763$ year; (b) Mode No. 2 with decay time $\tau = 0.749$ year; (c) Mode No. 3 with decay time $\tau = 0.746$ year; (d) Mode No. 4 with decay time $\tau = 0.735$ year. The mode amplitudes over the rest of the world are negligible. (Figures originally generated in the paper by coauthor GRN © *Tellus A*, permission not required.) (Wu and North (2007). Reproduced with permission of Taylor and Francis.)

modes. Figure 9.13a corresponds to $n = 1282$, $\tau = 0.03277$ year. Times for the other panels (Figure 9.13b–d) are listed in the figure caption. These modes were selected because they form a family of modes connected with the Eurasian continent. Figure 9.14 shows four modes associated with the North American continent. The modal shapes in both figures show the familiar "drumhead" patterns of eigenmodes for either wavelike or diffusive-like systems.

9.6.1 Statistical Properties of TDMs

We have already remarked that the TDMs cannot correspond to EOFs because EOFs are mutually orthogonal, whereas the TDMs are skewed. If the mixed layer were infinitely

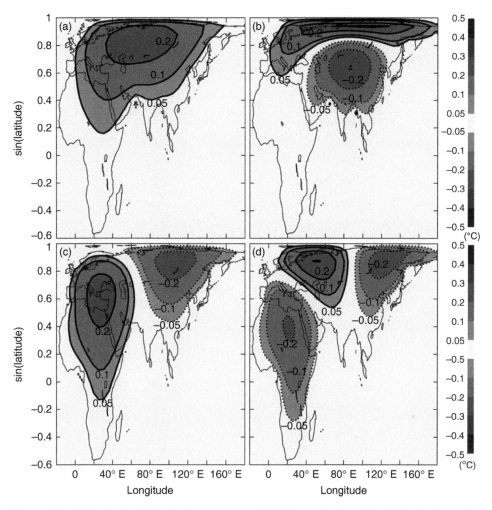

Figure 9.13 (a) Mode No. 1282 with decay time $\tau = 0.03277$ year; (b) Mode No. 1287 with decay time $\tau = 0.01561$ year; (c) Mode No. 1283 with decay time $\tau = 0.02329$ year; (d) Mode No. 1286 with decay time $\tau = 0.01642$ year. The mode amplitudes over the rest of the world are close to zero. (Wu and North, 2007. Reproduced with permission of Taylor and Francis.)

deep, there would be no response to space–time white noise over the oceans and this would force the TDMs in that limit to be mutually orthogonal.[4] Looking at the last two figures, one is tempted to think along these lines at least for heuristic purposes.

To get an idea about the statistical properties of TDMs refer to (9.50) and (9.51). First, note that the TDMs are dynamical modes. This means that if certain spatial mode patterns are present in the forcing, only those mode patterns will be found in the response. Similarly, if certain temporal frequencies are present in the forcing, only those frequencies will be found in the response. This follows from the stationarity of the forcing $F(\hat{\mathbf{r}}, t)$ and its response $T(\hat{\mathbf{r}}, t)$. But the nonorthogonality of the solutions leads

---

4 In this case, the boundary condition at the shorelines would be that the temperature (anomaly) has to be zero.

248 | *9 Perturbation by Noise*

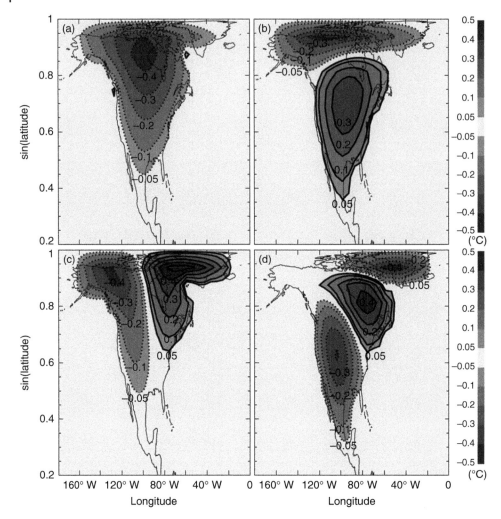

Figure 9.14 Examples of North American family modes: (a) Mode No. 1284 with decay time $\tau = 0.02031$ year; (b) Mode No. 1289 with decay time $\tau = 0.01191$ year; (c) Mode No. 1294 with decay time $\tau = 0.00952$ year; (d) Mode No. 1301 with decay time $\tau = 0.00747$ year. The mode amplitudes over the rest of the world are close to zero. (Wu and North, 2007. Reproduced with permission of Taylor and Francis.)

to a very peculiar property, namely, the dynamical modes are correlated in time. EOFs are uncorrelated, but EOFs are not dynamical decay modes. Recall that this would have been the case for the uniform planet where the TDMs are orthogonal and coincide with the EOFs (this holds as well for the case of the infinitely deep mixed layer).

Another interesting property is that the functions $\phi_n(\hat{\mathbf{r}}) \equiv \sqrt{C(\hat{\mathbf{r}})}\psi_n(\hat{\mathbf{r}})$ are mutually orthogonal.

Notes for Further Reading

There are many books on stochastic processes. For an introduction, the nicely written and inexpensive book by Bulmer (1979) covers the principles of statistics including calculus. Some math techniques for statistics are in the classic by Cramér (19th printing in 1999). The book by Cramér and Leadbetter (1995) serves as a good introduction to stochastic processes including vector processes. A useful handbook to consult is that of Gardiner (1985). Electrical engineering books often have good coverage of stochastic methods, e.g., Gardner (1989); and the more comprehensive, Papoulis (1984).

Exercises

9.1 Let us consider a time-dependent energy balance model forced by a sinusoidal forcing:

$$C\frac{\partial T(\hat{r}, t)}{\partial t} - D\nabla^2 T(\hat{r}, t) + A + BT(\hat{r}, t) = F\, e^{i\omega t}, \tag{9.52}$$

where A, B, C, D, and F are all constants.

(a) Let the solution (temperature) of the energy balance model be in the form

$$T(r, t) = \sum_{k,l} T_{k,l} Y_k^l(\phi, \theta) e^{i\omega t}. \tag{9.53}$$

Set up the energy balance model for the temperature field given above and derive the temperature field.

(b) How do the amplitude and the phase of the temperature field depend on the frequency ω of forcing?

9.2 Let us consider a time-dependent energy balance model forced by noise:

$$C\frac{\partial T(\hat{r}, t)}{\partial t} - D\nabla^2 T(\hat{r}, t) + A + BT(\hat{r}, t) = F(\hat{r}, t), \tag{9.54}$$

where A, B, C, and D are all constants.

(a) Let the solution (temperature) of the energy balance model be in the form

$$T(\hat{r}, t) = \sum_j \sum_{k,l} T_{k,l}^j Y_k^l(\phi, \theta) e^{i 2\pi j t}. \tag{9.55}$$

Set up the energy balance model for the temperature field given above and derive the temperature field.

(b) Find the phase lag of the temperature field with respect to the noise forcing.

9.3 Find the equilibrium solution of the energy balance model forced by an impulsive radiative forcing:

$$C\frac{\partial T(\hat{r}, t)}{\partial t} - D\nabla^2 T(\hat{r}, t) + A + BT(\hat{r}, t) = F(\hat{r}, t), \tag{9.56}$$

where the impulsive radiative forcing is given by

$$F(\hat{r}, t) = \begin{cases} 0, & \text{for } t < 0, \\ F_0, & \text{for } t \geq 0. \end{cases} \tag{9.57}$$

Note that A, B, C, D, and F_0 are all constants.

9.4 Let us consider a time-dependent energy balance model forced by an impulsive radiative forcing:

(a) Show that the Fourier transform of the impulsive radiative forcing is given by
$$\mathcal{F}(F(\hat{\mathbf{r}}, t), t \to \omega) = \int_{-\infty}^{\infty} F(\hat{\mathbf{r}}, t) e^{-i2\pi\omega t} \, dt = \frac{F_0}{2}\left(\delta(\omega) - \frac{i}{\pi\omega}\right). \quad (9.58)$$

Hint: $F(\hat{\mathbf{r}}, t) = \begin{cases} 0, & \text{for } t < 0 \\ F_0, & \text{for } t \geq 0 \end{cases} = \lim_{a \to 0} \begin{cases} 0, & \text{for } t < 0, \\ F_0 e^{-at}, & \text{for } t \geq 0. \end{cases} \quad (9.59)$

(b) Given the result in Part (a), determine the solution in the form
$$T(\hat{\mathbf{r}}, t) = \sum_{\omega} \sum_{k,l} T_{k,l}^{\omega} Y_k^l(\phi, \theta) e^{i2\pi\omega t}. \quad (9.60)$$

(c) In the limit of $\omega \to 0$, show that the solution is of the form
$$\lim_{\omega \to 0} T(\hat{\mathbf{r}}, t) = \frac{F_0}{B}\left(\frac{1}{2} + t\right) - \frac{A}{B}. \quad (9.61)$$

(d) Determine the lag of the solution as a function of frequency ω. Then, show that the time-dependent solution approaches the equilibrium solution in Exercise 9.3.

9.5 (a) Let $F(\hat{\mathbf{r}}, t)$ be spatially white noise forcing. Show that its expansion coefficients satisfy
$$\langle F_{nm} F_{n'm'}^* \rangle = \sigma_F^2 \delta_{nn'} \delta_{mm'}. \quad (9.62)$$

(b) If the noise forcing is white both spatially and temporally, show that the expansion coefficients of $F(\hat{\mathbf{r}}, t)$ satisfy
$$\langle F_{nm}^{\omega} F_{n'm'}^{\omega' *} \rangle = \sigma_F^2 \sigma_F^2 \delta_{nn'} \delta_{mm'} \delta_{\omega\omega'}, \quad (9.63)$$

which is the desired relationship.

9.6 Let us consider a time-dependent energy balance model for anomalous temperature due to a noise forcing:
$$C \frac{\partial T(\hat{\mathbf{r}}, t)}{\partial t} - D\nabla^2 T(\hat{\mathbf{r}}, t) + A + BT(\hat{\mathbf{r}}, t) = F(\hat{\mathbf{r}}, t), \quad (9.64)$$

where $F(\hat{\mathbf{r}}, t)$ is white both in space and time (see Exercise 9.5).

(a) Using spherical harmonics and Fourier basis functions, obtain the solution of the energy balance model forced by spatially and temporally white noise forcing.

(b) Determine the spectral density function of the temperature response.

(c) Show that
$$\langle T(\hat{\mathbf{r}}, t) T(\hat{\mathbf{r}}', t') \rangle = \sum_{\omega} \sum_{n} \frac{2n+1}{4\pi} \sigma_T^2(n, \omega) P_n(\hat{\mathbf{r}} \cdot \hat{\mathbf{r}}'). \quad (9.65)$$

On the basis of the expression above, derive an expression for spatial covariance $\langle T(\hat{\mathbf{r}}, t) T(\hat{\mathbf{r}}', t) \rangle$ and contemporaneous spatial variance at a point $\langle T(\hat{\mathbf{r}}, t) T(\hat{\mathbf{r}}, t) \rangle$, which is essentially an integration of the spectrum $\sigma_T^2(n, \omega)$ with respect to n and ω.

(d) Show that

$$\langle T(\hat{\mathbf{r}}, t) T(\hat{\mathbf{r}}, t') \rangle = \sum_n \frac{2n+1}{4\pi} \frac{\sigma_F^2 \tau_n}{2 B^2 \tau_0^2} e^{-|\tau|/\tau_n}, \tag{9.66}$$

where

$$\tau_0 = C/B, \quad \tau_n = \frac{\tau_0}{1 + n(n+1)(D/B)}, \quad \text{and } \tau = t - t'. \tag{9.67}$$

9.7 Let us consider the solution of an energy balance model forced by an arbitrary but steady forcing:

$$-D\nabla^2 T(\hat{\mathbf{r}}) + BT(\hat{\mathbf{r}}) = h(\hat{\mathbf{r}}). \tag{9.68}$$

(a) Using the Green's function method, show that the solution of the energy balance model above is given by

$$T(\hat{\mathbf{r}}) = \int_{\Omega'} G(\hat{\mathbf{r}}, \hat{\mathbf{r}}') h(\hat{\mathbf{r}}') \, d\hat{\mathbf{r}}', \tag{9.69}$$

where the Green's function $G(\hat{\mathbf{r}}, \hat{\mathbf{r}}')$ is the solution of the equation

$$-D\nabla^2 G(\hat{\mathbf{r}}, \hat{\mathbf{r}}') + BG(\hat{\mathbf{r}}, \hat{\mathbf{r}}') = \delta(\hat{\mathbf{r}} - \hat{\mathbf{r}}'). \tag{9.70}$$

(b) Obtain the Green's function in Part (a) in terms of spherical harmonics.
(c) Using the Green's function obtained in Part (b), determine the solution of the given energy balance model.

9.8 Let us consider an energy balance model forced by $F(\hat{\mathbf{r}}, t)$:

$$C(\hat{\mathbf{r}}) \frac{\partial T(\hat{\mathbf{r}}, t)}{\partial t} - \nabla \cdot (D(\hat{\mathbf{r}}) \nabla T) + A + BT(\hat{\mathbf{r}}, t) = F(\hat{\mathbf{r}}, t). \tag{9.71}$$

Note that the parameters $C(\hat{\mathbf{r}})$ and $D(\hat{\mathbf{r}})$ are now functions of position. Recast the given energy balance model in a spectral form using spherical harmonics and Fourier functions as basis sets in space and time, respectively.

9.9 Consider an energy balance model forced by $F(\hat{\mathbf{r}}, t)$:

$$C(\hat{\mathbf{r}}) \frac{\partial T(\hat{\mathbf{r}}, t)}{\partial t} - \nabla \cdot (D(\hat{\mathbf{r}}) \nabla T) + A + BT(\hat{\mathbf{r}}, t) = F(\hat{\mathbf{r}}, t), \tag{9.72}$$

where the parameters $C(\hat{\mathbf{r}})$ and $D(\hat{\mathbf{r}})$ are now functions of position. Assume that

$$T(\hat{\mathbf{r}}, t) = \psi(\hat{\mathbf{r}}) e^{-\lambda t}. \tag{9.73}$$

(a) Rewrite the homogeneous form of the energy balance equation above in terms of $\psi(\hat{\mathbf{r}})$ by using the assumed form of the solution. The resulting equation should be in the form of an eigenvalue problem. Discuss the orthogonality properties of the resulting eigenfunctions.
(b) Determine the solution of the given energy balance model in terms of the eigenfunctions derived in Part (a).

10

Time-Dependent Response and the Ocean

Understanding and estimating the evolving temporal response of the system due to a time-dependent forcing are key problems in climate theory. In particular, the layers of air, land, and water have an effective heat capacity that can delay the response to a time-dependent stimulus. The column of air above land for forcing frequencies in the annual cycle range involves only a fraction of the atmospheric column's heat capacity. This effect can delay the warmest day of the year from the day of maximum heating by up to a month. Over open ocean, the same delay can be a whole season or mathematically a quarter of a cycle. A quarter cycle delay turns out to be the maximum when the effective heat capacity is large enough. The reason for the delay difference is that for the column of air, the relaxation time is small compared to the period of the periodic forcing. But the mixed-layer of the ocean has an effective heat capacity that leads to a radiative relaxation time of several years, depending on the depth chosen for its thickness. In a simple linear system of the seasonal cycle, this corresponds to a high frequency forcing (period of 1 year compared to a relaxation time of several years). Land surface response is probably confined to a meter or so into the soil and this, together with the air column, has a relaxation time of the order of a month when forced at roughly annual frequencies. A meter or two below the surface, say in a cave, we find the mean temperatures independent of season – the ground above filters out signals of period shorter than a few months. Lower frequencies tend to penetrate deeper with an effective depth inversely proportional to the square root of the frequency. We will explain this in Section 10.2 where a pure diffusive vertical heat transport is employed (valid in homogeneous soil and perhaps in the upper layers of the ocean).

A good opening example is the response to a sudden spike of forcing such as the negative spike from the dust veil following a volcanic eruption. This example is often used in linear analysis and is the so-called impulse/response function or the temporal Green's function.[1] A second example is the response to an instantaneous doubling of CO_2. This forcing is proportional to the discontinuous Heavyside step function (0 for $t < 0$; 1 for $t \geq 0$). In addition, we would like to see how the system responds to a periodic forcing. The latter has application to the seasonal cycle, the solar cycle, and the response to white noise forcing according to which amplitudes of the Fourier components of the forcing are spread with variance evenly across all frequencies

---

1 Many textbooks on mathematical or engineering physics are available that discuss the Green's function technique, for example, Arfken and Weber (2005).

Energy Balance Climate Models, First Edition. Gerald R. North and Kwang-Yul Kim.
© 2017 Wiley-VCH Verlag GmbH & Co. KGaA. Published 2017 by Wiley-VCH Verlag GmbH & Co. KGaA.

(see Chapters 2 and 9). The damped-diffusion dynamics acts as a low-pass filter, yielding a response more concentrated in the lower frequency bands. The (simplified) forcing most directly bearing on the climate change problem is the response to a linear ramp forcing in time, beginning with a "cold start" at $t = 0$. As discussed in Chapter 2, this occurs when the concentration of carbon dioxide is increasing exponentially starting at $t = 0$. The forcing is linear in time because the forcing due to the greenhouse effect (Chapter 4) is nearly logarithmic in the concentration of CO_2 (see Section 4.10). It is an observed fact that the CO_2 concentration is increasing at a rate of 0.5% year^{-1}. But other greenhouse gases are also increasing at about the same rate, and it is conventional to take the forcing to be one that is increasing at a rate of 1.0% year^{-1}, which yields a doubling time of about 70 years. Many general circulation climate model results are presented for this prototype experiment. In fact, the so-called *transient climate sensitivity* is the amount of global average temperature change from the onset to the time of doubling. All of the models treated in this chapter are governed by systems of equations that are linear and with time-independent coefficients. These exercises constitute an approach to linear systems through probing with some representative forcings followed by examination of the system's response characteristics.

In this chapter, we examine a variety of models, each treating thermal participation of different levels of the ocean in a hierarchy of complexity (a nice early review of transient models is given by Harvey and Schneider, 1985). Most of the models in the chapter are for a planet covered completely with ocean, the exception being one (see Section 10.6) where the full two-dimensional land–sea geographical surface distribution is considered. The simplest model is one which only considers the mixed layer of the ocean. This is the layer that is stirred by wind stresses at the ocean–air interface. Its depth depends on the wind stress and the static stability of ocean at a particular point. It tends to be deeper in the Southern Hemisphere. According to the same factors, the depth of the mixed layer depends on season and location, but we will mostly be concerned with global averages and we will use different depths from one exercise to another (mainly because of the variety of depths used by different authors in the literature). Using time averages of a month or so, this idealization should hold reasonably well. A similar scheme is often used in common atmospheric general circulation model (GCM) experiments involving the seasonal cycle.

As with virtually every chapter in this book, we remind the reader that the models studied are highly idealized. The ocean models considered here are far from realistic from the point of view of lateral heat transport, but their level of simplicity is roughly commensurate with the energy balance models (EBMs) introduced so far in the text. As with the other chapters, the models have the advantage that the reader can solve or at least understand the steps involved in detail. There are phenomenological coefficients (i.e., they are fitted to observations), but nothing is hidden.

10.1 Single-Slab Ocean

We begin with a couple of crude but surprisingly helpful exercises with the single-slab ocean. First is the response to a Dirac delta function $F(t) = -g_0 \delta(t)$, then a step-function forcing.

10.1.1 Examples with a Single Slab

Here we have in mind the global surface temperature response to a prototype volcanic eruption. Such an eruption has to blast material vertically with sufficient thrust to populate the stratosphere with debris (hygroscopic gases such as sulfur oxides or nitrous oxides can attract and adhere water molecules to form electrolyte solutions of sulfuric and sulfurous acid, also nitrous and nitric acids – the resulting aerosol particles reflect sunlight back to space). The light-reflecting debris can remain in the stratosphere for several years before coagulating to form larger aerosol particles and settling out (with some help from stratospheric circulation). The aerosol particles form a thin layer, inducing a negative forcing that cools the planet. The heating can be taken here as a negative delta function with its spike at time $t = 0$. Such a pulse causes a quick depression of the surface temperature (ideally globally distributed, but often the homogenization takes less than a year). The depressed temperature leads to a radiation imbalance and is followed by an exponential-like recharge that depends on the time constant of the Earth–atmosphere–ocean system. The recovery is long compared to the duration of the negative spike of forcing. The behavior of the temperature anomaly can be described by an energy balance equation:

$$\frac{dT}{dt} + \frac{T}{\tau} = -\frac{g_0}{C}\delta(t), \tag{10.1}$$

where $\tau = C/B$, C is the heat capacity of a column of water whose horizontal cross section is 1 m², whose thickness is the depth of the mixed layer (typically 50–100 m), and $B \approx 2.00$ W m^{-2} K^{-1} is Budyko's coefficient of the surface temperature for the outgoing radiation to space – it is sometimes referred to as the *radiation damping coefficient*. The portion of the heat capacity due to the atmospheric column is small compared to that of the mixed layer and is therefore neglected. This notation has been used often in previous chapters. To solve the ordinary differential equation, we multiply through by the integrating factor, $e^{t/\tau}$ and rearrange to obtain

$$\frac{d}{dt}\left(e^{t/\tau} T(t)\right) = -\frac{g_0}{C} e^{t/\tau} \delta(t). \tag{10.2}$$

Next we integrate each side from 0− to t, noting that $T(0-) = 0$, where 0− means the value just infinitesimally below 0:

$$e^{t/\tau} T(t) = -\frac{g_0}{C}, \tag{10.3}$$

and finally, we have

$$T(t) = \begin{cases} 0; & \text{if } t < 0, \\ -\frac{g_0}{C} e^{-t/\tau}; & \text{if } t \geq 0. \end{cases} \tag{10.4}$$

As t crosses the origin, the temperature abruptly falls by an amount $-\frac{g_0}{C}$. After the pulse at $t = 0$, the temperature "recharges" exponentially to zero as shown in Figure 10.1a. The mathematical steps for the adjustment from a flat forcing to a higher one of strength g_o are almost identical to those just described for the impulse/response. The results are shown in Figure 10.1b. These experiments with GCMs are standard practice (e.g., Donahoe et al., 2015).

Next we take up the ramp forcing. As already mentioned, the dependence of outgoing terrestrial radiation on CO_2 concentration is approximately logarithmic. Hence, if

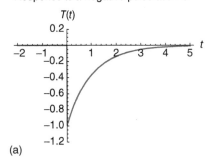
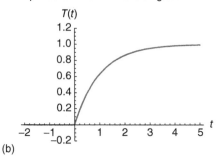

Figure 10.1 Illustration of how a global model whose surface consists of a single slab with strong vertical thermal conductivity, analogous to a mixed-layer ocean, responds to two idealized types of time-dependent forcing: the (negative) impulse and the step function. Shown are graphs of adjustment in the two examples. (a) Example of the recovery after a negative pulse such as a volcanic dust veil of short duration. In this case, the strength of the pulse $g_0/C = 1.0$ and $\tau = C/B = 1.0$. (b) Example of the adjustment of the global temperature from a flat forcing to a higher one of strength g_0. The values of the parameters are $g_0/C = 1.0$ and $\tau = C/B = 1.0$.

the amount of CO_2 is increasing exponentially in time, the increase in radiative forcing is approximately linear in time. In this section, we imagine the system to be in steady state with the exponentially increasing CO_2 being switched on at time $t = 0$. Atop the atmosphere of our all-ocean planet, we express this as

$$I = A + BT(t) - \gamma t H(t), \tag{10.5}$$

where t is time in years, the Heaviside step function is defined by $H(t) = 1, t \geq 0; 0$ for $t < 0$, and $\gamma \approx 4\,\text{W m}^{-2}/\tau_{\text{doub}}$ with τ_{doub} the equivalent doubling time for CO_2 (conventionally taken to be 70 years).

This case and its response to different radiative-imbalance forcings can be found in many papers (e.g., Kim et al., 1992, and Watts et al., 1994). This model will serve as a convenient benchmark for us to compare with more complicated models later in this chapter. The slab world is defined by its energy balance equation:

$$C_m \frac{\partial T_m(t)}{\partial t} + A + BT_m(t) = Q a_p + F_\uparrow + G(t), \tag{10.6}$$

where C_m is the slab heat capacity per unit horizontal area ($\text{J m}^{-2}\,\text{K}^{-1}$) and $F_\uparrow$ is the net flux density of heat (W m^{-2}) from the layers below and $G(t)$ is an external time-dependent heat source applied at the surface (such as γt in the ramp case). The slab is assumed to have a high thermal conductivity such that its vertical temperature profile is homogenized in a short time compared to the radiative relaxation time (months to years). In the single-slab model, $F_\uparrow = 0$—there is no responding medium below the mixed layer. The time constant for relaxation is $\tau_m = C_m/B \approx 5$ years, for a representative mixed-layer slab of about 80 m thickness.

The departure from steady state for $t > 0$ is the solution to

$$\frac{dT}{dt} + \frac{T}{\tau_m} = \frac{\gamma}{C_m} t, \tag{10.7}$$

where $\tau_m = C_m/B$. We have a first-order linear nonhomogeneous equation to be solved for $T(t)$. The solution will consist of two parts: a *homogeneous solution* and a *particular solution*

$$T(t) = T_{\text{homog}}(t) + T_{\text{part}}(t). \tag{10.8}$$

A satisfactory particular solution (found by trial and error) is given by

$$T_{\text{part}}(t) = \frac{\gamma}{B}(t - \tau_m), \tag{10.9}$$

which is the same as a $C_m = 0$ solution except for the time lag τ_m. The homogeneous solution is given by

$$T_{\text{homog}}(t) = c_1\, e^{-t/\tau_m}. \tag{10.10}$$

The integration constant c_1 is to be chosen so as to fit the initial condition, $T(0)$:

$$T(t) = \underbrace{\left(T(0) + \frac{\gamma}{B}\tau_m\right) e^{-t/\tau_m}}_{\text{homogeneous}} + \underbrace{\frac{\gamma}{B}(t - \tau_m)}_{\text{particular}}. \tag{10.11}$$

This partitioning of the solution is useful for physical insight (and for comparison later with more complex models where a similar trick will be used). Note that the particular solution as we have constructed it is the asymptotic form as t becomes large compared to τ_m. We could think of the particular solution as the *attractor*, as the homogeneous solution always decays away because of the way we have partitioned it. The homogeneous solution which depends on the initial condition $T(0)$ decays away in a characteristic time τ_m leaving only the particular or asymptotic solution. It is interesting that we have a familiar characteristic adjustment time τ_m for the solution to reach its asymptotic form (this adjustment structure and characteristic time will vary according to the model's complexity). The asymptotic form consists of a straight line corresponding to a slab with no heat capacity, but lagged by the characteristic time τ_m. These curves are illustrated in Figure 10.2. In the construction of the two terms, we have contrived to have the constant

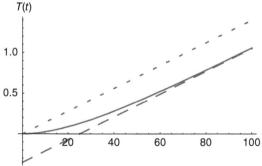

Figure 10.2 The response of a slab model's surface temperature to linear ramp heating (analogous to exponentially increasing CO_2 concentration). Shown are solutions to linear ramp forcing for zero slab thickness (short dashed) and for a finite thickness of the slab (solid) corresponding to a relaxation time (C_m/B) of 25 temporal units. The long-dashed asymptote intersects the origin at the value of τ_m. The long characteristic timescale is for visual convenience. Time is in units of $\tau_m = C_m/B$ and the temperature units are arbitrary.

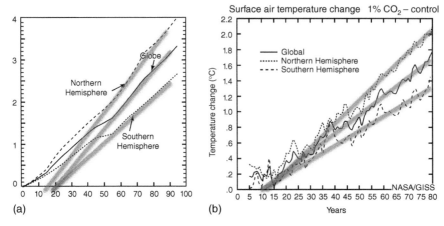

Figure 10.3 Illustration of how two GCMs (of the 1990s) respond to ramp forcing. (a) Coupled ocean–atmosphere simulation of global and hemispherical average temperature for ramp forcing. The vertical axis is temperature change in kelvins and the abscissa is in years. Taken from Manabe et al. (1991). ((Manabe et al., 1991). © American Meteorological Society. Used with permission.) (b) Another coupled ocean–atmosphere simulation of global and hemispherical average temperature for ramp forcing using the Goddard Institute for Space Studies model. The vertical axis is temperature change in kelvins and the abscissa is in years. Taken from Russell et al. (1999). Both models relax to the linear ramp (except for natural variability) in both hemispheres (Russell and Rind (1999). © American Meteorological Society. Used with permission.)

$-\frac{\gamma}{B}\tau_m$ to be canceled at $t = 0$ by the coefficient of the exponential in the homogeneous term. This decomposition involving the cancellation strategy holds in some of the more complex models that have a varying vertical structure below the surface.

Figure 10.3a shows the evolution of global and hemispherical surface temperatures based on an early version of the Geophysical Fluid Dynamics Laboratory (GFDL) coupled ocean–atmosphere GCM. After the transient dies out, one can trace a straight line to the time axis (shown as the heavy gray lines in the figure) to find lags of about 20 years for the global and Southern Hemisphere temperatures, while the Northern Hemisphere appears to have a lag closer to 15 years (Manabe et al., 1991). Other models, such as the Goddard Institute for Space Studies (GISS) coupled model of the same era, show similar behavior (Russell and Rind, 1999) as indicated by Figure 10.3b. In these experiments, the extrapolation is closer to 10 years for the lags. While these particular GCMs have evolved since these figures were produced, they indicate that the slab model concept of a decaying transient followed by the ramp of upward temperatures for large-scale surface temperature averages has some heuristic value.

10.1.2 Eventual Leveling of the Forcing

Next consider the increase of greenhouse gases to end at time t_0 followed by a leveling of the right-hand side of (10.7) to become $\gamma t_0 / C$ where we have dropped the subscript on the heat capacity of the slab to simplify the notation. For times earlier than t_0, the solution is the same as in the previous section if we continue to use the initial condition that $T(0) = 0$. After time t_0, the solution can be shown to be

$$T_{>t_0} = T(t_0)e^{-(t-t_0)/\tau_m} + \frac{\gamma t_0}{B}\left(1 - e^{-(t-t_0)/\tau_m}\right). \tag{10.12}$$

Figure 10.4 Single-slab model solutions to linear ramp forcing for slab thickness corresponding to 25 time units (as in Figure 10.2) until $t = t_0 = 50$ years (twice the relaxation time τ). After that, the forcing is flat at its ramp value $t = t_0$. The horizontal dashed line indicates the point ($t = 50$ years) when the ramp switches to flat forcing.

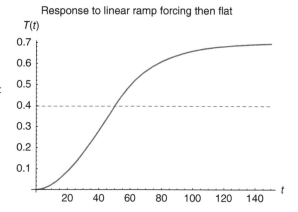

For $t \gg t_0, \tau$, we see that the solution finally settles down to the constant value of $\gamma t_0/B$, which is just the value of the equilibrium-to-equilibrium change in global temperature ΔT for a forcing of $\Delta F = \gamma t_0$. Note that the latter is the total accumulated forcing (proportional to the total log of CO_2 concentration) in the atmosphere.

Figure 10.4 shows a numerical example of what happens after the forcing ramp is flattened, starting at $t = 50$ years. The forcing has been raised to a new constant level, but because of the lag as seen in the second term of (10.11), the temperature has not caught up with the forcing. The temperature continues to rise until it reaches the equilibrium-to-equilibrium value associated with a change in forcing from $t = 0$ until $t = t_0$. The adjustment after the leveling begins is exponentially damped to the new level with the same time constant $\tau = \tau_m$.

The rise above the dashed line in Figure 10.4 is referred to as the *commitment*, because we would be committed to this much additional warming although we have zeroed the rate of rise of the forcing after time t_0.

10.2 Penetration of a Periodic Heating at the Surface

To gain some insight into how heat is transmitted in a layered medium, we next consider a (vertically) thermally diffusive medium. It may seem odd to consider the more complicated continuous medium before looking at multiple slabs, but the analysis is not difficult and it helps to understand how the finite slab layers work. We take up first the case of periodic heating at the surface, because the frequency analysis of such a process gives an idea how far down the heat is conducted before the next round of cyclical forcing is applied. As usual, we allow the cooling of the surface to obey $R_{IR}^\uparrow = A + BT$ (see Chapter 2). Homogeneous terrestrial soil is the perfect candidate for this problem. We use it as a prototype for the ocean. Initially, we imagine a semi-infinite slab of soil (or ocean) ($0 \geq z > -\infty$), hence $z = 0$ is at the surface and $z \to -\infty$ is deep in the soil (or ocean). For simplicity, we take the thermal conductivity k to be constant in space and time. For the temperature anomaly, $T(z, t)$ (the departure of the temperature from its steady-state form for no external heating), we can write

$$c\frac{\partial T}{\partial t} = k\frac{\partial^2 T}{\partial z^2}; \quad z \leq 0, \tag{10.13}$$

where c is the heat capacity of the medium (for seawater it is $4.18 \times 10^6 \,\mathrm{J\,m^{-3}\,K^{-1}}$; we ignore any density variation with depth), and k, with units $\mathrm{W\,K^{-1}\,m^{-1}}$, is the (macro)thermal conductivity. The term k/c has units of $\mathrm{m^2\,s^{-1}}$ or $8000\,\mathrm{m^2\,year^{-1}}$. The vertical component of the heat flux density $\mathbf{q}_z$ in $\mathrm{J\,s^{-1}\,m^{-2}}$ is given by

$$\mathbf{q}_z = -k\frac{\partial T}{\partial z}. \tag{10.14}$$

We will be imposing a sinusoidal forcing (heating) at the surface, $z = 0$, of $g_\omega\,e^{i\omega t}$ where ω is the angular frequency and $i = \sqrt{-1}$. We take the response temperature to be of the form $T(z,t) = T_\omega(z)e^{i\omega t}$.

At the top of the ocean, the boundary condition is

$$-k\frac{\partial T_\omega}{\partial z}\bigg|_{z\to 0-} = [BT_\omega - g_\omega]_{z\to 0-}. \tag{10.15}$$

Moreover, the lower boundary condition is that $T_\omega(z)$ must be finite as $z \to -\infty$.

Since the problem is linear with real, constant coefficients, we may take the real part of the solution we find at the end. The ratio of the imaginary part of the solution to the real part will be the tangent of the phase lag in radians as a function of z. Inserting this form, we find

$$i\omega c T_\omega(z) = k T_\omega''(z), \tag{10.16}$$

$$-kT_\omega'(z)|_{z\to 0-} = BT_\omega(z)|_{z\to 0-} - g_\omega, \tag{10.17}$$

where, in the last two equations, the prime indicates differentiation with respect to z. Next we assume the form

$$T_\omega(z) = \tilde{T}_\omega\,e^{az}, \tag{10.18}$$

where a is a coefficient (likely complex) to be determined. Substitute this into (10.16) and (10.17). After noting that

$$\sqrt{i} = (e^{\frac{\pi}{4}i}, e^{\frac{5\pi}{4}i}) = \pm\frac{i+1}{\sqrt{2}}, \tag{10.19}$$

we find

$$a = \pm(1+i)\sqrt{\frac{\omega c}{2k}}. \tag{10.20}$$

We dismiss the negative choice in the solution (10.20) because it would lead to unacceptable behavior as $z \to -\infty$. Finally, we arrive at the complex amplitude:

$$\tilde{T}_\omega = \frac{g_\omega}{ka + B}e^{az}. \tag{10.21}$$

Then, with $\alpha = \mathrm{Re}(a) = \mathrm{Im}(a) = \sqrt{\omega c/2k}$ and $|\alpha| = \sqrt{\omega c/k}$. The magnitude of the complex amplitude (after some algebra) is

$$|\tilde{T}_\omega(z)| = \frac{g_\omega\,e^{\alpha z}}{\sqrt{B^2 + \sqrt{2kc\omega}B + kc\omega}}, \tag{10.22}$$

where, in the last step, we substituted for α. Note that the z dependence is in the numerator, and it indicates that the warming signal is modulated by an exponentially decreasing function as $z \to -\infty$. The characteristic depth of heat penetration λ_ω is inversely proportional to the square root of the angular frequency of the driver.

$$\lambda_\omega = \sqrt{\frac{2k}{\omega c}} = \alpha^{-1}. \tag{10.23}$$

Note that the penetration depth is also proportional to the square root of the thermal diffusivity k/c. Low-frequency components can penetrate deeper than those with higher frequency. We can get an intuitive idea of how the heat is diffused from the surface toward the depths by looking at how a Gaussian distribution of diffusing substance spreads in a time t. The distance, d, of the one-standard-deviation width of a spreading Gaussian is $d \sim \sqrt{kt}$ (see Section 6.4). If we replace t by $2\pi/\omega$, we find

$$d \sim \sqrt{\frac{2\pi k}{\omega c}}. \tag{10.24}$$

At $z = 0$, we find that the amplitude squared (power or variance) can be written as follows:

$$|\tilde{T}_\omega(0)|^2 = \frac{g_\omega^2}{B^2 + \sqrt{2kc\omega}B + kc\omega}. \tag{10.25}$$

This form has more power at low frequencies and for large ω, it monotonically decays as $1/\omega$. Unlike the case of white noise, its integral over all frequencies is not bounded. In the statistics and engineering literature, it is known as "pink noise" ($1/\omega^s$ with $0 < s < 2$). In reality, the forcing (white noise) is never really constant as $\omega \to \infty$, but rather it must cut off at some finite frequency. In this book, we usually think of the white noise drivers as atmospheric weather which has a timescale of a few days. Note that if there is no cooling to space ($B = 0$), the power spectrum diverges as $\omega \to 0$. But note that the numerator, which is constant for $z = 0$, is proportional to $\exp((\omega c/2k)^{\frac{1}{2}}z)$ and hence, for $z < 0$, causes convergence for $\omega \to \infty$.

Figure 10.5 shows the real part of the complex amplitude $\tilde{T}_\omega(z)$: Figure 10.5a is the amplitude without the damping factor $e^{\alpha z}$ and Figure 10.5b shows the same but including the damping factor. In this example, $k/c = 6000 \text{ m}^2 \text{ year}^{-1}$ (i.e., below the surface) leading to a value of $\lambda_\omega \approx 44$ m. We can compute the phase lag by using $B = 2.0 \text{ W m}^{-2} (°C)^{-1}$.

In real soil, the medium is not usually homogeneous; hence, both k and the local heat capacity are z-dependent. Moreover, water in the column of soil can influence both of these parameters as well. Ignoring these complications for our purposes, we see that long timescales in the forcing can penetrate deeper in the soil. One might even posit that the effective heat capacity is proportional to the e-folding length scale, $1/\alpha$, which depends on the frequency ω. Heuristically, the penetration of the daily heating of the soil is much less than that of the annual cycle. Anyone who has toured a cave will find that the annual cycle of temperatures below ground is very small. We encounter the mean annual temperature instead. A similar effect can be noted with permafrost. Permafrost is mostly a few meters below the surface. For this reason, it is the mean annual temperature that matters at such depths. One could say the high-frequency components of the driving signal are filtered out by the vertical diffusion process.

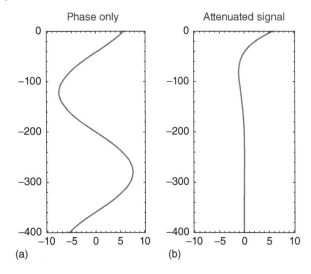

Figure 10.5 The nature of penetration into a thermally diffusive medium such as soil and roughly a mixed-layer ocean. Values of the parameters are chosen to be characteristic of the ocean with annual forcing at its surface: $k/c = 8000 \text{ m}^2 \text{ year}^{-1} = 2.54 \times 10^{-4} \text{ m}^2 \text{ s}^{-1}$, $c = 4.18 \times 10^6 \text{ J m}^{-3} \text{ K}^{-1}$, $B = 2.0$ $\text{J s}^{-1} \text{ m}^{-2} \text{ K}^{-1}$, $\Omega = 1.99 \times 10^{-7} \text{ s}^{-1}$, $g_\omega = 238 \text{ W m}^{-2}$, $\alpha^{-1} = 50.4 \text{ m}$. (a) The real part of the temperature amplitude, $T_\omega(z)$, as a function of z (in units of meters), without the exponential damping factor $e^{\alpha z}$. (b) The same but including the damping factor.

An application of a more refined ground-heat-conduction model with a knowledge of the nonhomogeneous properties can be used as a proxy for past surface temperatures (see, e.g., Pollack and Huang, 2000; NRC, 2006). Proxy data generated from preexisting boreholes are very noisy, but when many of them are averaged together, the warming signal is clear.

While the vertical transport of heat in the ocean is hardly a diffusive medium, some of the intuition gained from the soil case is helpful. The case of upwelling of ocean waters combined with thermal diffusion gives the best[2] vertical model as we will see in this chapter. Keep in mind that diffusion is a molecular process in which heat energy is transported by molecular collisions. Eddies in the fluid motion also transport heat energy and we imitate this transport mechanism by diffusion, but the real turbulent stirring process is more complicated than diffusion. However, in our ocean modeling slabs, diffusion and upwelling are appropriate models consistent with our overall strategy in this book.

10.3 Two-Slab Ocean

In this section, we consider a very simple two-slab model following the work of Gregory (2000)[3] who provides background and motivation for the model. This was also the same class of model used by Held *et al.* (2010) to infer the long-term behavior of global temperatures after CO_2 emissions are leveled or even stopped.

$$C_m \frac{dT_m}{dt} + BT_m + k(T_m - T_d) = +F(t), \tag{10.26}$$

---

2 Here "best" means the most consistent with our usual practice with EBMs. Serious oceanographers are likely to take exception to such a simple approach to dynamics of the ocean.
3 Gregory (2000) used rather different values for his slab thicknesses than we have done. His upper layer has thickness 150 m and his deeper layer is of the order of 2400 m. In our case, we choose a thin upper layer, taken to be 50 m and our deep slab layer is down to approximately the thermocline, 50 + 500 m. We have chosen these as illustrations, which demonstrate similar features.

$$C_d \frac{dT_d}{dt} + k(T_d - T_m) = 0. \tag{10.27}$$

We denote the mixed-layer temperature as T_m and the deeper-layer temperature as T_d. The equation is for the anomaly from equilibrium response to the forcing at the top of the ocean $F(t)$. The flux density of heat between the two layers is proportional to their temperature differences with an exchange coefficient k. We can introduce the two-component vector:

$$\vec{T} = (T_m, T_d)^T. \tag{10.28}$$

The homogeneous (differential equation) system can be expressed as follows:

$$\frac{d\vec{T}}{dt} + \mathcal{M} \cdot \vec{T} = 0, \tag{10.29}$$

with

$$\mathcal{M} = \begin{pmatrix} \dfrac{B+k}{C_m} & -\dfrac{k}{C_m} \\ -\dfrac{k}{C_d} & +\dfrac{k}{C_d} \end{pmatrix}. \tag{10.30}$$

Next, insert $\vec{T}(t) = \vec{T} e^{-\lambda t}$:

$$\text{Det}(\mathcal{M} - \lambda \mathbf{1}) = 0. \tag{10.31}$$

The eigenvalues $\lambda_{1,2}$ can be found by solving the resulting quadratic equation for λ. The inverse $\tau_{1,2} = 1/\lambda_{1,2}$ are the decay times for the two eigenvectors. Figure 10.6 shows the two relaxation times corresponding to the inverse of the eigenvalues as a function of the exchange coefficient k. When $k \to 0$, the two slabs decouple and the shorter timescale becomes that of the mixed layer alone, ~5.0 years, while that of the lower slab becomes indefinitely large. For large values of k, the larger root tends toward the relaxation time for both slabs taken together, while the shorter time tends to zero. The solution corresponding to the lower timescale becomes a singular solution, because if (10.27) is divided through by k, the time derivative term drops out, reducing the order of the system by one.

The actual solution for the homogeneous case can be written as a pair of exponentials that will decay according to the mix of the eigenvectors corresponding to the eigenvalues λ_1 and λ_2.

$$\begin{pmatrix} T_m(t) \\ T_d(t) \end{pmatrix} = C_1 \begin{pmatrix} \hat{E}_m^{(1)} \\ \hat{E}_d^{(1)} \end{pmatrix} e^{-t/\tau_1} + C_2 \begin{pmatrix} \hat{E}_m^{(2)} \\ \hat{E}_d^{(2)} \end{pmatrix} e^{-t/\tau_2}, \tag{10.32}$$

where $\hat{E}_m^{(1)}$ stands for the component of the unit eigenvector 1 (corresponding to λ_1) in the direction of T_m, and so on. The coefficients $C_{1,2}$ must be found from the initial conditions $T_m(0), T_d(0)$. We refer to the coefficients $C_{1,2}$ as the *eigenmode amplitudes*, as they indicate the strength of the *eigen patterns* represented by the unit eigenvectors $\hat{E}^{(1,2)}$. Once the coupling parameter k is fixed, the eigenvector components $\hat{E}_{m,d}^{(i)}$ are known constants. Note that the matrix $\mathcal{M}$ is not symmetric, and therefore the eigenvectors

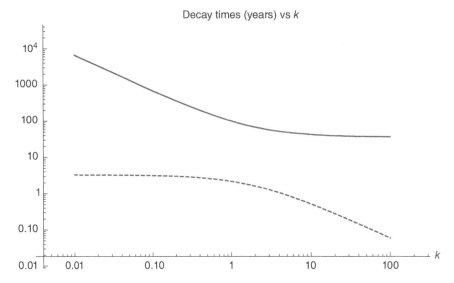

Figure 10.6 The behavior of the eigenvalues as a function of the coupling coefficient between the two slabs k. The two curves show how the characteristic times (inverses of the eigenvalues) of the two eigenmodes vary over a wide range of the coupling parameter k. For small values of k, only the upper layer is active, and the decay time is 3.31 years (that of a 50 m slab) as expected. The other eigenmode is very large for small k but eventually settles down to approach a value of 36.5 years for large k when the combination single slab has thickness 550 m.

are not orthogonal, but they still can be used as a basis set in two-dimensional space, provided they are not parallel. This means the temperature of each slab is a different linear combination of two exponential decays. For weak coupling, $k \sim 0$, this would be a short decay determined by the radiative decay time added to a component with long decay corresponding to the situation where there might be a lot of heat stored below that takes a long time to transfer to the upper layer where it can be radiated to space. This is the effect referred to as the *recalcitrant climate effect* (Held et al., 2010) that might occur after a long period of slow warming due to, say, greenhouse gases followed by a shutdown or even a leveling of the greenhouse gases. The recalcitrance is the long time required to release the heat from lower layers to the upper layer, where it can cool to space. If the coupling to lower layers is weak, the warming due to greenhouse gases does not reverse itself rapidly.

The eigenvalues are the roots of the quadratic equation derived from the null determinant $|\mathcal{M} - \lambda \mathbf{1}| = 0$.

$$\lambda_n(k) = (Bc_d + c_d k + c_m k)/2c_d c_m \qquad (10.33)$$
$$+ \frac{(-\delta_{n,1} + \delta_{n,2})}{2c_d c_m} \sqrt{(Bc_d + c_d k + c_m k)^2 - 4Bc_d c_m k}.$$

The eigenvectors are given by complicated expressions (but easily handled by MATHEMATICA as analytical expressions, and numerically as well). It is convenient to use (normalized) unit vectors, $\hat{\mathbf{E}}^{(1,2)}$. These unit vectors are not orthogonal. Hence, it is useful to introduce *reciprocal vectors*, $\mathbf{D}^{(1,2)}$, which[4] are defined by

$$\hat{\mathbf{E}}^{(1)} \cdot \hat{\mathbf{E}}^{(1)} = 1, \qquad \hat{\mathbf{E}}^{(2)} \cdot \hat{\mathbf{E}}^{(2)} = 1, \tag{10.34}$$

$$\mathbf{D}^{(1)} \cdot \hat{\mathbf{E}}^{(1)} = 1, \qquad \mathbf{D}^{(2)} \cdot \hat{\mathbf{E}}^{(2)} = 1, \tag{10.35}$$

$$\mathbf{D}^{(1)} \cdot \hat{\mathbf{E}}^{(2)} = 0, \qquad \mathbf{D}^{(2)} \cdot \hat{\mathbf{E}}^{(1)} = 0. \tag{10.36}$$

The eigenvectors have been normalized to have unit length, whereas the reciprocal vectors $\mathbf{D}^{(1,2)}$ are not unit vectors. Next consider the relationship of decay times τ_1 and τ_2 and the thermal coupling k. Figure 10.6 shows how the decay times (inverse of the eigenvalues) vary as a function of k. The inverses of the eigenvalues are the decay time constants for the eigenmodes. As $k \to 0$, one decay time is infinite and the other approaches 3.31 years, the decay times for the upper mixed-layer slab alone. In this limit, the corresponding eigenvectors are orthogonal as shown in Figure 10.7. The passage of $k \to 0$ is a singular limit because as the coupling goes to zero, the degree (dimension of the system) is reduced from two components to one. The result is that one of the timescales of the eigensystem tends to infinity.

But as k is increased to very large values, we see that the decay time of τ_2 becomes vanishingly small, while the other approaches 36.4 years, the decay time of a 550 m single, aggregated slab. In this limit, both slabs are tightly coupled and their entire mass is participating as a unit. As k increases, the eigenvectors become more skewed at an

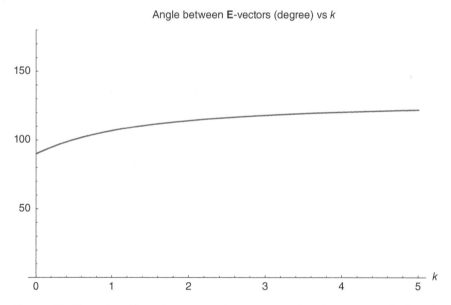

Figure 10.7 Illustration of how the two unit eigenvectors go from being perpendicular at $k = 0$ to a limiting larger angle of about 128°. Parameter values: $B = 2.00$; $C_m/B = 3.0$ years; $C_d/B = 30.0$ years.

---

4 Reciprocal or *dual* vectors are not found in many recent books. They are more commonly covered by tensor notation, for example, Arfken and Weber (2005), which would overcomplicate our treatment. An older but very valuable and inexpensive book is that of Wills (1958). The reciprocal basis set is a set of vectors used when the basis set of interest is nonorthogonal. As will be seen, reciprocal vectors allow us to project out the component amplitudes easily.

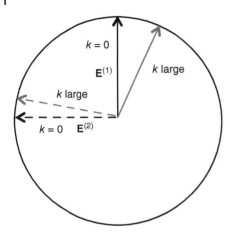

Figure 10.8 Diagram of the unit vectors $\hat{\mathbf{E}}^{(1)}$ and $\hat{\mathbf{E}}^{(2)}$. At $k = 0$, the unit vectors are orthogonal but as k increases, they change direction and are no longer orthogonal. Note that $\hat{\mathbf{E}}^{(2)}$ points in the negative direction in the left upper quadrant of the (m, d) plane.

angular separation of ~128° (see Figure 10.7). Figure 10.8 shows how the two unit vectors $\hat{\mathbf{E}}^{(1)}$ and $\hat{\mathbf{E}}^{(2)}$ rotate as k increases from 0 to large values. Note that $\hat{\mathbf{E}}^{(2)}$ points in the negative direction in the (m, d) plane.

For reference, we list the components of $\mathbf{D}^{(1,2)}$ in terms of the $\hat{\mathbf{E}}^{(1,2)}$:

$$D_m^{(1)} = \frac{-\hat{E}_d^{(1)}}{\det}; \quad D_d^{(1)} = \frac{\hat{E}_m^{(2)}}{\det}; \tag{10.37}$$

$$D_m^{(2)} = \frac{-\hat{E}_d^{(2)}}{\det}; \quad D_d^{(2)} = \frac{\hat{E}_m^{(1)}}{\det}; \tag{10.38}$$

and

$$\det(k) = \hat{E}_d^{(1)} \hat{E}_m^{(2)} - \hat{E}_m^{(1)} \hat{E}_d^{(2)}. \tag{10.39}$$

Note that $\det(k)$ is the cross product of $\hat{\mathbf{E}}^{(1)}$ and $\hat{\mathbf{E}}^{(2)}$. It vanishes if these two unit vectors are orthogonal, which is the case as $k \to 0$ (see Figure 10.7).

10.3.1 Decay of an Anomaly with Two Slabs

Consider an anomaly or initial condition where $T_m(0) = 1$ K and $T_d(0) = 0$ K. (Note that this is equivalent to the upper slab being subjected to a delta function heating at time $t = 0$.) This anomaly will decay according to (10.32) after the constants $C_{1,2}$ are determined by the conditions

$$C_1 \hat{E}_m^{(1)} + C_2 \hat{E}_m^{(2)} = 1; \quad C_1 \hat{E}_d^{(1)} + C_2 \hat{E}_d^{(2)} = 0. \tag{10.40}$$

A numerical example is shown in Figure 10.9 and the late part of the decay is shown in Figure 10.10. In this numerical example, the upper slab has thickness 50 m and the lower slab 500 m, as in the previous subsection. We chose a fairly large value of the coupling parameter $k = 2.5$ W m^{-2} for illustrative purposes. Given this value of k, the characteristic times (inverse of the eigenvalues) of the eigenmodes are 61.5 and 1.43 years. As the initial anomaly decays in the upper layer because of the fairly strong coupling, heat is transferred to the lower layer, which soon heats up to a warmer temperature than the upper layer. In this case, the cooling of the upper layer to the lower one is more efficient than the cooling to space. In weaker coupling (e.g., $k = 0.2$ W m^{-2}) scenarios, this does

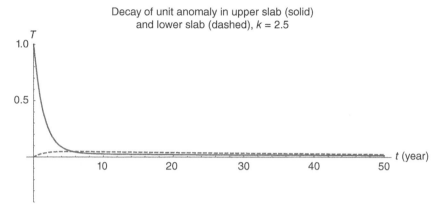

Figure 10.9 The decay of an initial anomaly of 1 K in the upper slab T_m. The temperature in the upper slab $T_m(t)$ is shown as the solid curve and the temperature in the lower slab $T_d(t)$ is shown as a dashed curve. Parameter values are $c_m/B = 3.31$ years, $c_d/B = 33.1$ years and $k = 2.5$ W m^{-2} K^{-1}. Slab thicknesses are 50 and 500 m. Given this value of k, the characteristic times of the two eigenmodes are 61.5 and 1.43 years.

Figure 10.10 The later portion of the decay of an initial anomaly of 1 K in the upper slab T_m. The temperature in the upper slab $T_m(t)$ is shown as the solid curve and the temperature in the lower slab $T_d(t)$ is shown as a dashed curve. Note that the upper slab has a substantial anomaly even after many e-folding times of the single uncoupled slab of the same thickness.

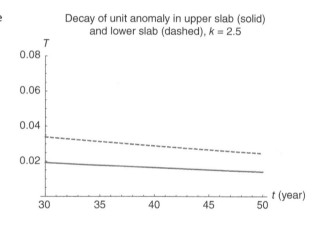

not happen until much later (~20 years). Figure 10.10 shows a magnification where we see that both slabs are at nearly 0.02 K, even after many characteristic time lengths of the shorter e-folding time. The lower level is still warmer than the upper even at 50 years after the initial anomaly is launched.

Using C_m as the heat capacity of the mixed layer, the decay time of an anomaly $\tau_{slab} = C_m/B$ is related to sensitivity for a one-slab model, as the equilibrium sensitivity to doubling CO_2 (see Chapter 2) is proportional to $1/B$. Here we might be including in B the feedbacks that are working at longer timescales, say, a few years. It is tempting to speculate that one could estimate the sensitivity by observing the decay time or equivalently the autocorrelation time. Aside from not having a good estimate of the slightly ambiguous term, C_m, we have to worry about the coupling to the slab below the mixed layer. The coupling constant k is similar to C_m, in that it is not well known. Figure 10.11 shows several decays for unit anomaly and various values of the coupling parameter k.

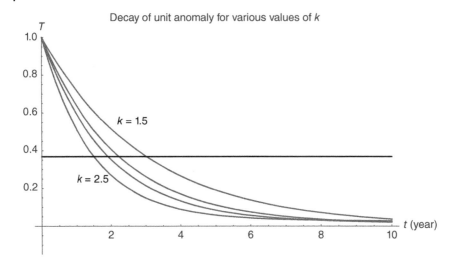

Figure 10.11 The decay curve from a unit anomaly of the upper layer for four values of the coupling k. The values of k and 1/e timescale values for the curves from the leftmost to the rightmost are (2.5, 1.48), (1.5, 1.91), (1.0, 2.22), (0.20, 3.01); the units of k are W m^{-1} K^{-1}, and those of time are years.

In the figure, there are four decay curves that look similar to exponentials (but consist of a linear combination of two of them), whose time constants can be read from Figure 10.6. From Figure 10.11, we see that for very weak coupling the decay time is close to 3.31 years, the time for an uncoupled single upper slab. But as k is increased, the decay time shrinks. The casual observer might think the sensitivity is less than that inferred by a value of $B = 2.00$. Lindzen and colleagues (Lindzen, 1994; Lindzen and Giannitsis, 1998) have used simple models similar to[5] those in this chapter to show that, because of the unknown coupling parameter, we cannot get an unambiguous estimate of the sensitivity of climate. This seems to be borne out in our model as well.

10.3.2 Response to Ramp Forcing with Two Slabs

In this section, we consider the same model as in the previous section that has a mixed layer of thickness $h_m \approx 50$ m (whose effective *volumetric* heat capacity is C_m).[6] Below this thin layer lies another well-mixed slab whose bottom is at the thermocline. The thickness of the lower slab is $h_d \sim 500$ m with effective volumetric heat capacity C_d. This places the thermocline at about 550 m below the surface. In our model, there is no vertical heat transport below the thermocline. Several models of this genre were studied numerically by Harvey and Schneider (1985). The radiative decay time of the uncoupled mixed layer is taken to be 3.31 years and that of the lower slab to be 33.1 years. As before, let the two slabs transfer heat proportional to their temperature difference (warmer to cooler) with a coupling coefficient k. The surface temperature is $T(t)$ (taken to be the

---

5 Actually Lindzen's model is a bit more complicated and perhaps more realistic. It has a mixed layer slab and below it a thermally diffusive model down to the thermocline with upwelling of cool water transmitted below it. The two-slab model in our chapter takes the lower portion as a well-mixed slab.
6 It is convenient to consider the volumetric specific heat of water (4.18×10^6 J m^{-3} K^{-1}) rather than the mass specific heat, as in our treatment, the variation of water density with depth is neglected.

same as that of the mixed layer) with the usual heat flux densities $A + BT(t)$ and Qa_p. The governing equations are (10.26) and (10.27).

Next consider obtaining a solution for ramp forcing $F(t) = \gamma t$ applied to the upper layer only. The problem can be stated as follows:

$$\frac{d}{dt}\begin{pmatrix} T_m \\ T_d \end{pmatrix} + \mathcal{M}\begin{pmatrix} T_m \\ T_d \end{pmatrix} = \begin{pmatrix} \gamma t/c_m \\ 0 \end{pmatrix}, \tag{10.41}$$

with

$$\mathcal{M}\hat{E}^{(1,2)} = \lambda_{1,2}\hat{E}^{(1,2)}. \tag{10.42}$$

We can now use

$$\begin{pmatrix} T_m \\ T_d \end{pmatrix} = C_1(t)\begin{pmatrix} E_m^{(1)} \\ E_d^{(1)} \end{pmatrix} + C_2(t)\begin{pmatrix} E_m^{(2)} \\ E_d^{(2)} \end{pmatrix}. \tag{10.43}$$

We can now insert (10.43) into (10.41) to find

$$\dot{C}_1\hat{E}^{(1)} + \dot{C}_2\hat{E}^{(2)} + \lambda_1 C_1\hat{E}^{(1)} + \lambda_2 C_2\hat{E}^{(2)} = \begin{pmatrix} \gamma t/c_m \\ 0 \end{pmatrix}, \tag{10.44}$$

where the overdot denotes time differentiation. Next, project out the equations for $C_1(t)$ and $C_2(t)$ by multiplying from the left by $\mathbf{D}^{(1)}$, then $\mathbf{D}^{(2)}$ using (10.34)–(10.36). The result is two uncoupled ordinary differential equations, each of which is of the same form as the one-slab problem:

$$\dot{C}_1 + \lambda_1 C_1 = D_m^{(1)}\frac{\gamma t}{c_m}, \tag{10.45}$$

$$\dot{C}_2 + \lambda_2 C_2 = D_m^{(2)}\frac{\gamma t}{c_m}. \tag{10.46}$$

We can now solve each of these equations as in the solution of (10.7)–(10.11). The temperature in each slab can now be constructed for a given value of k by inserting $C_1(t)$ and $C_2(t)$ into (10.43). Thus each slab's transient will consist of linear combinations of exponentials with the characteristic timescales for the eigenmodes. Without solving the system explicitly, we can conclude that after the transients have been exhausted, the simulated temperature in each slab gets onto the straight-line ramp.

The procedure just derived for two slabs can be used to solve for n slabs by constructing the eigenvectors and their reciprocal vectors for the $n \times n$ matrix $\mathcal{M}$. Further, we can use the technique to solve for any forcing on the right-hand side of the governing equation. For example, one might consider white noise forcing from above the ocean surface generated by weather. In this case, we insert for the forcing $g_\omega e^{i\omega t}$ as we have done already in previous chapters.

10.4 Box-Diffusion Ocean Model

Lebedeff (1988) examined a box-diffusion model in which there is a deep layer below the mixed layer. The deep layer transports heat by the thermal diffusion mechanism.

We follow his derivation but omit some detail because of the technical details making use of the Laplace transform.[7]

The model has a finite-thickness mixed layer (slab) on top with a continuum below:

$$C_m \frac{\partial T_m}{\partial t} + \lambda T_m - F^\uparrow = F(t), \tag{10.47}$$

where T_m represents the departure from the steady state of the mixed-layer temperature, $C_m = ch$ and h is the thickness of the mixed layer (meters), c is the volumetric heat capacity of seawater (4.18×10^6 J m^{-2} K^{-1}), $F^\uparrow$ is the flux density of heat entering the surface layer from below. In this last equation, λ is used in place of B to include possibly other feedbacks such as ice-cap feedback.[8]

$$F^\uparrow = ck \left.\frac{\partial T_d}{\partial z}\right|_{z=0}, \tag{10.48}$$

where k is the (macro)thermal diffusivity of the ocean below the mixed layer (typically, in the range 1–10 m^2 year^{-1})—here the macro-diffusion is considered to be driven by random eddies. $T_d(z)$ is the departure of the ocean temperature below the mixed layer ($z \leq 0$) from its steady-state profile. We have denoted the vertical coordinate in the lower layer as z with the origin at the bottom of the mixed layer and having negative values below. The temperature in the lower levels is governed by

$$\frac{\partial T_d(z,t)}{\partial t} = k \frac{\partial^2 T_d(z,t)}{\partial z^2}. \tag{10.49}$$

The boundary conditions are specified by (10.48) and that

$$T_d(z=0, t) = T_m(t). \tag{10.50}$$

We take the "bottom" of the ocean to be insulating. A null heat flux at the bottom of the layer at $z = -h$ can be expressed as

$$\left.\frac{\partial T_d}{\partial z}\right|_{z=-h} = 0. \tag{10.51}$$

Lebedeff's solution for step-function forcing is

$$T_m(t; \kappa) = 1 + \frac{1}{\eta - 1/\eta} \left[e^{t\eta^2} \operatorname{erfc}(\eta \sqrt{t})/\eta - \eta \operatorname{erfc}(\sqrt{t}/\eta) \right], \tag{10.52}$$

where[9]

$$\eta = \frac{1}{2\kappa} + \sqrt{\frac{1}{4\kappa} - 1}; \quad \kappa = \frac{\lambda h}{ck}. \tag{10.53}$$

---

[7] The Laplace transform is a useful technique for solving linear differential systems, especially in cases where the initial conditions are known. Descriptions can be found in Arfken and Weber (2005) as well as in many books on mathematical methods in engineering and physics. A useful compilation of transforms and their inverses can be found in Abramowitz and Stegun (1964, available on line at http://people.math.sfu.ca/cbm/aands/intro.htm).

[8] We have usually taken a value of B at approximately 2.00 W m^{-2}. It would be less if some other positive feedback were included – water vapor feedback is presumably already included in B.

[9] The error function is defined as $\operatorname{erf}(x) \equiv \frac{2}{\sqrt{\pi}} \int_0^x e^{-t^2} dt$ and the complementary error function is $\operatorname{erfc}(x) \equiv 1 - \operatorname{erf}(x) = \frac{2}{\sqrt{\pi}} \int_x^\infty e^{-t^2} dt$. Both functions are included in MATLAB and MATHEMATICA.

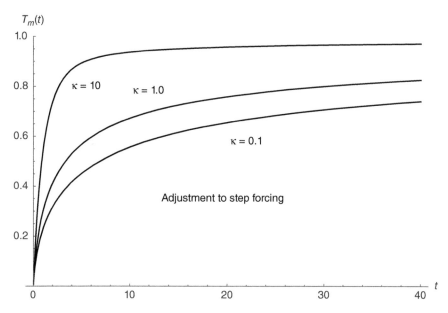

Figure 10.12 Graphs of Lebedeff's solutions to step-function forcing versus time for various values of the dimensionless parameter k. Dimensionless time, t, is in units of the mixed layer radiation relaxation time (nominally, a few years). The dimensionless parameter κ is inversely proportional to the thermal diffusivity k.

The formula for $T_m(t; \kappa = \frac{1}{4})$ is not defined, but Lebedeff provides a solution for this value of the constant, $\eta = 2$. Plots for $k = 0.1$, 1.0 and 10 are shown in Figure 10.12. The formula for κ tells us that it is the ratio of two timescales: the time for a particle to diffuse across a distance of the mixed-layer depth to the radiative relaxation time for an uncoupled mixed layer: $h^2/k : hc/\lambda$. In the numerator, the distance h is diffused in accordance with $\sqrt{kt}$. We discussed the spread of diffused heat in Section 6.4. Readers are referred to Lebedeff (1988) for more discussion and details of his solution. In his paper, he also derives simplified forms for long-term behavior of the solutions. Note that we have reversed the labels k and κ in Lebedeff's paper to be consistent with our notation elsewhere in this chapter.

10.5 Steady State of Upwelling-Diffusion Ocean

An upwelling-diffusion (UD) ocean is yet another highly simplified model, but slightly more realistic than the purely diffusive one. It dates back to early works of Robinson and Stommel (1959) and Walter (1966). Many others have worked on this class of models as well (e.g., Hoffert et al., 1980; Harvey and Schneider, 1985; Morantine and Watts, 1994; Watts et al., 1994). One very attractive feature of this model is that a thermocline-like feature comes out naturally in the solution to the steady-state problem. The level of realism of this model is pretty low (for an up-to-date survey of ocean circulation theories, see, e.g., Huang, 2013). We adopt the UD model in the same spirit as other schematic approximations throughout this book.

The thinking behind the UD model is that throughout the ocean, eddies transport heat down gradient from the warm water above to the colder waters below. In addition cold water in the abyss is constantly supplied by downwelling of cold, dense waters in the polar regions. These volumes of colder and more saline volumes of water sink to the bottom or in some cases intermediate levels, spreading out horizontally mostly across the entire world ocean floor. This supply of cold water and its upward displacement of water above is the reason for the upwelling everywhere. It can also suggest us to make our lower boundary condition at $z \to -\infty$ be at a temperature of $\sim 0\,°C$ or zero difference from the equilibrium solution when we deal with anomalies.

The local rate of change of thermal energy in the mixed layer is given by the equation:

$$ch\frac{\partial T}{\partial t} + A + BT - cwT + ck\frac{\partial T}{\partial z} = Qa_p + F(t), \quad z = 0; \tag{10.54}$$

where c is the specific heat of seawater (in per volume units), $ch = C_m$ is the effective heat capacity per unit horizontal area of the mixed layer, whose thickness is h, $F(t)$ is an external forcing, and as in Chapter 2, $A + BT$ is the infra-red outgoing radiation to space, Qa_p is the absorbed solar radiation flux density. Below the mixed layer is the ocean interior where the energy changes are governed by

$$h\frac{\partial T}{\partial t} = k\frac{\partial^2 T}{\partial z^2} - w\frac{\partial T}{\partial z}, \quad z < 0, \tag{10.55}$$

where k and w are called the vertical diffusivity and the upwelling velocity, respectively. Note that $z = 0$ is taken to be at the bottom of the mixed layer. In this formulation k/c has dimensions length^2/time (m^2 year^{-1}) with typical values 2000–6000 m^2 year^{-1} The upwelling velocity per unit heat capacity w/c has dimensions length/time (m year^{-1}) with typical values a few m year^{-1}. These parameters suggest a vertical length scale $\ell \sim \frac{k}{w}$ ranging from 400 to 1000 m and a timescale $\tau \sim \frac{k}{w^2} \sim 250$ years. The lower boundary condition is:

$$T \to 0 \quad \text{as } z \to -\infty. \tag{10.56}$$

In obtaining the solution first consider the steady-state solution for the UD model with $\frac{\partial}{\partial t} \to 0$. Equation (10.55) yields after one integration

$$\frac{dT}{dz} - \frac{w}{k}T = 0, \tag{10.57}$$

where we have set the integration constant to zero in order to satisfy the boundary condition at $z \to -\infty$. Using $\ell = \frac{k}{w}$ we find

$$T(z) = T_0 e^{z/\ell}; \quad z < 0. \tag{10.58}$$

This very simple model of the world ocean leads us to a thermocline whose characteristic depth is ℓ below the mixed layer.

If the CO_2 is suddenly doubled, the asymptotic solution (long times) is

$$\Delta T_p = -\frac{\Delta A}{B}e^{z/\ell}, \quad t \to \infty, \tag{10.59}$$

where the subscript "p" stands for particular solution. The transient solution will be uniform over the all-ocean planet subjected to step-function forcing. Its solution is presented by Morantine and Watts (1994). They begin by finding the impulse/response function via a Laplace transform method. The steps in the derivation are more complicated than necessary for this book, but can be found in their paper.

10.5.1 All-Ocean Planetary Responses

One concern with the box-diffusion model raised by Hoffert and Flannery (1985) is that due to the response after long times, the entire column of water is heated during ramp forcing, whereas the UD model suggests that only the (effective) mass of water above the thermocline participates in the long-term (asymptotic) solution. The reason for this is that the upwelling cool water presses against the downward diffusing warm water and a match is found at the thermocline. We can find the so-called impulse/response function (also called *Green's function*) by finding the response to a forcing spiked at the origin, the Dirac delta function, $\delta(t)$. As mentioned in the last sections, this is equivalent to finding the decay of an initial anomaly in the mixed layer.

Watts *et al.* (1994) summarize several models including the pure diffusion model, the UD model, and a UD model that includes land masses in much the same way as the beach-ball model of Chapter 8. In this elegant survey, they used the Laplace transform method introduced by Lebedeff (1988) to provide analytical solutions. Here we show one example, the UD model for an all-ocean planet. The solution for this model is given by the following formula (Equation 9 of Morantine and Watts, 1994):

$$\frac{\theta(\tau)}{R_0/\lambda} = a\frac{t_u}{t_m}\left(\frac{e^{(a^2-0.25)\tau}\operatorname{erfc}(a\sqrt{\tau})}{(a-b)(a^2-0.25)}\right)$$
$$- b\frac{t_u}{t_m}\left(\frac{e^{(b^2-0.25)\tau}\operatorname{erfc}(b\sqrt{\tau})}{(a-b)(b^2-0.25)}\right) \quad (10.60)$$
$$- \frac{0.5t_t\operatorname{erf}(0.5\sqrt{\tau})}{t_u-t_t} + \frac{t_u-0.5t_t}{t_u-t_t},$$

where $\theta(\tau)$ is the departure of the mixed-layer temperature from its initial steady-state value, τ is time normalized by the upwelling timescale $\tau \equiv t/t_u$ with t_u defined below. The denominator of the left-hand side of the last equation consists of $R_0(t)$, the radiative forcing of the Earth–atmosphere system (e.g., the imbalance due to a sudden change [step function] in CO_2), λ is the climate sensitivity parameter,[10] typically approximately 2.0 W m^{-2} K^{-1}. The climate sensitivity parameter is the inverse of the change in global average temperature per unit of radiative forcing expressed in units, W m^{-2} K^{-1}. The special functions in the formula are the error function, erf(x), and the complementary error function, erfc(x) defined in Section 10.4. Other terms are defined below starting with $\tau = t/t_u$ and

$$a = 0.5t_t/t_m + \sqrt{(0.5t_t/t_m + 0.5)^2 - t_u/t_m},$$
$$b = 0.5t_t/t_m - \sqrt{(0.5t_t/t_m + 0.5)^2 - t_u/t_m},$$
$$t_u = k/w^2 = \text{upwelling timescale},$$
$$t_t = \rho c k u / \lambda w = \text{thermocline timescale},$$
$$t_m = C/\lambda = \text{mixed-layer timescale}. \quad (10.61)$$

An example of the response to step-function forcing is shown in Figure 10.13.

---

[10] Watts and Morantine used a value of 2.2 W m^{-2} K^{-1}, both their value and the one usually used in this book ignore ice feedback which would lower the value of λ.

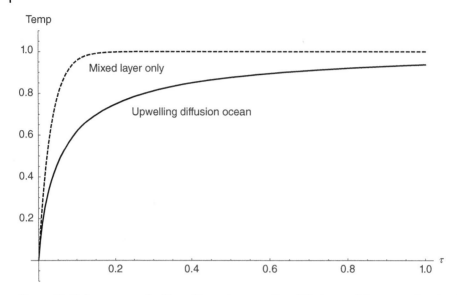

Figure 10.13 Response to the Watts–Morantine upwelling diffusion model to a step-function forcing. The ordinate is a normalized response in units of the asymptotic solution. The abscissa is time in units of the mixed-layer-only relaxation time, C/λ. The upper curve (dashed) is the response to a mixed-layer ocean and the lower curve (solid) is for an upwelling-diffusion model. The parameters are $D = 100$ m; $k = 3000$ m² year⁻¹; $w = 4$ m year⁻¹, $\lambda = 2.2$ W m⁻² K⁻², D here means the depth of the mixed layer. The time axis in units of the mixed layer response time $C/\lambda = 6.02$ years.

10.5.2 Ramp Forcing

Watts and colleagues also solve the problem with ramp forcing for the global ocean (no land as with the step-function forcing in the previous section). They approach the problem in much the same style as we have earlier in this chapter. The particular solution for the mixed-layer temperature responding to a forcing γt is

$$T_p(t) = \frac{\gamma}{B}(t - \tau_m - \tau_\ell), \tag{10.62}$$

where $\tau_m = C_m/B$, $\tau_\ell = C_\ell/B$, $\ell = k/w$, and $C_\ell = c\ell$ with c in per unit volume units. In its asymptotic (in time) form, the lag behind the response with no ocean or atmosphere is the sum of the lag for the mixed layer and the lag associated with the effective heat capacity of the water between the thermocline and the surface. The transient solution is much more complicated, involving a mixture of exponentials, error functions, and so on. Note the very long adjustment time in Figure 10.13 of the UD ocean compared with the mixed-layer-only case. The transient or homogeneous solution contains two timescales, τ_m and $\tau_m + \tau_\ell$. Morantine and Watts (1994) also point out in their discussion that naturally induced changes in w (the response to which they give solutions) can cause changes in the global surface temperature that are as large as the global warming signal.

10.6 Upwelling Diffusion with (and without) Geography

The study by Kim *et al.* (1992) takes the past studies by Morantine and Watts (1990) and others (e.g., Wigley and Schlesinger, 1985; Lebedeff, 1988; Morantine and Watts, 1994) to include the land–sea geography in the sense of the previous chapters of this book, namely, the local surface is characterized by its effective heat capacity. The approach is similar with the starting point being the trial of the particular solution:

$$T_p = \alpha(t - \tau(\hat{\mathbf{r}}) + z/w)e^{z/\ell}, \tag{10.63}$$

where $\tau(\hat{\mathbf{r}})$ is a lag that depends on the position on the globe. It turns out that $\ell = k/w$, where ℓ can be thought of as the depth of the thermocline minus the depth of the mixed layer. By insertion of this last form into the governing equation, we find that $\tau(\hat{\mathbf{r}})$ satisfies

$$-\nabla \cdot \left(\frac{D(x)}{B}\nabla \tau(\hat{\mathbf{r}})\right) + \tau(\hat{\mathbf{r}}) = \tau_m(\hat{\mathbf{r}}) + \tau_\ell(\hat{\mathbf{r}}). \tag{10.64}$$

Over ocean, the terms on the right-hand side sum up to the radiative relaxation time for the entire thermocline. Over land, these terms add up to the relaxation time of an effective column of air. These quantities are known from the land–sea geography. This inhomogeneous equation has the form of the steady-state EBM, but with $\tau(\hat{\mathbf{r}})$ as dependent variable. The dependent variable is the response to the source terms on the RHS. It is damped by the $B\tau(\hat{\mathbf{r}})$ term and smeared out by the diffusion term. Through this process of filtering the geography is entered. The solution for lag $\tau(\hat{\mathbf{r}})$ will be large over oceans, small over land, but the hard edges are removed over a length scale of the order of $\sqrt{D/B}$. Figure 10.14a shows the lag field $\tau(\hat{\mathbf{r}})$ for parameter values $w = 4\,\text{m year}^{-1}$ and $k = 4000\,\text{m}^2\,\text{year}^{-1}$. As expected, the lag is large over the oceans and small over land masses. The features are smoothed by the diffusion operator in accordance with the local length scale $\sqrt{D/B}$. Figure 10.14b shows a contour map of the asymptotic form for large values of time after the transients have died out, with the same parameters employing a deep thermocline of about 1100 m. Kim *et al.* (1992) used this deeper thermocline depth in these experiments to enhance the land–sea contrast of the signal.

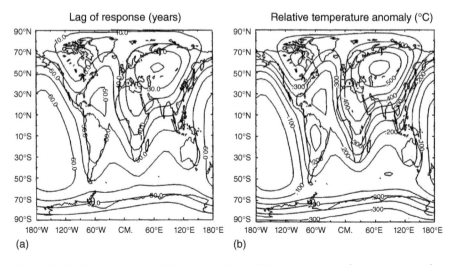

Figure 10.14 (a) Contour map of the lag function $\tau(\hat{\mathbf{r}})$ for $w = 4\,\text{m year}^{-1}$ and $k = 4000\,\text{m}^2\,\text{year}^{-1}$ (thermocline at 1100 m). (b) Contour map of the temperature field for the particular solution, $T_p(\hat{\mathbf{r}}, t)$ for the same parameters. These maps are the response without the transient terms to ramp forcing $F(t) = \gamma t$. (Kim *et al.* (1992). Reproduced with permission of Kim.)

10 Time-Dependent Response and the Ocean

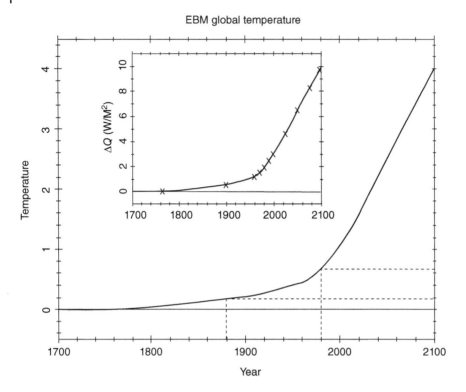

Figure 10.15 Global average temperature anomaly for a "business as usual" scenario with $w = 4$ m year^{-1} and $k = 2000$ m^2 year^{-1}. Note that in this case the thermocline is at 600 m (mixed layer plus the upwelling diffusion component depths $= 100$ m$+\ell$). This simulation includes the land–sea geography of the previous figure. (Kim et al. (1992). Reproduced with permission of Kim.)

Figure 10.15 shows the global average temperature for a case where the more shallow thermocline is at 600 m. This figure illustrates that the scenario needs to start from rest in the middle 1700s because of the long timescales associated with the transient relaxation times. Figure 10.16 shows the anomaly response including both transient and particular solutions.

10.7 Influence of Initial Conditions

During the adjustment to the particular solution, there is a potentially long period during which the initial conditions are important because of the amount of ocean water that has to be heated. The common thread in all of these solutions is the separation of the homogeneous or transient solution from the particular or long-term solution. Recalling (10.11), we see that in order to make the solution consistent at $t = 0$, the coefficient c_1 in (10.10) has to contain a term that will cancel the lag term in the particular solution for the right-hand side to be equal to $T(0)$. This same cancellation trick comes up again in the UD ocean model for the step function as well as in the ramp-heating cases. In these cases, the initial condition must involve the profile in the deep ocean as well as just its value at the surface (see Figure 10.17).

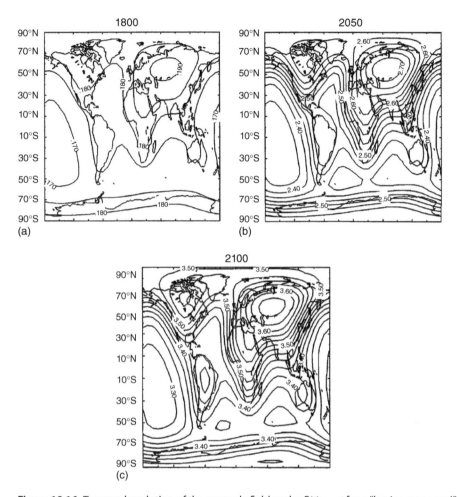

Figure 10.16 Temporal evolution of the anomaly field under $F(t) = \gamma t$ for a "business as usual" scenario. In this case, the thermocline is at 600 m (mixed layer plus the upwelling diffusion component depths=100 m+ℓ). (Kim et al. (1992). Reproduced with permission of Kim.)

In the case of the all-ocean planet, $\tau(\hat{\mathbf{r}})$ is a constant and from the terms in (10.63) one must cancel the exponential profile (of water mass) which has coefficient $\tau(\hat{\mathbf{r}})$ and the factor $(z/w)e^{z/\ell}$. Each of these terms will have a time constant of the order of the relaxation time of the whole thermocline (approximately few decades). As mentioned above, Morantine and Watts (1990) show that this has a complicated but closed-form analytical solution. The transient solution will also contain the timescale of the mixed layer alone, $\tau_m \sim$ a few years.

10.8 Response to Periodic Forcing with Upwelling Diffusion Ocean

In this section, we sketch the response of the UD Ocean to a periodic forcing at the surface. The approach is similar to that in Section 10.2, the difference being the inclusion

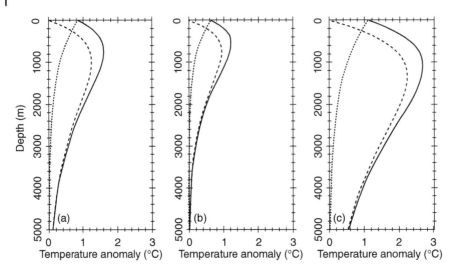

Figure 10.17 Vertical shapes of terms in the homogeneous solution that have to decay temporally in order to satisfy the initial conditions before assuming the asymptotic forms (particular solution) for various parameter values. (a) For $w = 4$ m year^{-1}, $k = 4000$ m^2 year^{-1}; (b) for $w = 4$ m year^{-1}, $k = 3000$ m^2 year^{-1}; (c) for 3 m year^{-1}, $k = 4000$ m^2 year^{-1}. Dotted lines are for $\alpha\tau(\hat{r})e^{z/\ell}$, dashed lines are for $\alpha(z/w)e^{z/\ell}$, and the solid lines are for the sum of the two. Aside from the response linear in time, the solid line essentially is a particular solution for $F(t) = \gamma t$. (Figure copied from Kim et al. (1992). Reproduced with permission of Kim.)

of upwelling. The complete details (including land–sea geography) can be found in Kim and North (1992). This problem is important because it allows us to compute the spectral density of this model system under white noise forcing. We summarize the procedure here for the all-ocean planet. Toward the end, we quote some results from published papers that include geography.

The governing equations are the same as in the last section except that the forcing is now periodic in time. The procedure is similar to that in Section 10.2, but the upwelling term complicates the analysis. As in that chapter, we are interested in the steady-state periodic solution as opposed to the transient solutions which die out after several characteristic timescale intervals. We begin with the all-ocean planet. The details of this section are from Kim and North (1992). As in Section 10.2, we substitute for $T(z,t)$:

$$T(z,t) = \tilde{T}_\omega e^{i\omega t + (\eta + i\zeta)z}, \tag{10.65}$$

where ω is the angular frequency of the periodic forcing at the surface, and η and ζ are the real and imaginary parts of the depth dependence in the exponential. After inserting this form into the governing equations and the boundary conditions, we find a relation between the parameters η, ζ, and the angular frequency ω:

$$i\omega h + w(\eta + i\zeta) = k(\eta + i\zeta)^2, \tag{10.66}$$

or equivalently:

$$\omega h = -w\zeta + 2k\eta\zeta; \tag{10.67}$$
$$0 = -w\eta + k\eta^2 - k\zeta^2. \tag{10.68}$$

At this point, it is convenient to normalize the variables to characteristic lengths (nominally, the depth of the thermocline):

$$\eta^* = \eta/(w/k); \quad \zeta^* = \zeta/(w/k). \tag{10.69}$$

Following this normalization, we drop the asterisks for simplicity. Then the previous equations become

$$-\zeta(1 - 2\eta) = \frac{\omega k h}{w^2} \equiv \tilde{\omega}, \tag{10.70}$$

$$-\eta + \eta^2 - \zeta^2 = 0. \tag{10.71}$$

Solving for η, we have a quartic equation with four roots for a given $\tilde{\omega}$:

$$g(\eta) = 4\eta^4 - 8\eta^3 + 5\eta^2 - \eta = \tilde{\omega}^2. \tag{10.72}$$

Recall that in the case with no upwelling ($w = 0$), the penetration decreases as the driving frequency is increased ($\eta \propto \omega^{-1/2}$). Figure 10.18 shows us that when the frequency (squared) is zero, the penetration depth is at $\eta = 1.0$ (the thermocline). As the frequency is increased (dotted line) from 0.5 to 3.0, the penetration depth changes from 1.25 to about 1.5 times the thermocline depth. In other words, the upwelling term induces quite a different penetration structure. This result that the effective depth is close to the thermocline might have been anticipated by the lag found in this model for ramp forcing in (10.62) where the thermocline relaxation time is added to the relaxation time of the mixed layer.

In the paper by Kim and North (1992), the model is developed with full geography as in the ramp-warming case of Section 10.6. In that paper, the seasonal cycle is updated with the new UD ocean and the model is subjected to white noise in space as well as time. The

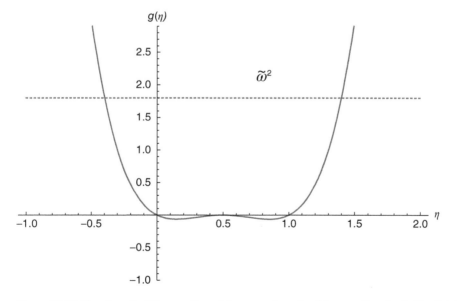

Figure 10.18 The characteristic equation $g(\eta)$ versus η in units of thermocline depth. The flat dotted line is for the normalized frequency squared. (See also Kim et al. (1992). (© American Geophysical Union, with permission).)

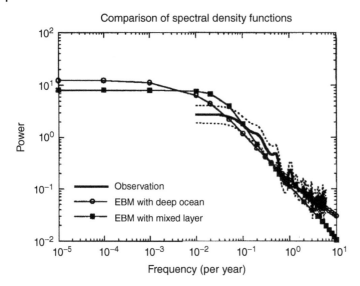

Figure 10.19 Spectral density functions of the global average temperatures: thick line, observations (smoothed using a Parzen window); dashed curves, 90% confidence band of the smoothed spectra; thin line with circles, response to the (upwelling diffusion model) coupled energy balance model with noise forcing, white in space and time; thin line with solid squares, linear (the ocean is mixed-layer-only) energy balance model response to a noise forcing that is white in space and time. Each spectral density function is normalized such that the area under the curve between the frequency of 0 and 6 is unity. The frequency 6 (2-month period) is the Nyquist frequency for the observational data. Both models include the full land–sea geography. (Figure taken from Kim and North (1992). (© American Geophysical Union, with permission).)

response to the noise forcing is shown in Figure 10.19 in the form of the spectral density. In constructing model-generated spectral densities, the strength of the forcing variance is arbitrary. We adjusted that strength to match the spectral density of the observations in the frequency range from a few decades to a period of a few months. Here we gain some insight especially when we consider low frequencies. At low frequencies, the models only differ by about a factor of 2, the UD model having more power as expected than the mixed-layer-only model. The UD model has a penetration of heat only just below the thermocline even at these very low frequencies. If upwelling is turned off in the infinitely deep ocean model, the power in this graph would be unbounded at lowest frequencies. This is a genuine difference between these two model structures.

10.9 Summary and Conclusions

In this chapter, we have considered a variety of oceanic models that are suitable for energy balance climate model (EBCM) studies. By moving through a list of models, each more complicated than the earlier one, we were able to uncover a number of features of the models that are important for determining whether they might be useful for a particular application. The simplest of these is the single-slab model. It satisfies the most elementary first-order linear ordinary differential equation with purely exponential-type solutions when it is unforced. There is a single timescale for this model, the radiative relaxation time of the slab. If found out of equilibrium, the solution returns to equilibrium with this time constant. If the model is forced at the top with white temporal noise, it reveals a spectral density identifiable as classical red noise.

Before going to multiple slabs, we treated the vertically continuous model with thermally diffusive transport of heat energy. If this is the only mechanism of vertical heat transfer and if the ocean is infinitely deep, we find some rather peculiar properties. The first is that the equilibrium solution is uniform in the vertical in disagreement with observations where a cooling of water from the surface to the thermocline is observed at a few hundred to a thousand meters. Another curious property of this model is that the depth of the penetration is strongly dependent on frequency of the surface heating. The penetration is inversely proportional to the square root of the frequency. Hence, low frequencies can penetrate very deep into the ocean, giving rise to a likelihood of very large power in the spectral density at low frequencies. In fact, if the radiation to space is made very small, the spectral density diverges at low frequencies. At high frequencies, there is another catastrophe, namely, the spectral density is of the pink-noise type ($\propto 1/\omega$), which means it is not integrable – not as bad as white noise, but still not finite.

Numerous investigators have known of this problem for many years. The logical cure that does not defy the EBCM framework and spirit seems to be to introduce an upwelling term. The basis for such a radical idea is that cool saline waters in the polar regions downwell water that is cold compared to the waters above. This dense water slips under the bulk of the abyssal waters, gradually lifting the waters above at a rate of a few meters per year. This is the so-called UD model. But before taking it up, we showed how models with two layers work. In this case, the layers are coupled with flux densities that are proportional to the temperature difference. To be consistent with the discussion above, we introduced an upwelling term as well. The solutions to this class are somewhat like the single-slab model in that it is often convenient to separate the transient solutions from the particular solution, which can often be identified as an asymptotic form for large time. In this way, the problem has two parts. The transient solutions tend to die away, leaving the asymptotic particular solution to emerge after all the transients have died out. In the case of several slabs, there are time constants that can be related to each slab, but as they are coupled, the time constants emerge as the inverses of eigenvalues. Even if the coupling of the slabs is weak, the long timescales of the deep tend to linger in the adjustment times, but we will find that the upwelling term limits the deep eigenvalues to some extent.

The most commonly desired product of interest throughout the model hierarchy is the response to a ramp-like forcing that would result from an exponential increase of CO_2. In this case, up and down through the hierarchy, one finds that after the transients have died down, the temperatures in the response fall on a straight-line ramp. The response temperatures lag behind the situation where there is no thermal inertia at the surface. The lags can often be guessed by knowing the finite slabs or by solving for the continuum solutions. We note one difference right away: the UD model leads to lags in the asymptotic forms that suggest that the heat capacity of the thermocline might dominate. This is in contrast to the case where the upwelling term is omitted. In that latter case, the contribution of the very deep layers are more important. In the better-behaved UD models, the asymptotic solutions are not too hard to extract from the equations. The transient solutions are more difficult to obtain, especially if one seeks analytical forms. Often, as we have seen in the UD models, the adjustment process to new conditions involves warming waters well below the surface and this can take a very long time. Is the time taken for adjustment onto the ramp tens or hundreds of years? Modern coupled GCMs suggest that it may be only a few decades until we are well into the ramp regime.

Once we have learned how to generate the solutions for the UD model for the all-ocean planet, it is possible to extend it to the case where the land–sea geography is specified as in the earlier chapters of this book. Solutions are of course dependent on parameter choice and these are not uniquely determined (the nemesis of all climate models great and otherwise). One easily sees that in the ramp-like scenarios, the land masses lead the oceans and the lags are shorter on land as well, both results agreeing with data and intuitively satisfying. The spectral density of these UD models driven by white noise in space and time are interesting. They exhibit power at low frequencies, somewhat higher than comparably generated solutions with a mixed layer. But the power at very low frequencies is not much higher than one might expect. This interesting feature appears to be limited to the lack of deep penetration of power in these models (contrasting with a pure diffusion model). If true, this could mean that our current period of global warming might be forced (as opposed to natural variability driven by noise at the surface) as many investigators seem to believe.

Exercises

10.1 (a) Determine the complete solution of a one-dimensional EBM with a sudden impulse forcing

$$\frac{dT}{dt} + \frac{T}{\tau} = F_0 \delta(t), \tag{10.73}$$

where $\delta(t)$ is the Dirac delta function.
(b) What is the initial temperature $T(t = 0+)$ right after the forcing is applied?
(c) How long does it take for the initial temperature change to be reduced to an e^{-1} (1/e) level?

10.2 (a) Obtain the solution of a one-dimensional EBM with a step forcing

$$\frac{dT}{dt} + \frac{T}{\tau} = \frac{\gamma}{C} H(t). \tag{10.74}$$

Here $\tau, \gamma,$ and C are constants, and the Heaviside step function is defined by

$$H(t) = \begin{cases} 0, & \text{for } t < 0, \\ 1, & \text{for } t \geq 0. \end{cases} \tag{10.75}$$

10.3 (a) Obtain the solution of a one-dimensional EBM with a ramp forcing

$$\frac{dT}{dt} + \frac{T}{\tau} = \frac{\gamma}{C} t, \quad t \geq 0, \tag{10.76}$$

where $\tau, \gamma,$ and C are constants. Assume that $T(t = 0) = 0$.
(b) Determine the lag between the solution and the linear forcing as time approaches infinity.
(c) Let us further assert that the forcing levels at a constant value $F_0 = \gamma t_0 / C$ at time $t = t_0$. Find the complete solution of the problem.

10.4 Let us consider a one-dimensional EBM on top of a diffusive ocean with an infinite depth. Inside the diffusive ocean, temperature changes according to the heat diffusion equation

$$C\frac{\partial T}{\partial t} = k\frac{\partial^2 T}{\partial z^2}, \qquad (10.77)$$

where C is heat capacity and k is thermal diffusivity. At the top of the ocean, energy balance is described as

$$A + BT + k\frac{\partial T}{\partial z} = Q, \quad \text{at } z = 0 \text{ (ocean surface)}, \qquad (10.78)$$

where we essentially ignored the atmospheric column and assumed that the atmospheric temperature is identical with the surface temperature of the ocean.
(a) Set up an equation together with the boundary condition for anomalous temperature driven by additional sinusoidal forcing $Q = Q_\omega e^{i\omega t}$.
(b) Find the solution of the problem in Part (a).

10.5 Let us consider a two-layer EBM in the form

$$C_m \frac{dT_m}{dt} + BT_m + k(T_m - T_d) = F(t), \qquad (10.79)$$

$$C_d \frac{dT_d}{dt} - k(T_m - T_d) = 0, \qquad (10.80)$$

where the subscript "m" and "d" denote mixed layer and deeper layer, respectively.
(a) Set up the equation as a system of coupled linear equations.
(b) Let us assume a homogeneous solution of the linear system in Part (a) in the form $\vec{T} = \vec{A}\, e^{-\lambda t}$. Determine the parameters λ.
(c) Plot the decay timescale (inverse of λ) and the angle between two eigenvectors as a function of the coupling parameter k. Use $\ln k \in [-2, 2]$ (equivalently, $k \in [0.01, 100]$).

10.6 Let us consider an energy balance system for a diffusive ocean with a mixed-layer at the top:

$$\frac{\partial T_d(z,t)}{\partial t} = k\frac{\partial^2 T_d(z,t)}{\partial z^2}, \quad -h < z < 0, \qquad (10.81)$$

$$C_m \frac{\partial T_m(z,t)}{\partial t} + BT_m(z,t) - F^\uparrow(t), \quad z = 0, \qquad (10.82)$$

where

$$F^\uparrow = ck\left.\frac{\partial T_d(z,t)}{\partial z}\right|_{z=0} \qquad (10.83)$$

is the flux entering the mixed layer from below. Further, the boundary condition is written as

$$\left.\frac{\partial T_d(z,t)}{\partial z}\right|_{z=-h} = 0. \qquad (10.84)$$

Derive the solution for a step forcing $F(t) = F_0$ for $t \geq 0$.

10.7 Let us consider an energy balance equation with a slab ocean:

$$C\frac{dT}{dt} + A + BT = Qa_p + \gamma t,$$

where $C = 9.7$ W year m^{-2} °C^{-1} represents the heat capacity of the mixed-layer ocean which is approximately 70 m deep and γt is a ramp forcing. Find the complete solution.

10.8 Let us consider an ocean whose interior temperature is governed by a diffusive process:

$$\frac{\partial T}{\partial t} = K\frac{\partial^2 T}{\partial z^2}, \quad -h < z < 0,$$

$$\frac{\partial T}{\partial z} = 0, \quad z = -h.$$

(a) Explain the physical meaning of the boundary conditions.
(b) Obtain the solution of the problem.

10.9 Let us consider a UD ocean model in the form

$$\frac{\partial T}{\partial t} = K\frac{\partial^2 T}{\partial z^2} - w\frac{\partial T}{\partial z}, \quad -h < z < 0,$$

where w is the upwelling speed.
(a) Calculate the vertical heat flux density for this model.
(b) Boundary conditions for anomalous temperature are given by

$$C\frac{\partial T}{\partial t} + BT + C_0\left(K\frac{\partial T}{\partial z} - wT\right) = \gamma t, \quad z = 0,$$

$$K\frac{\partial T}{\partial z} - wT = 0, \quad z = -h,$$

where the ocean is assumed to be insulated from below. Non-dimensionalize the governing equation using $z^* = \ell z$ and $t^* = \tau t$.
(c) Find the complete solution of the problem.

10.10 Let us consider an UD ocean coupled with a two-dimensional EBM as a top boundary condition:

$$\frac{\partial T(\hat{\mathbf{r}}, z, t)}{\partial t} + w\frac{\partial T(\hat{\mathbf{r}}, z, t)}{\partial z} = K\frac{\partial^2 T(\hat{\mathbf{r}}, z, t)}{\partial z^2}, \quad z \leq 0,$$

$$C\frac{\partial T}{\partial t} + BT - \nabla \cdot (D\nabla T) - wC_0 T + KC_0\frac{\partial T}{\partial z} = F(r, t), \quad z = 0,$$

$$T \to 0, \quad \text{as } z \to -\infty,$$

$$T(\hat{\mathbf{r}}, z, t = 0) = T_0(\hat{\mathbf{r}}, z),$$

where $\hat{r}$ represents a point (θ, ϕ) on the sphere, C is local heat capacity per unit area, D is local horizontal diffusion coefficient in the atmosphere and the mixed layer, B is the slope of the best linear!it for the infrared emission to space, w is upwelling speed, K is vertical heat diffusion coefficient, C_0 is the heat capacity of seawater per unit volume, and $F(r, t)$ is radiative forcing.

(a) Show that thermal flux at the bottom of the mixed layer is given by

$$F^\uparrow = wC_0 T - KC_0 \frac{\partial T}{\partial z}.$$

(b) Let us consider a ramp forcing $F(r, t) = \gamma t H(t)$, where $H(t)$ is the Heaviside step function. Let us consider the particular solution of the problem in the form

$$T^{(p)}(\hat{r}, z, t) = \alpha(\gamma t - \tau(\hat{r}) + z/w)e^{\delta z}.$$

Show that the lag $\tau(\hat{r})$ satisfies

$$B\tau(\hat{r}) - \nabla \cdot (D\nabla \tau(\hat{r})) = C(\hat{r}) + C_0(\hat{r})/\delta,$$

where $\alpha = \gamma/B$ and $\delta = w/K$

(c) Let us consider the radiative forcing in the form

$$F(\hat{r}, t) = F_f(\hat{r})e^{2\pi i f t}.$$

Let us consider the solution in the form

$$T(\hat{r}, z, t) = R_f(\hat{r})\exp(2\pi i f(t - \tau(\hat{r})))\exp((\eta + i\zeta)z)$$
$$= T_f(\hat{r})e^{2\pi i f t}\, e^{(\eta + i\zeta)z},$$

where $\tau(\hat{r})$ is the position-dependent temporal lag of the solution, η and ζ are constants determining the vertical structure of the solution, and $T_f(\hat{r})$ is a complex-valued amplitude. Determine η and ζ as a function of f.

11

Applications of EBMs: Optimal Estimation

11.1 Introduction

An *estimator* is an algorithm making use of an observation or combination of observations in order to provide a useful estimate of some system parameter such as the global average temperature. In order to conduct such a procedure, we have to make a *statistical model* of our process. Once we have a statistical model in place, we can examine it and the estimation procedure to learn several things about the underlying system parameter. Usually, the estimator is a random variable that has some probability distribution (*pdf*) due to errors in the measurement process or perhaps sampling error. The following are some questions of interest: (i) does the mean of the pdf of the estimator coincide with the actual value of the system parameter? If it does, we say the estimator is *unbiased*. (ii) What is the variance of the estimator? The square root of this variance is the standard deviation or *root mean square* error or *RMS* error.

EBMs can be of service in some estimation problems. The usefulness of an EBM for a particular application comes in two forms:

1. Many estimation problems can be evaluated with the assistance of general circulation model (GCM) simulations. In this chapter, we use the EBM to show how a few of these work in the simpler context. In many cases with the EBM, we are able to solve the problem analytically or nearly so. This puts aside the issue of whether the more realistic model actually has a solution or whether it is over-fitted, and so on. In this application, we can focus on the estimation process from its beginning to its end without letting the details or mathematical transgressions cloud the picture. The main point is the *understanding* of the estimation process in a simple and more heuristic context. It might be that the lessons learned can be applied with much more complicated models.
2. There are some estimation problems where it might not be feasible to solve the problem in the more complicated models because of lack of computing or data resources. For example, how many fully coupled GCMs have a 10 000 year control run with which to assess the low-frequency statistical parameters of natural variability? How can our intuition about such problems be enhanced? To paraphrase a statement by John Maynard Keynes, "it might be better to have a rough idea of the truth than a very precise [or satisfying] answer which is wrong."

In a typical case, there may be many unbiased estimators for a given problem, often combining lots of measurements, such as areal or temporal averages. Think of the estimate of the global average temperature. We might take gauge (point) measurements from a number of different stations or perhaps other observing systems (e.g., satellites). We suspect that if we simply average the data, then these averages might form an unbiased estimate of the global average temperature. We might ask whether a straight arithmetic average is actually the best unbiased estimator in terms of its RMS error. Perhaps some nonuniform weighting would improve the estimate. This general class of problems is the subject of this chapter. We will apply the method with the help of the EBM to two different examples: Section 11.3 on estimating the global average temperature; Section 11.4 on detecting faint deterministic signals (such as the greenhouse warming or episodic cooling by volcanic dust veils) in the climate system. We start with a simple problem involving two imperfect observations of a heat reservoir.

11.2 Independent Estimators

Consider estimating the temperature of a reservoir with two devices. Let the estimators $\widehat{T}_1$ and $\widehat{T}_2$ be unbiased; that is, $\langle \widehat{T}_1 \rangle = \langle \widehat{T}_2 \rangle = T$ where T is the true temperature, and $\langle \cdot \rangle$ means ensemble average.

The individual estimators are assumed to be of the form $\widehat{T}_i = T + \varepsilon_i$; where the errors $\varepsilon_i, i = 1, 2$ are assumed to be random variables taking on different values in each realization of the measurement process. The errors or noise are assumed to have mean zero when considered over a large number of trials: $\langle \varepsilon_i \rangle = 0$, and the covariances of the errors are given by $\langle \varepsilon_i \varepsilon_j \rangle = \sigma_i^2 \delta_{ij}$; $i, j = 1, 2$. The previous expression states that the errors of the separate devices are assumed to be uncorrelated and that their individual variances are given by σ_1^2 and σ_2^2. We assume that these characteristics of the errors are known beforehand. Our task is to take one realization of the measurement process and obtain an optimal estimate of the true reservoir temperature. We wish to make maximal use of the data collected from each device in an appropriate linear combination. The question is, what is the appropriate weighting to assign to each measurement? We form the estimate

$$\widehat{T} = \alpha \widehat{T}_1 + (1-\alpha)\widehat{T}_2, \tag{11.1}$$

where α is a weight to be adjusted to make the mean square error (MSE) the minimum. The estimator $\widehat{T}$ is clearly unbiased if the individual estimators are. We can form the MSE for the measurement as

$$\varepsilon^2 = \langle (\widehat{T} - T)^2 \rangle \tag{11.2}$$
$$= (\sigma_1^2 + \sigma_2^2)\alpha^2 - 2\alpha\sigma_2^2 + \sigma_2^2. \tag{11.3}$$

The latter is a quadratic in α and is shown in Figure 11.1 for a choice of $\sigma_1^2 = 2\sigma_2^2 = 1$. The point of this figure is that the MSE is rather insensitive to the choice of α so long as it is near its optimum value. This is an important point to be stressed later in the climate signal detection exercises.

The minimum of the quadratic above is easily found, and it yields the familiar and very important result:

$$\widehat{T}_{\text{opt}} = \frac{1}{\eta^2}\left(\frac{\widehat{T}_1}{\sigma_1^2} + \frac{\widehat{T}_2}{\sigma_2^2}\right) \tag{11.4}$$

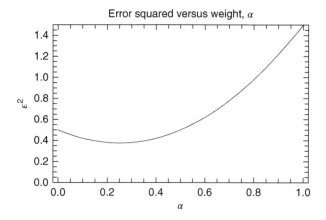

Figure 11.1 Error squared ε^2 versus weighting α for two unbiased estimators.

where

$$\eta^2 = \frac{1}{\sigma_1^2} + \frac{1}{\sigma_2^2}. \tag{11.5}$$

The result is easily generalized to include K independent unbiased estimators:

$$\widehat{T}_{\text{opt}} = \frac{1}{\eta^2} \sum_{k=1}^{K} \frac{\widehat{T}_k}{\sigma_k^2} \tag{11.6}$$

and

$$\eta^2 = \sum_{k=1}^{K} \frac{1}{\sigma_k^2}. \tag{11.7}$$

An interesting way of expressing these last results is

$$\widehat{T}_{\text{opt}}^{(N)} = \mathbf{\Gamma} \cdot \widehat{\mathbf{T}}, \tag{11.8}$$

with the column vector $\widehat{\mathbf{T}} = (\widehat{T}_1, \ldots, \widehat{T}_K)$ and

$$\Gamma_{kk'} = \frac{1}{\eta^2} \frac{\delta_{kk'}}{\sigma_k^2}. \tag{11.9}$$

This last form gives us a convenient way of viewing the optimal estimation procedure in the form of an *optimal filter* of the raw data. The filter loads each observation with a weight inversely proportional to its individual error variance. The factor $1/\eta^2$ assures the normalization necessary for unbiasedness.

After some algebra, it can be shown that the optimal error variance is just

$$\varepsilon_{\text{opt}}^2 = \frac{1}{\eta^2} = \frac{1}{\sum_k \frac{1}{\sigma_k^2}}. \tag{11.10}$$

This shows that adding another device always improves the signal-to-noise ratio indicator no matter how poor its quality.

The derivation presented above required that the individual errors be uncorrelated with one another. If this were not so, the coordinate axes could simply be rotated to the principal axes of the error or noise covariance matrix. Then the entire formalism goes through as before except in the rotated coordinate system. In climatology, this is the transformation to the empirical orthogonal function (EOF) basis set, which we will return to in later sections.

In two dimensions, this is easily spelled out explicitly. Let the covariance matrix of the noise be given by

$$\mathbf{C}_{ij} = \langle \varepsilon_i \varepsilon_j \rangle. \tag{11.11}$$

Taking the noise to be distributed bivariate normally, the contours of equal probability of occurrence of pairs of values of $(\varepsilon_1, \varepsilon_2)$ are given by (see Thiébaux, 1994):

$$\frac{T_1^2}{\sigma_1^2} - 2\rho \frac{T_1 T_2}{\sigma_1 \sigma_2} + \frac{T_2^2}{\sigma_2^2} = \text{constant}, \tag{11.12}$$

which is an ellipse in the (T_1, T_2) plane. In this two-dimensional case, we can find an angle θ to rotate the coordinate axes through, such that the principal axes of the ellipse coincide with new coordinate axes (T_1', T_2'). In the new coordinate system, T_1' and T_2' are uncorrelated. In the case of K dimensions, the figure is an ellipsoid in the K-dimensional space and a simple length-preserving rotation can also be used to find the appropriate coordinate system. This rotation of the coordinate axes is familiar in data analysis as the transformation from spatial coordinates to the EOF basis set.

The simple derivation of optimal weighting of independent estimators is familiar to many researchers. The result is very intuitive. We simply weight each estimator inversely according to its individual error variance.

11.3 Estimating Global Average Temperature

We will follow the method of Shen *et al.* (1994). The global average $\overline{T}_\tau(t)$ changes over time and it is smoothed over an interval τ centered at t (it is a *running average*):

$$\overline{T}_\tau(t) = \frac{1}{4\pi} \int_{4\pi} T_\tau(\hat{\mathbf{r}}, t) d^2\Omega, \tag{11.13}$$

where Ω refers to solid angle, $\hat{\mathbf{r}}$ is a unit vector originating at the Earth's center and pointing to a location on the sphere, and

$$T_\tau(\hat{\mathbf{r}}, t) = \frac{1}{\tau} \int_{t-\tau/2}^{t+\tau/2} T(\hat{\mathbf{r}}, t') dt'. \tag{11.14}$$

An anomaly at point $\hat{\mathbf{r}}$ is defined by

$$\Delta T_\tau(\hat{\mathbf{r}}, t) \equiv T_\tau(\hat{\mathbf{r}}, t) - \langle T_\tau(\hat{\mathbf{r}}, t) \rangle \tag{11.15}$$

and

$$\Delta \overline{T}_\tau(t) \equiv \overline{T}_\tau(t) - \langle \overline{T}_\tau(t) \rangle; \tag{11.16}$$

and by definition, $\langle \Delta \overline{T}_\tau(t) \rangle = 0$.

The global average temperature at time t can be estimated for a given fixed network of stations $\{\hat{\mathbf{r}}_j, j = 1, 2, \ldots, N_{\text{net}}\}$:

$$\hat{\bar{T}}_\tau(t) = \sum_{j=1}^{N_{\text{net}}} w_j T_\tau(\hat{\mathbf{r}}, t), \tag{11.17}$$

and to ensure unbiasedness:

$$\sum_{j=1}^{N_{\text{net}}} w_j = 1. \tag{11.18}$$

Our estimator is

$$\hat{\bar{T}}_\tau(t) = \frac{1}{4\pi} \int_{4\pi} w_{\text{net}}(\hat{\mathbf{r}}, t) T_\tau(\hat{\mathbf{r}}, t) d^2\Omega, \tag{11.19}$$

where

$$w_{\text{net}}(\hat{\mathbf{r}}) \equiv 4\pi \sum_{j=1}^{N_{\text{net}}} w_j \, \delta(\hat{\mathbf{r}} - \hat{\mathbf{r}}_j), \tag{11.20}$$

and the MSE is given by

$$\varepsilon^2 = \langle (\bar{T}_\tau - \hat{\bar{T}}_\tau)^2 \rangle. \tag{11.21}$$

After multiplying the factors together,

$$\varepsilon^2 = \int_{4\pi} d^2\Omega \int_{4\pi} d^2\Omega' \left[\frac{1}{(4\pi)^2} - \frac{2}{4\pi} \sum_{i=1}^{N_{\text{net}}} w_i \delta(\hat{\mathbf{r}} - \hat{\mathbf{r}}_i) \right.$$

$$\left. + \sum_{i,j=1}^{N_{\text{net}}} w_i w_j \delta(\hat{\mathbf{r}} - \hat{\mathbf{r}}_i) \delta(\hat{\mathbf{r}}' - \hat{\mathbf{r}}_j) \right] \rho_\tau(\hat{\mathbf{r}}, \hat{\mathbf{r}}') \tag{11.22}$$

and we have introduced the temporally smoothed covariance

$$\rho_\tau(\hat{\mathbf{r}}', \hat{\mathbf{r}}'') = \langle T_\tau(\hat{\mathbf{r}}', t) T_\tau(\hat{\mathbf{r}}'', t) \rangle. \tag{11.23}$$

To choose the optimal weighting coefficients, we need to use the method of Lagrange multipliers (e.g., Arfken and Weber, 2005). We minimize the function

$$J[\mathbf{w}] = \varepsilon^2[\mathbf{w}] - 2\Lambda \left[\sum_{j=1}^{N_{\text{net}}} w_j - 1 \right], \tag{11.24}$$

where 2Λ is a Lagrange multiplier.

Next, we take partial derivatives

$$\frac{\partial J}{\partial w_i} = 0, \quad i = 1, \ldots, N_{\text{net}} \tag{11.25}$$

and

$$\frac{\partial J}{\partial \Lambda} = 0. \tag{11.26}$$

After inserting the expression for ε^2 and rearranging, we find

$$\sum_{j=1}^{N_{\text{net}}} w_j \rho_\tau(\hat{\mathbf{r}}_i, \hat{\mathbf{r}}_j) - \Lambda = \frac{1}{4\pi} \int_{4\pi} \rho_\tau(\hat{\mathbf{r}}, \hat{\mathbf{r}}_i) d^2\Omega, \quad i = 1, \ldots, N_{\text{net}}; \quad \sum_{i=1}^{N_{\text{net}}} w_i = 1. \tag{11.27}$$

Our result is similar to the N thermometer case of the last section. The problem here is that the temperatures at one location $\hat{\mathbf{r}}$ and another $\hat{\mathbf{r}}'$ are correlated. To proceed, we need to know how to find variables that are not correlated. Thus the next section introduces EOFs.

11.3.1 Karhunen–Loève Functions and Empirical Orthogonal Functions

We begin by noting that $\rho_\tau(\hat{\mathbf{r}}', \hat{\mathbf{r}}) = \rho_\tau(\hat{\mathbf{r}}, \hat{\mathbf{r}}')$ is a real symmetric function.[1] Such functions play an important role in mathematical analysis. Consider, for example, the kernel of the integral in the eigenvalue problem[2]:

$$\int_{4\pi} \rho_\tau(\hat{\mathbf{r}}, \hat{\mathbf{r}}')\psi_n(\hat{\mathbf{r}}')\mathrm{d}^2\Omega' = \lambda_n\psi_n(\hat{\mathbf{r}}). \tag{11.28}$$

This equation is in the form of a Stürm–Liouville system (introduced in Chapter 7). The functions $\psi(\hat{\mathbf{r}})$ are the eigenfunctions for integer index $n : n = 1, \ldots, \infty$ and the $\lambda_n > 0$ are the (real and positive) eigenvalues. Properties of Stürm–Liouville systems can be found in most books on mathematical methods for physicists and engineers (e.g., Arfken and Weber, 2005 and later editions). Additional properties include the orthogonality relation:

$$\int_{4\pi} \psi_n(\hat{\mathbf{r}})\psi_m(\hat{\mathbf{r}})\mathrm{d}^2\Omega = \delta_{mn}, \tag{11.29}$$

and the completeness relation:

$$\sum_{n=1}^{\infty} \psi_n(\hat{\mathbf{r}})\psi_n(\hat{\mathbf{r}}') = \delta(\hat{\mathbf{r}} - \hat{\mathbf{r}}'). \tag{11.30}$$

The first of these tells us how to expand any reasonably well-behaved function[3] on the sphere into these basis functions. The functions which are the eigenfunctions of the covariance kernel (11.28) are called *the Karhunen–Loève functions* (K-LFs). They form a convenient basis set into which many useful decompositions might be derived.

Consider developing the function $G(\hat{\mathbf{r}})$ into a series of these basis functions:

$$G(\hat{\mathbf{r}}) = \sum_{n=1}^{\infty} G_n\psi_n(\hat{\mathbf{r}}). \tag{11.31}$$

The coefficients G_n can be calculated from (11.29):

$$G_n = \int_{4\pi} G(\hat{\mathbf{r}})\psi_n(\hat{\mathbf{r}})\mathrm{d}^2\Omega. \tag{11.32}$$

The completeness relation (11.30) assures that any well-behaved function on the sphere can be represented in the series.

---

1 The Wikipedia article, *Karhunen–Loève Theorem*, is particularly good on the K–L functions.
2 Note the similarity of the eigenvector equation for a finite-dimensional symmetric matrix.
3 Here, *reasonably well behaved* includes functions with discontinuities and discontinuous derivatives. For a rigorous treatment of convergence in Fourier-like series, see Körner (1989).

Using these relations, we find that a function such as $\rho_\tau(\hat{\mathbf{r}}, \hat{\mathbf{r}}')$ can be represented as

$$\rho_\tau(\hat{\mathbf{r}}, \hat{\mathbf{r}}') = \sum_{n=1}^{\infty} \psi_n(\hat{\mathbf{r}}) \lambda_n \psi_m(\hat{\mathbf{r}}'). \tag{11.33}$$

We can insert this last expression into (11.22) to obtain a matrix equation for the MSE, $\varepsilon^2(\mathbf{w})$. Inspection of the result reveals that the problem can be cast into the form of a filter through which data in the form of correlation information can be entered. Shen et al. (1994) proceed by expanding each EOF, $\psi_n(\hat{\mathbf{r}})$, into a spherical harmonic[4] series:

$$\psi_i(\hat{\mathbf{r}}) = \sum_{n=1}^{n_{\text{trunc}}} \sum_{m=-n}^{n} a_{nm}^{(i)} Y_n^{(m)}(\hat{\mathbf{r}}), \tag{11.34}$$

where n_{trunc} is a truncation level, typically set at spherical harmonic degree 11 or 15 in the experiments to be described.

Let us now ask why optimal weighting helps. The key is the understanding of how the annual averaged surface temperature data are correlated from one location to another. This was examined in an important paper by Hansen and Lebedeff (1987). Figure 11.2

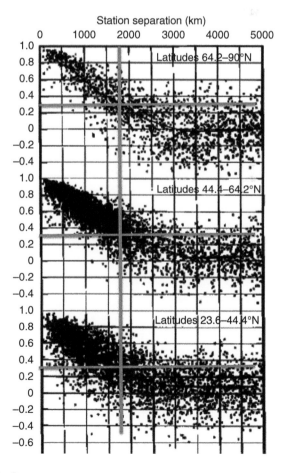

Figure 11.2 Spatial autocorrelation diagrams for annually averaged temperatures at stations separated by distances s. The solid lines are averages of the scattered points. The vertical and horizontal gray lines indicate the 1/e point and their values on the abscissa. Note that in the polar and mid-latitudes, the correlation lengths are about 1500 km. (Hansen and Lebedeff (1987). © American Meteorological Society. Used with permission.)

---

4 Spherical harmonics are discussed in Chapter 8.

shows scatter diagrams of the correlation of the annual averaged surface temperatures in different latitude belts (the figure concentrates on the polar and mid-latitude belts; tropical spatial autocorrelations are not well defined, but in general are much longer). The correlations (on the average) fall off with separation distance to a level 1/e at about 1500 km.[5]

The tropical temperatures do not follow this scheme. The reason is that in polar and mid-latitudes, the weather with autocorrelation times of the order of a few days drives the surface temperature field, which, for large scales, has an autocorrelation time of the order of weeks to a month over land and much longer over ocean. The conditions are right for the Langevin approximation used in Chapter 9. However, in the tropics, there is no weather noise and the dynamics smear out heat very quickly via direct circulations rather than in a kind of thermal diffusion.

We follow Shen et al. (1994) here to show that a relatively small number of gauges or sites (~ 64) are needed to achieve pretty good accuracy for estimating the global average temperature. Figure 11.3 shows the MSE for several gauge configurations: 4×4, 6×4, 9×7, a 63 station well-dispersed gauge network used by Angell and Korshover (1983); and finally a 20×10 array. The EOFs (or K–LFs) were computed as indicated above using a spherical harmonic basis truncated at degree 11 using the data itself, and using the noise-forced 2-D model of Chapter 9. Figure 11.3 shows a graph of the MSE versus the number of EOF modes retained (M) (the number of terms retained in (11.33)). This

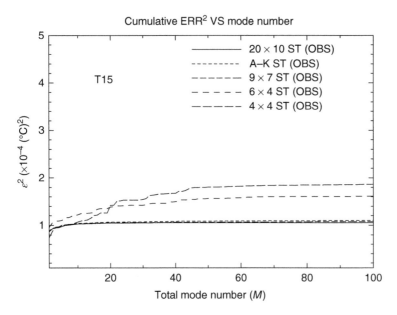

Figure 11.3 The mean squared error for estimation of the global average temperature as a function of the total number of spherical harmonic modes retained. (Shen et al. (1994). © American Meteorological Society. Used with permission.)

---

5 Longer time averages lead to longer autocorrelation lengths until an upper limit is reached: in the middle latitudes this can be over 3000 km. See North et al. (2011).

latter determines the dimension of the matrix problem. Note in the figure that the MSE levels off at a particular value of M for each network configuration. Coarser networks require more modes than dense networks.

To be complete, we must mention that the Shen *et al.* (1994) paper implicitly assumes that there is no power beyond the truncation level of the spherical harmonic expansion in the data. This is not quite true. But given the snugness of the fit in the figures, this might not be a bad approximation when the data are smoothed by time averaging (Figures 11.4 and 11.5).

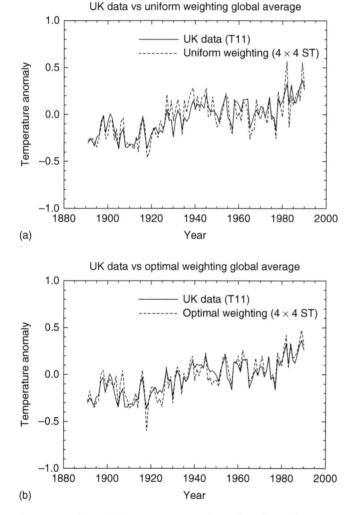

Figure 11.4 Plots of the temperature estimate based on only 16 geometrically symmetrically located gauges (dashed line) together with the best estimate of the data from the UK CRU (Climate Research Unit) data set. The EOFs used in the optical weighting were based on the UK data. (a) Uniform weighting. (b) Optimal weighting. (Shen *et al.* (1994). © American Meteorological Society. Used with permission.)

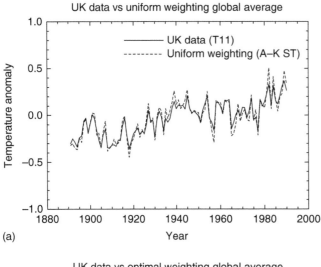

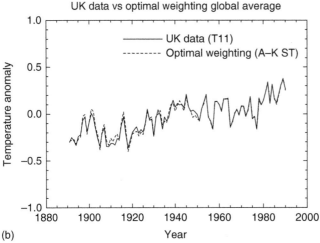

Figure 11.5 Similar to Figure 11.4 except for the 63 well-dispersed Angell–Korshover network. Note that the agreement is nearly perfect for the optimal weighting case. (Shen *et al.* (1994). © American Meteorological Society. Used with permission.)

11.3.2 Relationship with EBMs

So what does this have to do with EBMs? The answer lies in the functional form of the spatial autocorrelation functions in Figure 11.2. We can compare with Figure 11.6, where the annually averaged data are from Siberia. The autocorrelation length in the latter figure is about 50% larger than in Figure 11.2. Both are based on annually averaged data. The reason for the longer autocorrelation length is that the latter figure is over a land mass, while the data from Hansen and Lebedeff are mixed over land and ocean. Ocean correlation lengths for annually averaged data are shown in Figures 9.8 and 9.9. From Figure 11.7, we can see that over land, where $\omega\tau \sim 2\pi/12 \sim 0.5$, the autocorrelation length $r/\lambda \sim 1.8$, whereas over ocean, where $\omega\tau \sim 2\pi \times 5 \sim 30$, and thus $r/\lambda \sim 0.5$.

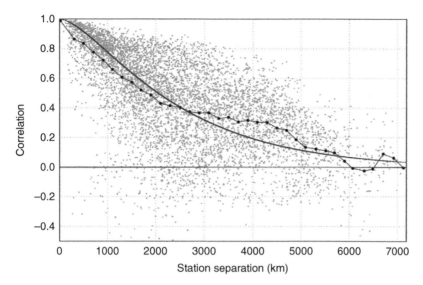

Figure 11.6 Scatter diagram of spatial autocorrelation data from eastern Siberia. The data were annually averaged. The solid curve is based on the well-known model form $rK_1(r)$, where r is the separation of stations in kilometers, $K_1(\cdot)$ is the modified Bessel function of the second kind and of degree unity (see Arfken and Weber, 2005), and the broken line is the average over the sample estimates. (North *et al.* (2011). © American Meteorological Society. Used with permission.)

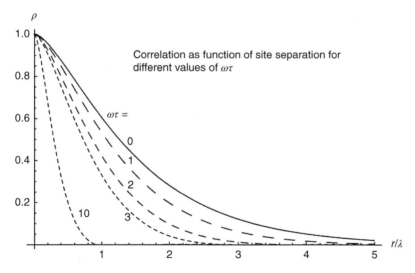

Figure 11.7 Theoretical (noise-forced EBM-based) spatial autocorrelation for different angular frequencies ω. The relaxation time for the surface is τ (typically, a few years for a mixed-layer model) and λ is the characteristic length scale for $\omega = 0$. Over land, the characteristic length ($2\pi/12 \sim 0.5$) is roughly twice the low-frequency limit over ocean ($2\pi \times 5 \sim 30$). (North *et al.* (2011). © American Meteorological Society. Used with permission.)

Hence, the autocorrelation length for all-land areas is large, while that over ocean is smaller, as suggested in both data and models of Figures 9.8 and 9.9.

11.4 Deterministic Signals in the Climate System

The problem of detecting climate signals in the noisy background was first advanced by Hasselmann (1979), but also see Hasselmann (1993, 1997). In this section, we take on the problem of detecting a faint signal in the noisy background of the climate system using a two-dimensional EBM. By signal, we mean a deterministic pattern in space–time—a response to a forced energy imbalance. The noise here is the natural variability as in our noise-forced EBMs of Chapter 9 or the natural variability in a GCM simulation. The statistical model we have in mind is that the data are a linear sum of signal and noise:

$$\mathbf{D}(\hat{\mathbf{r}}, t) = \alpha \mathbf{S}(\hat{\mathbf{r}}, t) + \mathbf{N}(\hat{\mathbf{r}}, t). \tag{11.35}$$

This would be true in the EBM world of a linear-sampled diffusive model driven by stationary random noise. It appears to be true also for large GCMs if $\alpha \mathbf{S}(\hat{\mathbf{r}}, t)$ is small enough. We have included a coefficient α in front of the signal because we usually want to *estimate* the strength of the signal. In many applications, we know the space–time shape of the signal, but we do not know how strong it is. Often, the strength of such a signal depends on feedback factors that are only poorly known. We begin with a single signal and its characterization. To begin our discussion consider a sinusoidal wave in one dimension:

$$S(t) = s^{(+)} \frac{1}{\sqrt{2}} \cos \omega t + s^{(-)} \frac{1}{\sqrt{2}} \sin \omega t. \tag{11.36}$$

The coefficients $s^{(+)}$ and $s^{(-)}$ determine the amplitude and phase of the wave. We can characterize the signal as a vector in a two-dimensional space:

$$\mathbf{S} = s^{(+)} \mathbf{e}_1^{(+)} + s^{(-)} \mathbf{e}_1^{(-)}, \tag{11.37}$$

where $\mathbf{e}_1^{(\pm)}$ are orthogonal unit vectors in the plane. The above result can, of course, be generalized to any number of dimensions, if, for example, the signal is composed of many harmonics. For example,

$$S(t) = s_0 + \sum_{n=1}^{n=N} \left(s_n^{(+)} \frac{1}{\sqrt{2}} \cos n\omega_0 t + s_n^{(-)} \frac{1}{\sqrt{2}} \sin n\omega_0 t \right). \tag{11.38}$$

Then the vector representation of $S(t)$ is

$$\mathbf{S} = \left(s_0; s_1^{(+)}, s_1^{(-)}; s_2^{(+)}, s_2^{(-)} \cdots \right) \tag{11.39}$$

or

$$\mathbf{S} = \sum_{n=0}^{N} \left(s_n^{(+)} \mathbf{e}_n^{(+)} + s_n^{(-)} \mathbf{e}_n^{(-)} \right). \tag{11.40}$$

Hence, if the signal is composed of N harmonics, there will be $2N + 1$ coefficients representing the amplitude and phase of each, with the exception of the zero-frequency harmonic which has no phase.

11.4.1 Signal and Noise

The additive noise can also be decomposed into frequency components (dropping the awkward (+) and (−) superscripts)

$$\mathbf{N} = n_1 \mathbf{e}_1 + n_2 \mathbf{e}_2. \tag{11.41}$$

In this case, the components n_1, n_2 are random variables and uncorrelated. (If they were correlated, we would rotate the axes to new coordinates such that there is no correlation; that is, in climate, we would use the EOF basis set.) For each realization of the process, a new value of n_1 and n_2 must be drawn from a distribution function and the distribution of n_1 and that of n_2 are independent. If the same frequency component of noise is added to that of the signal, we can write

$$\mathbf{D} = (s_1 + n_1)\mathbf{e}_1 + (s_2 + n_2)\mathbf{e}_2, \tag{11.42}$$

where we have used $\mathbf{D}$ to indicate "data." It is worth noting that if the noise process is a stationary time series, the noise from one frequency component to another is uncorrelated. Hence, in this simple case, no rotation is required.

In all the applications that follow, we assume the signals are linearly added to one another and to the natural variability background (the "noise"). We are now in a position to form some estimators of interesting quantities. For example, an unbiased estimator of s_1 is simply $\mathbf{e}_1 \cdot \mathbf{D}$, as $\langle n_1 \rangle = 0$ and thus $\langle \mathbf{e}_1 \cdot \mathbf{D} \rangle = s_1$.

11.4.2 Fingerprint Estimator of Signal Amplitude

A common problem in climatology is that we know the waveform of the signal (in the above example, the frequency and phase) but want to know its strength. In other words, we know the direction of the signal vector (indicated by the unit vector $\mathbf{e}_S \equiv \mathbf{S}/|\mathbf{S}|$). An unbiased estimator of $S \equiv |\mathbf{S}|$ is

$$\widehat{S}_{\text{rf}} = \mathbf{e}_S \cdot \mathbf{D} \tag{11.43}$$
$$= e_{S1} D_1 + e_{S2} D_2 \tag{11.44}$$
$$= \frac{s_1 D_1}{S} + \frac{s_2 D_2}{S}, \tag{11.45}$$

where the subscript "rf" indicates "raw fingerprint." In other words, we find the length of the component of the data vector which lies along the direction of $\mathbf{S}$. The raw fingerprint estimator has an MSE

$$\varepsilon_{\text{rf}}^2 = \frac{s_1^2 \sigma_1^2 + s_2^2 \sigma_2^2}{s_1^2 + s_2^2}. \tag{11.46}$$

The raw fingerprint method is very easy to implement and has attracted some users. On the other hand, it does not take advantage of the fact that σ_1^2 and σ_2^2 might be quite different. Hence, we might want to weigh the information from the two component directions optimally. The way to do this is presented in the next section.

11.4.3 Optimal Weighting

Consider a two-dimensional case in which we do know the direction of the signal vector ($\mathbf{e}_S$) and the angle θ_1 that it makes with the $\mathbf{e}_1$ axis. Then we can write for the component of **S** along the $\mathbf{e}_1$ axis:

$$s_1 = |\mathbf{S}| \cos \theta_1$$
$$= S \cos \theta_1$$
$$= S\mathbf{e}_1 \cdot \mathbf{e}_S, \tag{11.47}$$

or

$$S = \frac{s_1}{\mathbf{e}_1 \cdot \mathbf{e}_S}. \tag{11.48}$$

If there were no noise, we could calculate S by first obtaining its component in the $\mathbf{e}_1$ direction from data, then dividing by the direction cosine of the known signal vector and the $\mathbf{e}_1$ axis. This means we can form an unbiased estimate of S:

$$\widehat{S}^{(1)} = \frac{\mathbf{e}_1 \cdot \mathbf{D}}{\mathbf{e}_1 \cdot \mathbf{e}_S} \tag{11.49}$$

(note that $\langle \widehat{S}^{(1)} \rangle = S$). Hence, the data vector is to be projected along the 1-axis and inversely weighted by the direction cosine of the signal vector to the 1-axis. This unbiased estimator of S has an error variance of

$$\varepsilon_1^2 = \frac{\sigma_1^2}{(\mathbf{e}_1 \cdot \mathbf{e}_S)^2}. \tag{11.50}$$

But we have many statistically independent unbiased estimators of S, one for each component direction. The problem has been reduced to the same one as the thermometers in the reservoir analyzed at the beginning of this section. Hence, the optimal estimator of S is

$$\widehat{S}_{\text{opt}} = \frac{1}{\gamma^2} \sum_{k=1}^{K} \frac{(\mathbf{e}_S \cdot \mathbf{e}_k)(\mathbf{e}_k \cdot \mathbf{D})}{\sigma_{k^2}}, \tag{11.51}$$

$$= \left\{ \frac{1}{\gamma^2} \sum_{k=1}^{K} \frac{(\mathbf{e}_S \cdot \mathbf{e}_k)\mathbf{e}_k}{\sigma_{k^2}} \right\} \cdot \mathbf{D}, \tag{11.52}$$

with

$$\gamma^2 = \sum_{k=1}^{K} \frac{(\mathbf{e}_S \cdot \mathbf{e}_k)^2}{\sigma_{k^2}}. \tag{11.53}$$

In the last expression for $\widehat{S}_{\text{opt}}$, we show the data vector **D** factored out to emphasize that the procedure is a linear operation or projection of the data vector; hence, the term *filter*. Each term in the expression for the filter is an independent estimator for the signal's component along EOF$_k$, and each estimator is inversely proportional to the variance λ_k, as this variance is the eigenvalue of the corresponding EOF.

The matrix form of (11.53) will occur later:

$$\gamma^2 = \mathbf{e}_S \cdot \left\{ \sum_{k=1}^{K} \frac{\mathbf{e}_k \mathbf{e}_k}{\lambda_k} \right\} \cdot \mathbf{e}_S. \tag{11.54}$$

The form γ^2 is a kind of indicator of the signal-to-noise ratio squared. The numerator of each term is the projection the signal onto the EOF ($\mathbf{e}_k$) squared. The denominator is the variance corresponding to the natural variability of that EOF mode. Presumably, this series converges, but one must consider that the denominator will decrease toward zero because the EOFs are ordered by the magnitude of the eigenvalues. On the other hand, the numerator should tend toward zero as well, as the projection of the signals on the EOFs should diminish as the mode indices increase. The convergence will depend on the problem being investigated. In the case of signal detection in the climate system, we will see that the convergence is satisfactory.

Let us recall a few key assumptions. First and foremost, we assumed the linear additivity of the signal and the noise. This is likely to hold for weak signals that we expect in climate change problems. We have used the principal component directions of the natural variability to formulate the problem from the beginning; that is, we chose the coordinate axes to be the principal axes of the covariance ellipsoid of the noise vector. We had to assume knowledge of the direction of the signal waveform, and this had to be based on a model estimate itself. Our job is to estimate its strength given this information.

The quantity γ^2 is an *a priori* measure of the quality of the procedure, as for a signal strength of unity, the signal-to-noise ratio is squared. We can use γ^2 as computed with models to tell which vector components are most important in the estimation problem without really invoking the data. This is very important as we can use our climate models to condition our choice of the subspace within which we can make a reliable estimation of signal strength without involving the data (cheating).

Consider the error involved in the use of imperfect models in constructing the filter. The first type of error is in choosing an incorrect fingerprint. In the present context, this means the vector $\mathbf{e}_S$ has the wrong direction in the state space. An equivalent statement is that the direction cosines $\mathbf{e}_S \cdot \mathbf{e}_i$ are incorrect. The single constraint is that the squares of the direction cosines must add up to unity. An incorrect fingerprint can lead to a bias in the estimation of the signal strength. For this reason, it is well to find ways to eliminate aspects of the model-predicted signal which may lead to incorrect signal waveform prediction. This could be done by eliminating certain subspaces of the state space, but this is probably not a good approach as the EOFs are very irregular functions over the globe and it is not easy to relate these shapes to the areas that we know are weak in signal generation. Instead, it might be better to mask off certain regions on the globe, such as the polar regions where we know the models perform poorly. Once we have masked off certain areas (with tapered edges), we completely redo the problem including the EOFs on the newly masked planet. We do not pursue this possibility further in this book.

Another type of error comes from the optimal weights as generated from models. This type of error is less egregious than error in the signal waveform. Since the estimator is composed of K independent estimators which are assumed to be unbiased, the weighting does not introduce a bias. If erroneous, they can lead to a suboptimal estimator. In addition, they can lead to an underestimation of the theoretical MSE (γ^2). It turns out that as the minimum in the MSE as a function of the weights is the minimum of a multidimensional parabolic surface (actually, the intersection of this parabolic surface with the plane $\sum_i w_i = 1$), the MSE is not sensitive to the exact choice of the weights ($\Delta \varepsilon^2 \sim (\Delta W)^2$).

11.4.4 Interfering Signals

Following North and Wu (2001), four signals have been identified for climate signal detection. These are **A**, the cooling due to atmospheric aerosols; **G**, the greenhouse warming signal; **V**, the volcanic dust veil episodic cooling; and **S**, the solar cycle. Consider the case of two signals. If the unit vectors describing them are not orthogonal, we have to do some additional filtering. Suppose the signals **G** (the greenhouse gas signal) and **A** (the aerosol particle signal) are turned off. We want estimates of the amplitudes of **S** (the solar change signal) and **V** (the volcanic dust veil signal). Let us start with **S**. If the direction of **S** and interfering signal **V** (i.e., their space–time patterns or fingerprints) are known, we can obtain independent estimates of their strengths by estimating the components of each which are perpendicular to the other. For example, consider the component of **S** which is perpendicular to **V**:

$$\mathbf{S}_{\perp V} = (\mathbf{1} - \mathbf{e}_V \mathbf{e}_V) \cdot \mathbf{S}, \tag{11.55}$$
$$= \mathbf{S} - \mathbf{e}_V (\mathbf{e}_V \cdot \mathbf{S}), \tag{11.56}$$

where $\mathbf{e}_V \equiv \mathbf{V}/|\mathbf{V}|$ is a unit vector along **V**. Hence, using this projection procedure (operator), we can now proceed to estimate the strength of $\mathbf{S}_{\perp V}$ and therefore find the strength of **S**, as $S = |\mathbf{S}_{\perp V}|/\sqrt{1 - (\mathbf{e}_S \cdot \mathbf{e}_V)^2}$. The problem, of course, is that $\mathbf{S}_{\perp V}$ will be shorter than **S** with a corresponding loss of performance (signal-to-noise ratio $= S \cdot \mathbf{e}_V$) in the procedure.

We can now use the same procedure to find the length of $\mathbf{V}_{\perp S}$ and therefore the length of **V**. As a consistency check, we could then proceed to look at the parallel components if each signal, which, in principle, are now known.

$$\mathbf{S}_{\parallel V} = \mathbf{e}_V \cdot \mathbf{S}, \tag{11.57}$$
$$\mathbf{V}_{\parallel S} = \mathbf{e}_S \cdot \mathbf{V}. \tag{11.58}$$

It is of interest to know the angle between **S** and **V**,

$$\theta_{S,V} = \arccos(\mathbf{e}_S \cdot \mathbf{e}_V). \tag{11.59}$$

If the two signals are orthogonal to one another, there is no interference. If there is a significant alignment or anti-alignment of the two signals, there will be trouble discriminating between them. This condition is known in multiple regression as *collinearity*. If the length and direction of the interfering signals are both known, we have an unbiased estimator of the length of $\mathbf{S}_{\perp V}$, which when divided by $\sqrt{1 - (\mathbf{e}_S \cdot \mathbf{e}_V)^2}$ becomes an unbiased estimator of S. We can optimally combine this with the independent estimate based upon the parallel component which can be found by first subtracting (the known) $\mathbf{V}_{\parallel S}$ from the data stream.

Some interesting examples of the angles between signal vectors are given in North and Stevens (1998) for a narrow band of eight discrete frequencies centered at a period of one decade. In Table 11.1, we see that most of the combinations of the four signals (in the narrow frequency band used by North and Stevens) the vectors are reasonably perpendicular except for **G** and **A**. This latter is hardly a surprise as these two vectors clearly are nearly anti-collinear expressions of linear global warming from the greenhouse effect and a similar linear cooling effect due to aerosols.

Table 11.1 Angles between possible pairs of signal vectors.

Vector pair	Angle (°)
S · G	77.9
S · V	88.0
S · A	101.0
V · A	84.2
V · G	93.8
G · A	153.3

North and Stevens (1998).

11.4.5 All Four Signals Simultaneously

We can cast the problem in the following form:

$$T_m^{\text{data}} = \sum_{s=1}^{4} \alpha_s S_m^{(s)} + N_m, \tag{11.60}$$

where the subscript m is an index running over all space–time points in the record. For instance, in the published papers North and Stevens (1998), and North and Wu (2001), the number may be 100 years (of annual averages) × 36 sites (see Figure 11.8). North and Wu (2001), used several different space–time combinations.

The problem has been discretized for $i = 1, 2, \ldots, 36$ stations, and $j = 1, \ldots, 100$ time steps (see Figure 11.8). We have introduced the notation $S^{(s)}(\hat{\mathbf{r}}_i, t_j)$, $s = 1, 2, 3, 4$ for the four signals. The α_s are the four unknown coefficients that are to be estimated from the data stream. $N(\hat{\mathbf{r}}_i, t_j)$ is a Gaussian random field denoting the so-called natural variability. In order to build a set of statistically independent estimators of the α_s, we use space–time EOFs (we will refer to them as EOFs from here instead of Karhunen–Loéve functions). These are the eigenvectors of the space–time 3600 × 3600 covariance matrix:

$$\mathbf{K}_{mm'} = \langle N_m N_{m'} \rangle. \tag{11.61}$$

The angular brackets here imply an infinite-member ensemble average. Since we obtain these EOF basis vectors from very long runs of GCMs (or stochastic EBMs), we can assume the sampling errors in taking these averages are negligible. The eigenvector problem is posed as follows:

$$\sum_{m'} \mathbf{K}_{mm'} \psi_{m'}^{(k)} = \lambda_k \psi_m^{(k)}, \tag{11.62}$$

where $\psi_m^{(k)}$ is the kth eigenvector and λ_k is the corresponding (positive and real) eigenvalue. In what follows, we assume the $\psi_m^{(k)}$ and the λ_k are not random numbers because of the large number of realizations in determining them. The first step is to expand all

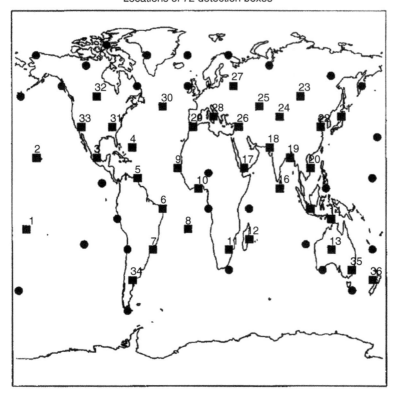

Locations of 72 detection boxes

Figure 11.8 Black squares showing the 36 stations used by North and Stevens (1998). Each of the 36 10° × 10° detection boxes comprised of four 5° × 5° boxes from the Climate Research Unit (UK) data set, each of which has 1200 months of data (1894–1993). These boxes were chosen based on where there was sufficient data, spatial sampling was maximized, and correlation between boxes was minimized. The sites designated by black disks were added by North and Wu (2001). These latter each contain 50 years of data. (North and Wu (2001). © American Meteorological Society. Used with permission.)

quantities in (11.60) into the eigenvectors.

$$N_m = \sum_k \tilde{N}^{(k)} \psi_m^{(k)}; \quad \tilde{N}^{(k)} = \sum_m N_m \psi_m^{(k)}, \tag{11.63}$$

$$S_m^{(s)} = \sum_k \tilde{S}_s^{(k)} \psi_m^{(k)}; \quad \tilde{S}_s^{(k)} = \sum_m S_m^{(s)} \psi_m^{(k)}, \tag{11.64}$$

$$T_m^{\text{data}} = \sum_k \tilde{T}_k^{\text{data}} \psi_m^{(k)}; \quad \tilde{T}_k^{\text{data}} = \sum_m T_m^{\text{data}} \psi_m^{(k)}. \tag{11.65}$$

To summarize, in what follows, the $S^{(k)}, S_m^{(s)}, \psi_m^{(k)}$, and λ_k are not random variables. Because of sampling error in the actual data record, the quantities $N_m, \tilde{N}^{(k)}, T_m^{\text{data}}, \tilde{T}_k^{\text{data}}$ are random variables. The $\tilde{N}_k^{(k)}$ are zero mean, normally distributed variates representing natural climate variability with the property

$$\langle \tilde{N}^{(k)} \tilde{N}^{(k')} \rangle = \lambda_k \delta_{k,k'}, \tag{11.66}$$

which means that when referred to the EOF basis set, the $\tilde{N}^{(k)}$ are uncorrelated from one component to another. After multiplying (11.60) through with $\psi_m^{(k)}$ and summing over m we arrive at

$$\tilde{T}_k^{\text{data}} = \sum_{s=1}^{4} \alpha_s \tilde{S}_s^{(k)} + \tilde{N}^{(k)}. \tag{11.67}$$

The last equation indicates that the equations for $k = 1, \ldots$ are statistically independent of one another. We would like to make estimates of the strength coefficients as a function of the number of EOFs retained, K. We can make this into a standard regression model by first normalizing the errors to white noise:

$$\epsilon = \frac{\tilde{N}^{(k)}}{\sqrt{\lambda_k}} = \check{N} = \check{T}_k^{\text{data}} - \sum_{s=1}^{4} \alpha_s \check{S}_s^{(k)}, \tag{11.68}$$

where the $\check{}$ implies that the variable is divided by $\sqrt{\lambda_k}$. Now we form the MSE and minimize it with respect to α_s. What we understand by "mean" here is

$$\langle (\cdot)_k \rangle_K = \frac{1}{K} \sum_{k=1}^{K} (\cdot)_k; \tag{11.69}$$

$$\sum_{k=1}^{K} \mathcal{M}_{ss'} \alpha_{s'}^{(K)} = \langle \check{S}_s^{(k)} \check{T}_k^{\text{data}} \rangle_K, \tag{11.70}$$

with

$$\mathcal{M}_{ss'} \langle \frac{\tilde{S}_s^{(k)} \tilde{S}_{s'}^{(k)}}{\lambda_k} \rangle K = \frac{1}{K} \sum_{k=1}^{K} \frac{\tilde{S}_s^{(k)} \tilde{S}_{s'}^{(k)}}{\lambda_k}. \tag{11.71}$$

Now we can invert the matrix to obtain our estimator:

$$\hat{\alpha}_s^{(K)} = \sum_{s'=1}^{4} (\mathcal{M}^{-1})_{ss'} \langle \check{S}_{s'}^{(k)} \check{T}_k^{\text{data}} \rangle_K. \tag{11.72}$$

It is important to realize that K has to be larger than or equal to 4, otherwise the matrix $\mathcal{M}$ will not have an inverse. As expressed in the conventional notation of unnormalized variables,

$$\hat{\alpha}_s^{(K)} = \sum_{s'=1}^{4} (\mathcal{M}^{-1})_{ss'} \sum_{k=1}^{K} \frac{\tilde{S}_{s'}^{(k)} \tilde{T}_k^{\text{data}}}{\lambda_k}. \tag{11.73}$$

This last is our optimal estimator of the four signal strengths. The above derivation is equivalent to a multiple regression model with four unknown coefficients. The value of our approach or decomposition is that it provides the solution as a function of the number of EOFs retained in the analysis as we will see later in the numerical example from North and Wu (2001). If the estimates are stable as the value of K is increased, we have more confidence in the procedure. This might not always be the case as some of the series leading up to K terms have λ_k in the denominator. This is a problem because the eigenvectors (EOFs) and their eigenvalues are traditionally arranged in descending order as a function of k. Hence, λ_k is likely to approach zero as $k \to \infty$. This is always a problem in optimal estimation as can be seen in the simple case of K thermometers.

11.4.6 EBM-Generated Signals

A persistent problem in detection and attribution problems is to accurately characterize the signals. In GCM studies, one can run many realizations of perturbed runs with only a single forcing applied, then average across the realizations to obtain the signal fingerprint in space–time. Presumably, the natural variability cancels out and one is left with the bare signal. In linear EBM studies, one can simply turn off the noise forcing and the signal will be evident (Figure 11.9). All of the four forcings can be superimposed because the problem is taken to be linear with no time-dependent coefficients (Figure 11.9d). This is the method used by Stevens and North (1996), North and Stevens (1998), and North and Wu (2001). The natural variability statistics can be gathered from long control runs from a GCM (usually a 1000 years or so) or EBM (Stevens calculated EOFs for a 10 000 year control run).[6]

Before proceeding, it is useful to show how the North and Wu (2001) signals compare with four realizations of a GCM (HadCM2) (dotted line in Figures 11.10–11.12) of the same era and observational data (light solid line in the same three figures) which include natural variability. In the same figures, the heavy solid line represents the evaluation of the EBM greenhouse signal (reduced by the factor $\alpha_G = 0.65$ to conform with our results shown in this section). Each box in the three figures represents one of the 36 boxes used

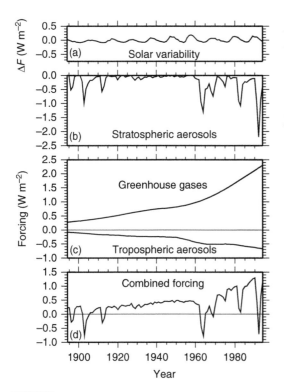

Figure 11.9 The four global average forcings. The abscissa is time in years. The ordinate for each panel is watts per meter square. (a) The solar cycle signal. (b) The stratospheric aerosol remaining in the stratosphere following volcanic eruptions. (c) The greenhouse and tropospheric aerosol forcing. (d) The sum of all four forcings globally averaged. (North and Stevens (1998). © American Meteorological Society. Used with permission.)

---

6 There is a huge literature on signal detection and attribution using GCMs for signals as well as natural variability. Each of the recent Reports of the International Panel on Climate Change (IPCC), *The Physical Science Basis* (Working Group I WGI) reports has a chapter on detection and attribution with many references to the recent literature. Copies of such reports are available at: http://ipcc.ch.

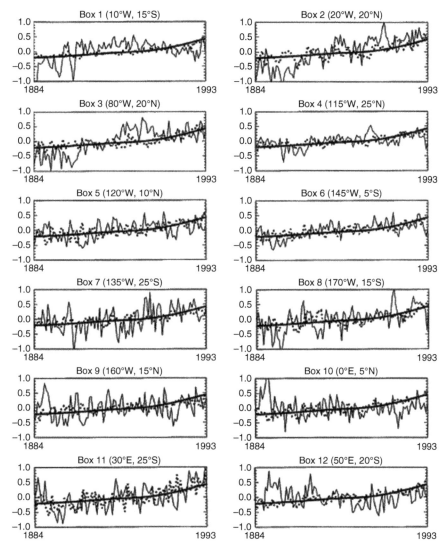

Figure 11.10 Each panel shows modeled and observed time series from a different observational site as indicated in Figure 11.8. The greenhouse gas signal from the EBCM (thick solid line) has been multiplied by 0.65 (in conformity with our detection results). The dotted line is an average across a four-member ensemble of HadCM2 forced by greenhouse gases (also multiplied by 0.65 to conform with our detection results). Observational data from Jones are shown by the thin solid line. (North and Wu (2001). © American Meteorological Society. Used with permission.)

in the analysis (Figure 11.8). Note the natural variability about the heavy black curve, indicating natural variability (see caption in Fig. 11.10). Also note the agreement of the (scaled) shape of the EBM curve versus the HadCM2 curve. Note also that if the average over the four realizations is taken to be the signal used in a detection study, there will still be a fair amount of noise in the signal pattern. Being satisfied with the space–time

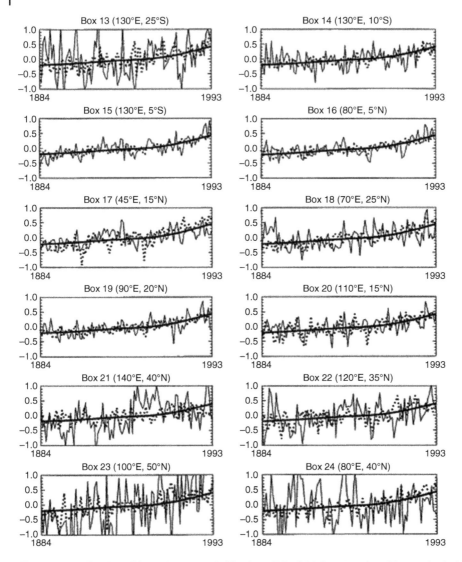

Figure 11.11 Continued from Figure 11.10. (North and Wu (2001). © American Meteorological Society. Used with permission.)

pattern as shown in the three figures, in what follows, we will use the EBM to generate the four signals.

The signals used in the North and Wu (2001) paper were taken from earlier work in a dissertation by Stevens (1997). We show here a few figures which illustrate the time dependence of the signals. Figure 11.9 shows the global average signal (actually the average over the 36 boxes shown as black squares in Figure 11.8. The topmost panel shows the faint solar signal, the next lower is the stratospheric aerosol signal from volcanic particles left after eruptions. The sharp dips going backward in time are Mt Pinatubo in 1992, then El Chichon, then Mt Agung. The next lower panel shows both the greenhouse

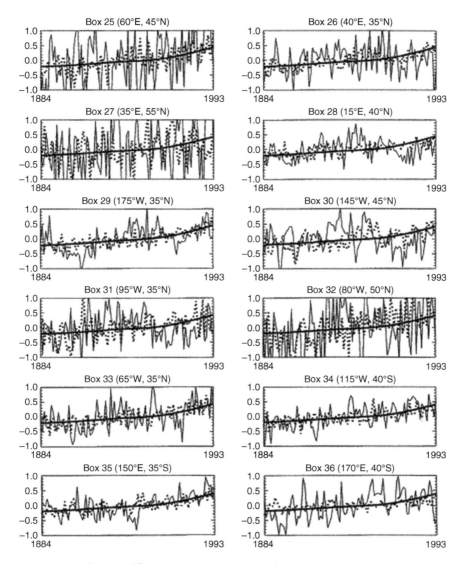

Figure 11.12 Continued from Figure 11.11. (North and Wu (2001). © American Meteorological Society. Used with permission.)

gas and tropospheric aerosol signals. Note the anti-collinearity of these two global signals, making it very difficult to discriminate between them in a detection scheme. The lowermost panel shows the time dependence of the sum of all four signals.

Figure 11.13 shows the same 36 stations. In this figure, signals are band-pass filtered in what Stevens called the "solar band," a frequency band straddling the frequency 1/10 year^{-1}. The squares of the real and imaginary parts of the Fourier frequency components are shown as columns above the stations. The amplitude is the square root of the sum of these two parts (site by site). Note that both of the imaginary parts are dominant over the real parts. This simply means that the phase lag is nearly $\pi/2$, as expected from a gradual temporal increase in the signal. The important thing about this figure is that the

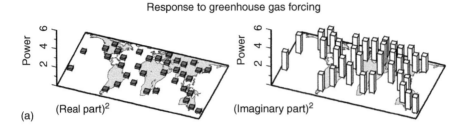

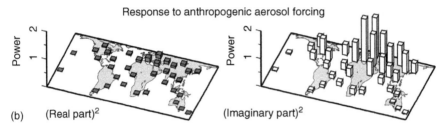

Figure 11.13 The real and imaginary parts squared for the Fourier component of the greenhouse gas forcing (a) and the tropospheric aerosol forcing (b). The Fourier frequency component is at a period of one decade. The imaginary part dominates in both upper and lower panels, suggesting that the phase lag is $\pi/2$. The important point is that there is considerable asymmetry between the two hemispheres, suggesting a good possibility of discriminating between the two signals. (North and Stevens (1998). © American Meteorological Society. Used with permission.)

two panels show a strong asymmetry between the hemispheres. This asymmetry should assert some discriminating power between the two signals and help us to distinguish one from the other in our detection process.

11.4.7 Characterizing Natural Variability

When Stevens selected the 36 boxes of Figure 11.8, he tried to space them so that there was as little correlation as possible. If there were no correlation, the 36-dimensional vector with unity in position one and zeroes elsewhere would be the first normalized EOF, and so on. This suggests that only a small rotation of the natural variability field will be necessary to generate the EOFs. It might also mean that the procedure will not be very sensitive to our choice of model-generated fields to use in our study. Nevertheless, North and Wu (2001) decided to use not only the EBM-generated EOFs but also to use several (one 1000 year run from the Max Planck Institute (ECHAM1/LSG), two 1000 year runs from different GFDL models, and one 1000 year run from the Hadley Centre (HadCM2)) GCM-generated sets. We could then compare them to see if there is much difference. Recall that even if we do not use the *best* set, we do not bias the result, but our estimate might be slightly suboptimal with respect to error variance with our estimators. That comparison will be indicated in the figures to follow in this chapter.

With four signals, it is best to recognize that the problem is equivalent to multiple regression for the signal amplitudes. The filter formalism we have used so far gets more complicated because we must make sure that for each of the four signals, the other three have no component parallel to that which is passed through and optimally weighted. On the basis of standard multiple regression analysis, the optimal estimator for a particular

$\hat{\alpha}_s$ is given by

$$\hat{\alpha}_s = \sum_{s'} \left[(S^T \cdot \mathbf{W} \cdot S)^{-1} \right]_{ss'} S^T \mathbf{W} T^{\text{data}}, \tag{11.74}$$

where $s, s' = (G, A, V, S)$. In (11.74), the quantity S is a diagonal matrix with diagonal entries given by the components of the signal vector. The matrix $\mathbf{W}_{mn} = \lambda_n^{-1} \delta_{mn}$ is the inverse space–time lagged covariance matrix of the natural variability in its EOF or diagonal form, $\mathbf{W} = \mathbf{K}^{-1}$, where $\mathbf{K}$ is the covariance matrix. It forms a metric tensor in space (Hasselmann, 1993; also see the Appendix of North and Wu, 2001).

The formalism leads to a similar expression to that of the single-signal case:

$$\gamma_{ss'}^2 = \sum_n \frac{S_{sn} S_{s'n}}{\lambda_n}. \tag{11.75}$$

The matrix $\boldsymbol{\Gamma}$ can be formed as the array of the $\gamma_{ss'}^2$,

$$\boldsymbol{\Gamma}_{ss'} = \gamma_{ss'}^2. \tag{11.76}$$

Then the covariance matrix of the estimators $\hat{\alpha}_s$ and $\hat{\alpha}_{s'}$ is just

$$\text{cov}(\hat{\alpha}_s, \hat{\alpha}_{s'}) = (\boldsymbol{\Gamma}^{-1})_{ss'}. \tag{11.77}$$

11.4.8 Detection Results

The results of the North and Wu (2001) paper are compactly summarized in chart form in Figure 11.14. This graphic shows the results for a total of five experiments, the asterisk indicating that the estimate is based on 20 tropical stations with 100 year records. The × symbol indicates the 36 stations and 100 years of data as in the previous figure. The ◊ symbol indicates that the experiment was for 43 stations—20 with 100 years record and 23 with only 50 years; the △ symbol indicates the experiment was conducted with 72 stations—36 with 50 years record and 36 with 100 years records; ○ is based on 72 stations with 50 years of data (1944–1993). The error bars in the figure represent a 90% confidence region. If an error bar reaches below the dotted line (zero) the corresponding α coefficient is not significant at the 90% level. The clusters labeled GFCLc, GFDLml, EBCM, ECHAM1/LSG, and HadCM2 are used to indicate that these experiments were conducted with the EOFs generated from long (usually 1000 years, but 10 000 years for the EBM).

We can examine each row individually to sort out the main features. The top row representing the detection of the solar cycle shows quite a few dips below the dotted line, indicating that in many experiments it was not significantly different from zero (i.e., the zero-amplitude hypothesis could not be rejected). As we scan the different natural variability choices, we see there is rather good consistency; further, it is the rightmost experiment with only 50 years of data that is most unstable, which is hardly surprising because temporally, less data have been included (less than five solar cycles). The greenhouse gas (second) row shows very tight error bars and the estimates of the G-signal strength seem very robust across the different experiments and across the different choices of natural variability. The volcanic signal (third row) shows wider error bars but with unanimous statistical significance. There are not that many volcanic events in the record. Finally, the fourth row showing the aerosol signal strength shows many overlaps

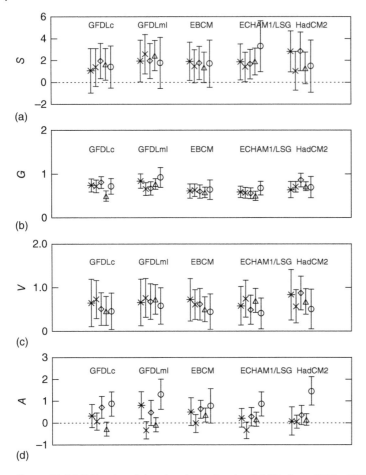

Figure 11.14 Estimates of signals using natural variability from GFDLc, GFDLml, EBM, MPI, and HadCM2, and using the EBM-generated signals for G, V, A, and S. This graphic also shows the results for a total of five experiments. The asterisk indicates that the estimate is based on 20 tropical stations with 100 year records. The × symbol indicates that 36 stations and 100 years of data as in the previous figure. The ◊ symbol indicates that the experiment was for 43 stations—20 with 100 year records and 23 with only 50 year records; the △ symbol indicates the experiment was conducted with 72 stations—36 with 50 year records and 36 with 100 year records; ○ is based on 72 stations with 50 years of data (1944–1993). (Figure from North and Wu (2001). (©Amer. Meteorol. Soc., with permission).)

with the zero line and very unstable results across all possible configurations. Each estimate seems to have very precise error bars, but there is strong dependence on all factors. We would have to conclude that no aerosol signal has been detected.

Next consider the ellipses in Figure 11.15 which represent the different elements of $\Gamma_{ss'}$, the covariance of the estimators for signals s and s'. Look first at the upper left corner. The collinearity of G and A are quite evident across all of the five ellipses in the box. An error on the small side of G is correlated with a similar small estimate of A. Note that all the ellipses intersect $A = 0$, meaning that it fails the significance test. Other figures in the diagram can be interpreted in a similar way.

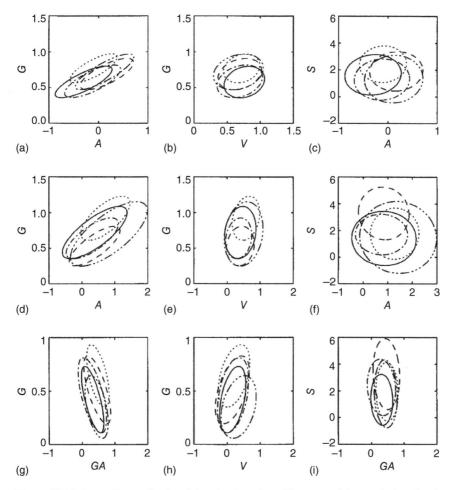

Figure 11.15 Error ellipses of pairs of signals, given five different model prescriptions for the natural variability: GFDLc (solid line); GFDLml (dotted line); MPI (dashed line); EBM (dashed-dotted line); HadCM2 (dashed-dotted-dotted line). Here (a), (b), and (c) are EBCM signals for 72 stations, 36 with 100 years of data, 36 with 50 years of data; (d)–(f) are EBM signals for 72 global sites all with 50 years of data; (g)–(i) are HadCM2 G and GA and EBM V and S signals for 72 global sites all with 50 years of data. (Figure from North and Wu (2001). (© Amer. Meteorol. Soc., with permission).)

11.4.8.1 Convergence

In this section, we examine the convergence of the estimation process. Figures 11.16 and 11.17 show the convergence results. The abscissa shows the number of space–time EOF modes included in the partial sum up to that many terms. The EOFs are arranged in order of descending variance (eigenvalue). We can see from (1a), (1b), (1c), and (1d) that all of the series converge. We already know that S and A are unstable statistically. Panels (1e) and (4e) show this by the irregularity of the amplitude estimate $\alpha_s; s = S, G$ as a function of EOF number. The estimate of α_A even drops below zero. On the other hand, the estimates of α_G and α_V are quite stable, both converging consistently to a value around 0.60.

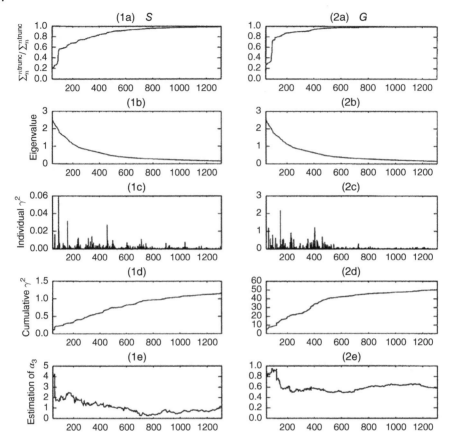

Figure 11.16 (1a)–(1e) For solar cycle (S), (2a)–(2e) for greenhouse gas (G), (3a)–(3e) for volcanic (V), and (4a)–(4e) for aerosol (A) in the case of the 36 global stations over 100 years. The space–time EOF modes are arranged in order of descending variance (EOFs from 10,000 year EBM control run). (1a)–(4a) The normalized cumulative fraction of variance of the signal, $\sum_n^{n_{trunc}} S_n^2$ with $\sum_n^{n_{all}} S_n^2 = 1$. (1b)–(4b) indicate the eigenvalue of each spatial–temporal mode; (1c)–(4c) indicate the contributions to SNR$^2 = \gamma_n^2 = S_{sn}^2$ from the individual EOF modes; (1d)–(4d) indicate the cumulative $\gamma^2 = \sum_n^{n_{trunc}} \gamma_n^2$. (1e)–(4e) The cumulative estimate of α including EOFs up to n_{trunc}. (Figure from North and Wu (2001). (©Amer. Meteorol. Soc., with permission).)

11.4.9 Discussion of the Detection Results

We suggest two reasons that the North and Wu (2001) study yields a somewhat lower estimate than expected amplitude for G and V as well as the near-zero amplitude for A. The Appendices of North and Wu (2001) contain a number of interesting tests of the detection program. For example, Figure 11.18 shows results of a Monte Carlo experiment using the EOFs from the 10,000 year run of the EBM for the natural variability (EOFs) and all four signals are included with $\alpha_s = 1$; $s = G, V, A, S$. Then 200 of the EBM 50 year runs are used as "data" inserted into the 72 data sites. In the figure, the error ellipses for 90% confidence are drawn along with the individual points representing each run. Note that each ellipse has (1, 1) at its center. In the left panel, the correlation of G and A is evident. In the right panel the orthogonality is expressed as virtually no correlation between the errors in V and G.

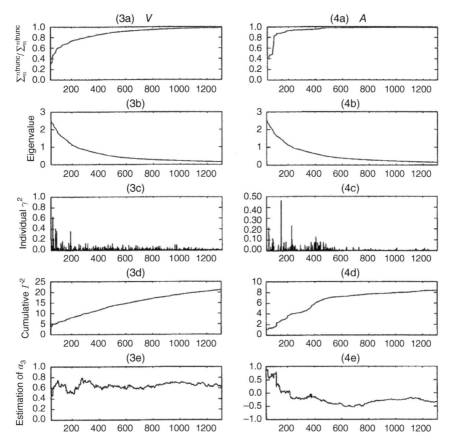

Figure 11.17 Continued from the previous figure. (Figure from North and Wu (2001). (©Amer. Meteorol. Soc., with permission).)

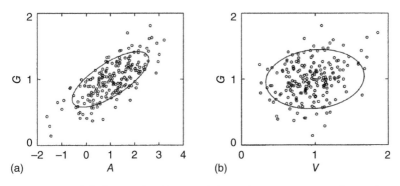

Figure 11.18 (left) Scatter plot of Monte Carlo studies and 90% error ellipse of detection studies for the pair of signal G–A for 72 boxes all with 50-yr (1944–93) observational data. In Monte Carlo studies, the artificial data is constructed by adding 200 50-yr EBCM control run and four EBCM signals S, G, V, and A. The truncated eigenmode is 500 in the 10 k yr control run of EBCM. (right) Same as (a) except for pair of signal G–V. (Figure from North and Wu (2001). (©Amer. Meteorol. Soc., with permission).)

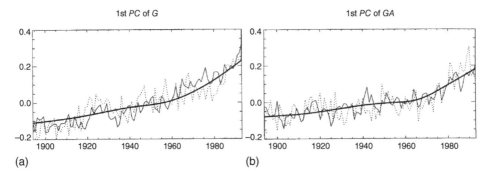

Figure 11.19 (a) First principal component time series of annual mean climate change signal for greenhouse-gas-only forcing from EBCM (heavy solid line), HadCM2 GCM (light solid line), and ECHAM4 GCM (dotted line); (b) same as (a) except for greenhouse-gas-plus-aerosol forcing. (Figure from North and Wu (2001). (©Amer. Meteorol. Soc., with permission).)

Another interesting test is shown in Figure 11.19. Here, the comparison of the time dependence of G with that of $G + A$ (V and S are omitted in the experiment), where the solid line is the EBM signal, the light solid line indicates the GCM HadCM2, and the dotted line represents the GCM ECHAM4. The EBM signal is very close to those of the two GCMs. One can notice a very slight difference between the two signals (panels (a) and (b)) with the aerosols making a slight bend in the $G + A$ curve. The statistical collinearity is evident in the global average curves. The only difference that can be used for discrimination between them must be in the inter-hemispheric difference (presumably there is higher A in the NH. Most GCM simulations have indicated a larger value of A that cancels part of a larger G. This latter would mean that the data point would lie in the upper right end of the ellipses in Figure 11.15a,d. In this case, the "equilibrium sensitivity" would be greater than the nominal 2.3 K that we assume for the EBM used in this book. We cannot rule out this case as it would lie within the 90% confidence area of those two panels of Figure 11.15.

Another possible reason for the discrepancy between the EBM detection study and that of many IPCC GCMs is that the EBM uses only a mixed-layer ocean in its EBM-generated signal. Compared to a deeper ocean coupling, this would make the EBM signal larger than the one which might have been generated from a coupled ocean–atmosphere GCM. The latter would hold down the signal from such a coupled model as we discussed in Chapter 10.

A final possible criticism of the North and Wu (2001) study lies in the fact that so many EOFs are used. Using very large numbers of EOFs often raises a red flag in statistical studies because their eigenvalues are necessarily close together and this means that they will "mix" (if the eigenvalues were close together, a linear combination of the associated eigenvectors is also an eigenvector). But this criticism is false. The (sample) PC time series associated with a particular EOF are orthogonal by construction and the EOFs form a good basis set.

Notes for Further Reading

Surface temperature data sets can be found at the following websites. In many cases, there are descriptions of how the data are processed and how the estimation procedure works:

1. The NASA/Goddard Institute for Space Studies (GISS) provides a very accessible digital as well as graphical (including maps) surface temperature data. http://data.giss.nasa.gov/gistemp/.
2. The Climate Research Unit (CRU) of the University of East Anglia provides both digital and colorful graphical information. They also provide a list of publications by Dr. Philip D. Jones and his colleagues. http://www.cru.uea.ac.uk/cru/data/temperature/.
3. The NOAA/National Climate Data Center (NCDC) website is a little more difficult to navigate, but it has the data. http://www.ncdc.noaa.gov/temp-and-precip/global-maps/.
4. NOAA also publishes online the pdfs of annual issues of the *Bulletin of the American Meteorological Society* on the *State of the Climate* for each year starting in 1991; there is also a decadal review covering 1981–1990. http://www.ncdc.noaa.gov/bams/past-reports.

Exercises

11.1 (a) Consider two measurements of surface temperature at a location. They are described as

$$\widehat{T}_1 = T + \varepsilon_1 \text{ and } \widehat{T}_2 = T + \varepsilon_2,$$

where T is the true temperature and ε_1 and ε_1 are measurement errors. Assume that measurements are unbiased, that is, $\langle \varepsilon_1 \rangle = \langle \varepsilon_2 \rangle = 0$, and error variance is given by $\langle \varepsilon_1^2 \rangle = \sigma_1^2$ and $\langle \varepsilon_2^2 \rangle = \sigma_2^2$. Further assume that the errors in the two measurements are uncorrelated, that is, $\langle \varepsilon_1 \varepsilon_2 \rangle = 0$. Show that an unbiased optimal estimator with the least MSE is given by

$$\widehat{T}_{opt} = \frac{1}{\sigma_1^2 + \sigma_2^2}(\sigma_2^2 \widehat{T}_1 + \sigma_1^2 \widehat{T}_2).$$

(b) Show that an unbiased optimal estimator for three independent unbiased measurements is given by

$$\widehat{T} = \frac{1}{\eta^2} \sum_{i=1}^{3} \frac{\widehat{T}_i}{\sigma_i^2}, \quad \eta^2 = \sum_{i=1}^{3} \frac{1}{\sigma_i^2}.$$

(c) For N independent unbiased measurements, the optimal unbiased estimator is obtained by minimizing the so-called error functional defined by

$$E^2 = \langle \varepsilon^2 \rangle - \Lambda \left(\sum_{i=1}^{N} \alpha_i - 1 \right),$$

where Λ is called a *Lagrange multiplier*. Show that the optimal unbiased estimator is given by

$$\widehat{T} = \frac{1}{\eta^2} \sum_{i=1}^{N} \frac{\widehat{T}_i}{\sigma_i^2}, \quad \eta^2 = \sum_{i=1}^{N} \frac{1}{\sigma_i^2}.$$

(d) Show that the error variance of the optimal estimator is given by

$$\langle \varepsilon^2 \rangle = \frac{1}{\eta^2}.$$

11.2 Consider two measurements, $(\widehat{T}_1, \widehat{T}_2)$, of which the errors are correlated. Assume that error covariance matrix is given by

$$\mathbf{C} = \{C_{ij}\} = \begin{pmatrix} 1.0 & 0.5 \\ 0.5 & 2.0 \end{pmatrix}.$$

(a) Find the principal axes for which two new measurements are uncorrelated.
(b) Show that the two measurements rotated according to the eigenvectors become uncorrelated.
(c) Determine the optimal unbiased estimator for $\widehat{T}$ based on the two measurements $(\widehat{T}_1, \widehat{T}_2)$.

11.3 As discussed in Exercise 11.2, two or more measurements on the surface of the Earth are typically correlated. Let us consider N measurements of length L of a variable $T(r,t)$ (say, temperature), $r = r_1, r_2, \ldots, r_N, t = 1, 2, \ldots, L$. Then, the covariance matrix of the N measurements is given by

$$\mathbf{C} = \{C_{ij} | i,j = 1, 2, \ldots, N\} = \{\langle T(r_i, t) T(r_j, t) \rangle | i,j = 1, 2, \ldots, N\}.$$

The eigenvalues and eigenfunctions of are determined by solving the Karhunen–Loève equation

$$C_{ij} \cdot \phi_j^{(n)} = \lambda_n \phi_i^{(n)},$$

where $\{\phi_i^{(n)} | i = 1, 2, \ldots, N\}$ is the nth eigenvector with corresponding eigenvalue λ_n. Then, the N measurements can be written as a unique linear combination of eigenvectors as

$$T(r,t) = \sum_n P_n(t) \phi_n(r), \quad \left\{ \phi_i^{(n)} | i = 1, 2, \ldots, N \right\}.$$

This procedure is called the *EOF* analysis. The unique amplitude time series, $\{P_n(t)\}$, is often called the it principal component time series; for this reason, this decomposition is also called the PCA (principal component analysis). The eigenfunctions are orthogonal to each other and eigenvalues are all positive, as the covariance matrix is real and symmetric. In addition, the eigenvectors (also called *EOF loading vectors*) and PC time series should satisfy the following:

$$\phi_n(r) \cdot \phi_m(r) = \frac{1}{N} \sum_{i=1}^{N} \phi_n(r_i) \phi_m(r_i) = \delta_{nm} \quad \text{(orthogonality)},$$

$$P_n(t) \cdot P_m(t) = \frac{1}{L} \sum_{t=1}^{L} P_n(t) P_m(t) = \alpha_n \delta_{nm} \quad \text{(uncorrelatedness)},$$

where α_n is a proportionality constant.

(a) Show that the covariance matrix, **C**, can be written as

$$C_{ij} = \langle T(r_i, t) T(r_j, t) \rangle = \sum_n \alpha_n \phi_n(r_i) \phi_n(r_j).$$

(b) Show that the EOF decomposition indeed satisfies the Karhunen–Loève equation and the proportionality constant $\alpha_n = \lambda_n$.

(c) Show how you can calculate the PC time series from the eigenfunctions of the covariance matrix.

11.4 Global average temperature is defined by

$$\overline{T}(t) = \int_\Omega T(r, t) \, d\Omega = \frac{1}{4\pi} \int_{-\pi}^{\pi} \int_0^{2\pi} T(\phi, \theta, t) \cos\theta \, d\phi \, d\theta,$$

where Ω denotes the surface of the Earth, and ϕ and θ represent longitude and latitude of a location r on the surface of the Earth. The factor 4π is introduced to properly scale the result.

(a) Let us consider the problem of estimating global average temperature based on a small number of samples on the surface of the Earth. Set up an optimal estimation problem based on samples $T(r_i, t), i = 1, 2, \ldots, N$.

(b) Solve the optimal estimation problem for the weights.

11.5 Consider a noise-forced one-dimensional energy balance model of the form

$$C \frac{dT}{dt} + BT = F_\omega e^{i\omega t}.$$

(a) Calculate the spectrum of T using the spectrum of noise forcing F.

(b) What is the nature of the spectrum of T when the model is forced by a white-noise forcing, that is, $\langle |F_\omega|^2 \rangle = \sigma^2$, regardless of the frequency ω? Plot the spectrum for ocean and land responses by using $\sigma^2 = 1$, $B = 2.0$ W m^{-2} (°C)$^{-1}$, $C_{\text{ocean}} = 10$ years, $C_{\text{ice}}/65$, $C_{\text{land}} = C_{\text{ocean}}/600$.

(c) Express the variance of the temperature response in terms of the spectrum of temperature and in terms of the spectrum of forcing.

(d) Calculate the autocovariance function for the temperature response when the model is driven by a white-noise forcing. Then, determine the e-folding timescale of response.

11.6 Consider a signal detection problem, where the normalized signal at two stations is given in the form

$$\vec{u} = u_1 \hat{e}_1 + u_2 \hat{e}_2, \quad \vec{u} \cdot \vec{u} = (u_1 \hat{e}_1 + u_2 \hat{e}_2) \cdot (u_1 \hat{e}_1 + u_2 \hat{e}_2) = u_1^2 + u_2^2 = 1,$$

where $(\hat{e}_1, \hat{e}_2)$ are orthogonal unit vectors. Actual data at the two stations are given by

$$\vec{D} = (\alpha u_1 + n_1) \hat{e}_1 + (\alpha u_2 + n_2) \hat{e}_2,$$

where α is the true strength of the signal and (n_1, n_2) are the natural variability at the two stations. Thus, the signal of constant strength is embedded amid randomly varying natural variability. Further, assume that

$$\langle n_1 \rangle = 0, \quad \langle n_2 \rangle = 0, \quad \langle n_1^2 \rangle = \sigma_1^2, \quad \langle n_2^2 \rangle = \sigma_2^2, \quad \langle n_1 n_2 \rangle = 0,$$

that is, natural variability at each station has mean zero and the random variability at the two stations are uncorrelated.

(a) One way to determine the signal strength is to project the normalized signal on the data set, that is, $\hat{\alpha} = \vec{u} \cdot \vec{D}$. Show that this is an unbiased estimator of the signal strength. What is the error variance of the signal strength α?

(b) Note that the estimator in Part (a) is not optimal, as it did not consider that the magnitude of natural variability differs at the two stations. One way to account for this is to estimate the signal strength at each station and weigh the two estimates optimally. Develop an optimal estimator based on this idea. Calculate the error variance and compare it with that in Part (a).

(c) Show for the signal and mutually uncorrelated noise background (natural variability) at N stations that

$$w_i = \frac{1}{\eta^2} \frac{u_i^2}{\sigma_i^2}, \quad \eta^2 = \sum_{i=1}^{N} \frac{u_i^2}{\sigma_i^2}, \quad \sigma_i^2 = \langle n_i^2 \rangle,$$

where the signal is determined by

$$\vec{S} = \alpha \sum_{i=1}^{N} u_i \hat{e}_1, \quad \hat{\alpha} = \sum_{i=1}^{N} w_i \frac{u_i D_i}{u_i u_i}.$$

11.7 Consider data consisting of four different temporally varying signals on top of natural variability:

$$T(m) = \sum_{n=1}^{4} \alpha^{(n)} S^{(n)}(m) + N(m), \quad m = (r, t), \quad r = 1, 2, \ldots, L, \quad t = 1, 2, \ldots, M,$$

where $S^{(n)}$ denote different signals, N represents natural variability, and $m = L \times M$ space–time points. Assume that the signals are not necessarily orthogonal to each other.

(a) Show that this problem can be recast in terms of EOFs of natural variability in the form

$$T_k = \sum_{n=1}^{4} \alpha^{(n)} S_k^{(n)} + N_k, \quad k = 1, 2, \ldots,$$

where k represents the EOF mode number.

(b) Show that an unbiased estimator for the signal strength can be derived from each EOF mode equation in Part (a) as

$$\hat{\alpha}_k^{(n)} = \sum_{n'} (M_k^{-1})_{nn'} S_k^{(n')} T_k.$$

12

Applications of EBMs: Paleoclimate

12.1 Paleoclimatology

There are a few areas in paleoclimatology that are particularly well suited for energy balance modeling. It is important to remember that EBMs as we present them in this book (except for those considering vertical structures in Chapters 3 and 4) only treat the surface temperature field. Precipitation and changing ocean circulations are also not included in these EBMs although they are clearly of interest in paleoclimatology. Fortunately, much of the data coming from empirical studies in paleoclimatology pertain to the surface temperature. Anything above the surface is out of bounds for us—it requires a much more sophisticated model to go above the surface or to deal with any transport phenomena such as the transport of water vapor above the boundary layer.

Before plunging into EBM applications directly, we present a very short summary of the Earth's climate history. One of the first problems we encounter is based on sound theoretical evidence of the lower brightness of the Sun at the time of the formation of the Earth and through its settlement into a planet with land and ocean surfaces. According to long-accepted astrophysical theory, the Sun was only about 70% as bright as it is today. Recalling our study of ice-cap models in Chapters 2 and 7, we see that the evolution of the Earth's climate was quite different from what these models based on a monotonically increasing solar brightness and no changes in atmospheric greenhouse gas might suggest. Bender (2013) and Feulner (2012) review the many theories of how we might avoid the "faint Sun paradox," according to which the planet could not possibly get to our present climate by a steadily increasing external forcing. The solution must lie in the atmosphere's changing composition—perhaps much more powerful greenhouse gas concentrations came to the rescue—but it would take substantial amount of forcing, perhaps tens of percent of the equivalent of solar brightness.

Ward and Kirschvink (2015) argue that there were incidences of "snowball Earths" as early as 2.2 Ga[1] (the so-called Huronian event). They also argue that these events might have been instrumental in the beginnings of life on the planet. This first massive glaciation might have lasted several hundred My. The event was coincident with a large

---

[1] The abbreviation Ga stands for billions of years before the present. The abbreviation Ma stands for millions of years before the present. We will use Gy and My to indicate intervals of billions or millions of years, respectively. Similarly for ka and ky with indicate thousands of years.

increase (several bars, one bar of pressure = that of one atmosphere) in O_2 in the atmosphere due to oxygenic photosynthesis. There might have been lots of methane in the atmosphere at that time and it might explain the hot planet preceding the ice event. The oxygenation of the methane might have cooled the planet sufficiently to cause the collapse to the snowball. Other possibilities include the increase of weathering (removal of CO_2) following accretion of large objects and collisional tectonics (Melezhik, 2006).

There is evidence suggesting that there was another interval when the planet was totally ice covered several times, later called the Neoproterozoic period (1 Ga to 542 Ma). These are also periods thought to include total ice cover (Kirschivink, 1992; Evans, 2000; Hoffman and Schrag, 2002) . There is a large literature on this subject; several of these papers tend to confirm the geological dating and resolve issues of global synchronicity for the ice cover. Among these are papers that present pretty sound evidence of tropical glaciers. Our elementary models suggest that, without wild changes in obliquity, it would be hard to have the tropics covered with ice without the whole globe so covered. These glaciations might have persisted for tens of My. The Neoproterozoic glaciations took place after the emergence of life on the planet and one wonders how organisms might have survived. Warren *et al.* (2002) propose that thin ice cover over the oceans might have allowed the passage of enough sunlight to support photosynthesis. Other papers have been concerned with the survival of multicellular organisms during such inhospitable conditions.

In the Paleozoic era ($\sim$ 500 to 200 Ma), there is evidence of numerous large-scale glaciations (but not snowballs). The continents were going through rearrangements over this and later periods on time scales of millions of years and these configurations of land/sea geography are likely to have played a role in the early ice ages. Figure 12.1 summarizes the geological time intervals to be used in the discussion.

At around 50 Ma the Earth began to gradually cool (geologically speaking) as shown in Figure 12.2. This figure, modified from the original found in Zachos *et al.* (2001), and a similar modification featured in Chapter 6 of Bender (2013) tell much about the climate over the last 65 Ma. Bender (2013) describes how the ^{18}O record from bottom-dwelling microbiota can be used to infer global temperatures. On the basis of Figure 12.2, we can trace a gradual descent from a much warmer ($\sim$12 °C warmer at around 50 Ma, called the *early Eocene thermal maximum*) planetary surface to that at present. The main governing factor was likely to be CO_2 decreases over this time. As time proceeds, the planet continues to gradually cool until the beginning of the Oligocene epoch ($\sim$34 Ma). At this point in time, there is a sharper rate of cooling which is attributed to the growth of continent-wide ice in Antarctica. From that point to the present, the planet has a large (south) polar ice sheet, with Greenland ice coverage likely occurring at the beginning of the Pliocene. The last million years or so (the Pleistocene) features large excursions in ice sheets, most of which occur in the Northern Hemisphere.

12.1.1 Interesting Problems for EBMs

- The first problem for EBMs has already come up in Chapter 2 in which we found a possible ice-covered Earth solution to our global energy balance equation. The operating curve showed that if the solar brightness was dropped by only a small percentage, the planet would plunge into a deep-freeze state. Of course, the percentage depends on the details of model parameter choices.

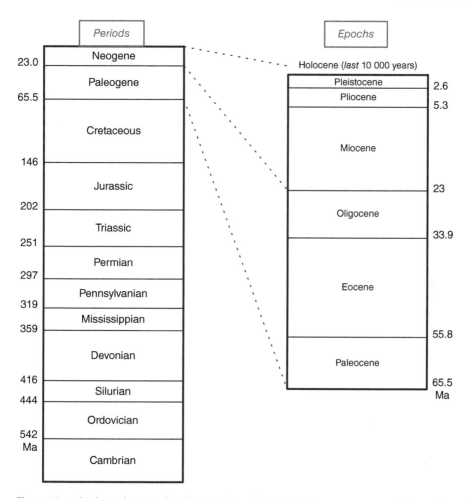

Figure 12.1 Geological timescale indicating boundaries separating eons, eras, periods, and epochs along with the times in Ma. Note that the durations of the intervals are not drawn to temporal scale in this chart. (Data from Bender (2013).)

- The second problem is the effect of the land–sea distribution over deep geological time scales where, through continental drift, the landmasses have continuously changed their configurations. It could be that summertime's warmest temperatures determine whether snow will linger over the warm season and allow the growth of an ice field and eventually a large-scale ice sheet. We know that the proximity to the poles and the shapes of landmasses can alter the summertime temperatures. Could EBMs tell us when ice sheets should have been present? The seasonal cycle is a key factor.
- The third concerns the most recent few tens of millions of years during which the Earth has shifted from a warmer state to one with large continental ice sheets, the most prominent being Antarctica and Greenland. What were the conditions for these and what might EBMs have to contribute?
- A fourth has to include the ice sheets that have waxed and waned over North America and Western Europe (but not over Siberia). Observations provide evidence suggesting

324 | *12 Applications of EBMs: Paleoclimate*

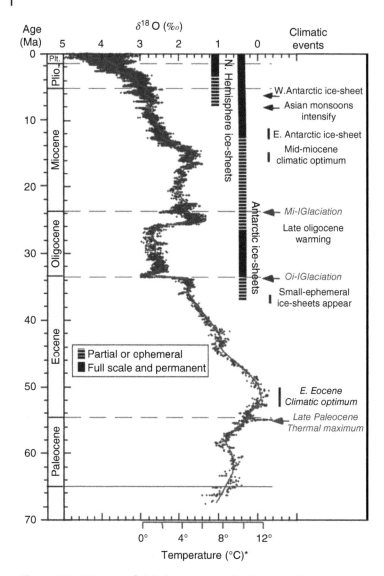

Figure 12.2 Estimates of global temperature (horizontal axis) versus time (vertical axis). The temperature is estimated from ^{18}O data collected from benthic foraminifera. (Zachos *et al.* (2001). Reproduced with permission of AAAS.)

that the glaciations exhibit a regularity and are in step with the theoretically calculated oscillations of orbital elements of our planet. Might EBMs be of use in understanding the glaciations over the Pleistocene (last few million years), and even the most recent 10 ka, (some authors use 12 ka) called the *Holocene*?

Not all problems are approachable, not to mention solvable, by EBMs. Atmospheric chemistry and ocean circulation play roles possibly as important as the changing land–sea configuration and the associated radiative-energy-balance effects on summertime temperatures. Sometimes, chemistry and aerosol effects can be inserted into an EBM, but even then the problem might not be solved or may not be interesting. For instance, if atmospheric CO_2 increased because of some geological process, the EBM contribution may be large, but it is not the important aspect of the problem (e.g., why did CO_2 increase?). But if the greenhouse gas and aerosol concentrations are given, it is straightforward to provide EBM solutions. Long-lived ocean current anomalies or surface temperature aberrations (e.g., a shift in the Gulf Stream) can be included by prescribing them, but as yet no one has found a way to include them dynamically in such a simple model. Unfortunately for the EBM aficionado, the use of a general circulation model (GCM) may be far more appropriate for such a forcing as they are likely to excite interesting quasi-stationary wave patterns that affect storm paths (all this is outside the scope of an EBM). As we found in Chapter 10, we can include some oceanic effects in EBMs, but to attempt the circulation of the ocean currents is well beyond the scope of simulation at this level of the model hierarchy. Perhaps the single, most important advantage for EBMs is to spell out how the seasonal cycle can play a role in various paleoclimate problems, as the seasonal cycle responds almost linearly to forcings due to changes in orbital elements, CO_2, and aerosol concentrations, as well as the geographical distribution of land and sea. These concepts have been a common theme in the many works of Crowley and his colleagues Hyde, Baum, and Short (for a list of others see the preface of this book.).

12.2 Precambrian Earth

The Earth settled into its present form with an atmosphere including distinctly defined ocean and landmasses a few billions of years ago. The earliest life forms appeared about 3 billion years ago (see Ward and Kirschivink, 2015). The Sun was not as bright as it is today. Gough (1981) describes the astrophysical problem as it relates to the history of the Sun's luminosity and its radius over the last 4.7×10^9 years that is consistent with arriving at its present conditions. Gough summarizes the increasing luminosity by the simple formula

$$L(t) = \frac{L_\odot}{1 + \frac{2}{5}\left(1 - \frac{t}{t_\odot}\right)}, \qquad (12.1)$$

where $L(t)$ is the luminosity as a function of time t; $L_\odot$ is the present luminosity; and $t_\odot$ is today. This result has not changed appreciably since Gough's paper was written (personal communication with Gough, 2012).

According to Figures 2.9 and 7.2, if the other factors (e.g., CO_2) are held fixed and given that the Sun was dimmer, the Earth should have been frozen over for billions of years before it might have jumped to an ice-free planet. Then it would have had to traverse the S-shaped control curve, jumping to the ice-free Earth, then cooling (Figures 2.9 and 7.2) to get back to our current climate. This is called the *faint Sun paradox*, and was first noted by Sagan and Mullen (1972). Kasting (2010) and Feulner (2012) discuss the problem and list entries in the literature that attempt to solve it (see also recent book Kasting (2014)). A popular proposal involved high concentrations of CO_2 (e.g., Owen *et al.*, 1979) but this one appears to be inconsistent with recent geological evidence according to Rosing *et al.* (2010). A major problem is that if the Earth were

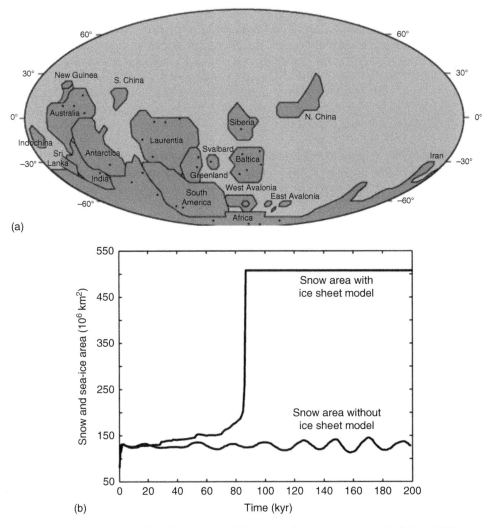

Figure 12.3 (a) Geography of the Neoproterozoic featuring the supercontinent Rodinia (∼1000 to 540 Ma). (b) Snow and sea-ice cover (10^6 km^2) as simulated in a 2-D seasonal EBM including orbital variations (from the Pleistocene). The lower wavy curve is for the EBM alone, and the upper curve shows the effects of including an ice-sheet model. (Hyde et al. (2000). Reproduced with permission of Nature.)

iced over, it is nearly impossible to see how to get its frozen surface melted without some extraordinary intervention in the energy balance. Thin cirrus clouds have been suggested as a solution, but they do not seem to increase the temperature enough to melt the ice and lower the very high albedo of the ice cover. Bender (2013) presents a recent review, devoting Chapter 2 of his book to the faint Sun problem. Bender (2013) agrees with Feulner (2012) that the problem remains unsolved.

In the interval 1 Ga to 542 Ma, there appear to be several glaciations that may have covered the entire Earth: the "snowball Earth" events. Crowley and Baum (1993) considered this problem with a number of experiments with the two-dimensional seasonal EBM that has been described in Chapter 8. The most interesting application of an EBM to these glacial events comes from Hyde et al. (2000). In this paper, experiments were conducted with the usual 2-D EBM, but also an ice sheet was included along the lines of that in Deblonde et al. (1992); Tarasov and Peltier (1997); as derived from the early theoretical model of Nye (1959). A large landmass concentrated at the pole in the Neoproterozoic (see Hyde et al., 2000, and Figure 12.3a). The model includes isostatic depression of the land under the heavy ice sheet. The Sun's brightness was taken to be 6% lower than today. Figure 12.3b shows a simulation with this coupled model for a period of 200 000 years. After about 20 ky, the ice sheet begins to grow in the coupled model, while the model with no ice sheet dynamics does not grow. The ripples are due to changes in the orbital elements. It seems that in the coupled model simulation, the ice mass relaxes under its own weight to spread laterally. If the ice at the terminus is thick enough, it can survive the summer ablation (provided new ice can be transported into the ablating area). As the ice spreads, the planetary albedo increases and eventually a threshold is crossed and the solution plunges to the ice-covered state. The authors experiment over a wide range of parameters such as precipitation rates (for which there is an empirical formula), ice sheet parameters, and so on. Such a self-spreading ice sheet may be essential for explaining the Pleistocene glaciations as well.

12.3 Glaciations in the Permian

Figure 12.4 gives an idea of how the continents begin to spread apart after the Permian. There is observational evidence for glaciations during the period 365–260 Ma, Permian. Chapter 5 in Bender (2013) is devoted to ice cover in the time frame 370 to 260 Ma (Glacial I), 325 to 310 Ma (Glacial II), and 300 to 285 Ma (Glacial III), as shown in the upper level in Figure 12.5 labeled "Overall Gondwana." Figure 12.6 shows the polar wander over the whole entire period of the three Glacials. The South Pole is in the neighborhood of Antarctica (during this period the pole wanders slowly (timescales of 20 My) over much of the lower part of the landmass in Figure 12.6 (Frank et al., 2008)). The glaciations and polar positioning of the landmasses is of interest to EBM modelers for two reasons: (i) because the ice–albedo effect is large, there is some evidence of rapid transitions to large ice cover. (ii) The positioning of the continents at high latitudes can have dramatic effects on summer temperatures, which if below freezing can lead to ice-cap growth.

12.3.1 Modeling Permian Glacials

Much of climate change over the long term (tens of millions of years) is governed by the greenhouse gas concentrations. This is not an especially interesting aspect of the dynamical behavior of EBMs. The dependence on CO_2 concentration is logarithmic (see (2.13) and surrounding discussion). When CO_2 concentration doubles, the global temperature will increase by 2–4 K (including feedbacks, and after equilibrium is attained). If the CO_2 concentration is halved, the temperature will correspondingly decrease by 2–4 K. Interesting problems for EBMs come up when there is a large change in the seasonal cycle, especially when a landmass moves near the pole. When the landmass is large and its center is over the pole, summers will be too hot to initiate or sustain an

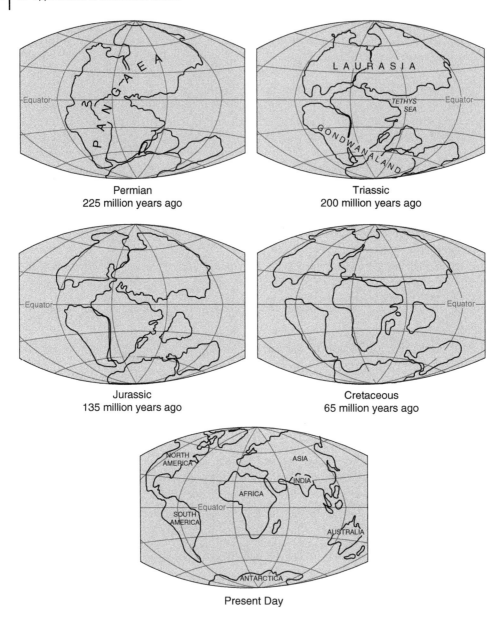

Figure 12.4 Geography of the continental configurations from the Permian (~225 Ma) to the present. (US Geological Survey, https://pubs.usgs.gov/gip/dynamic/historical.html.)

ice sheet. The action comes when a landmass is near the pole where the mean annual surface temperatures will be low enough for freezing to occur, but again the key is for the summer temperatures to be below freezing. This latter happens if the landmass edge is near the pole or if there is a field of broken landmasses near the pole. The presence of smaller-scale water-covered regions can moderate the summer temperatures, keeping them below freezing. As Pangea breaks up, the chances are good for this to happen.

12.3 Glaciations in the Permian | 329

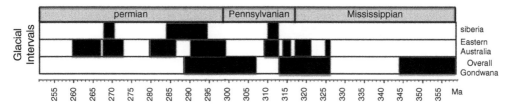

Figure 12.5 Glacial intervals in deep time. Shaded areas are indications of low sea level, inferring the presence of continental ice sheets. The inference of glacial intervals is drawn from sea level evidence (low sea level indicates large volumes of land-based ice). (Rygel *et al.* (2008). © American Meteorological Society. Used with permission.)

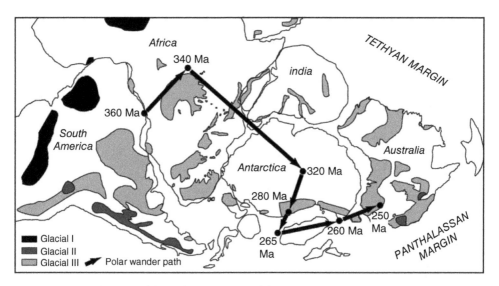

Figure 12.6 An indication of the shifting locations of the South Pole as it passes over portions of Africa, Antarctica, India, and Australia at different times as shown. There are three glacials indicated by the shading in the legend. (Frank *et al.* (2008). Reproduced with permission of Geological Society of America.)

A series of studies by Hyde *et al.* (1990) with a nonlinear (snow/ice–albedo feedback) two-dimensional EBM including a seasonal cycle consider the importance of continental size and positioning with respect to the poles. This study does not include ice-sheet dynamics. There has to be some land near the poles to initiate glaciation. However, the continent covering the poles cannot be too large, otherwise the summers will be too warm to allow snow to build up into an annual ice cover. When the continent is smaller, say 3000 km across, this smaller landmass will allow penetration of the mild maritime summer to suppress the seasonal cycle in the continental interior. Under these conditions, summers will be mild enough (staying below freezing) to allow the snow to build up, eventually leading to an ice field or glacier. The upshot is that the actual geography including shoreline configuration near the pole is important.

An example of conditions that lead to bifurcations (or tipping points) is shown in Figure 12.7 where Crowley and Baum (1993) considered the continental configuration near the South Pole around 300 Ma. In this study, they used the two-dimensional,

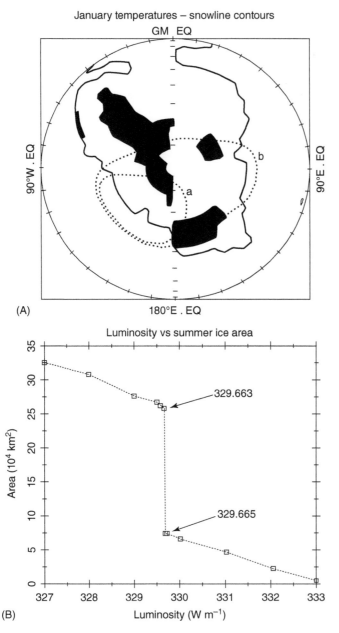

Figure 12.7 Results of a nonlinear 2-D seasonal EBM with ice feedback at 305 Ma. The model was iterated to find the steady solutions. (A) The heavy black line is the outline of the continent (Gondwanaland), the black-shaded areas are locations where data suggest ice sheets during this time. The dotted lines are simulated ice sheets: (a) is for the warmer TSI, and (b) is for an infinitesimally large value of TSI. (B) Shows the area of ice cover as a function of TSI, which is indicated on the abscissa as luminosity. Note the abrupt transition to a smaller ice cap as the TSI passes the bifurcation point. (Baum and Crowley (1991). Reproduced with permission of Wiley.)

seasonal EBM similar to the one discussed in Chapter 8. They examined the areas of ice cover for a continuous range of values of the solar luminosity (we have used the more modern term total solar irradiance, TSI ÷ 4) from 327 to 333 W m^{-2}. A sharp bifurcation is found as the ice cap suddenly decreases in area.

12.4 Glacial Inception on Antarctica

The EBM mechanism for the Antarctic glacial-inception case is based on the geological evidence that Antarctica was once joined to the landmasses of the present Indian subcontinent and present Australia, making the composite continent very "continental" (see Figure 12.8a). Note that the (present) Antarctic continent remains at the pole. Sometime between 80 and 20 Ma the landmass of present India separates from present Antarctica followed by the departure of present Australia (see Figure 12.8b). India proceeds equatorward and across to the collision that results in the present uplift known as the *Himalayan Mountains*. We are interested in the change leading to a much more maritime Antarctica, wherein the summers at the South Pole would be cooler (see Figure 7.10 and surrounding discussion). As can be seen in Figure 12.9, the summer temperature drops dramatically as Australia disassociates with the Antarctic landmass making each smaller than the combination and much more maritime. The polar temperature falls to about 2.5 °C, but no lower once the influence of the separated twin is no longer important. It is also possible that small fluctuations could cause

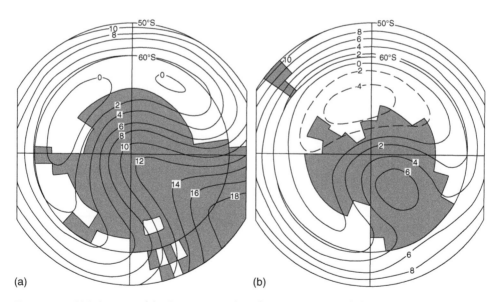

Figure 12.8 (a) Polar view of the SH continental configuration 60 Ma. Solid line contours indicate mid-January temperatures in degree celsius as simulated by the two-dimensional seasonal EBM (North et al., 1983). (b) Same as (a), except 20 Ma. In the interim, Australia shifts away from Antarctica, leaving the latter to be more maritime, thus moderating summer heating at the pole. Poleward of 60°N, the heat capacity of the model's surface is that of sea ice. (Crowley and North (1988). Reproduced with permission of AAAS.)

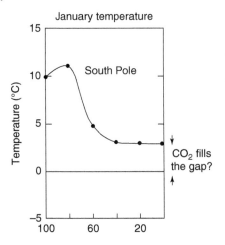

Figure 12.9 Temperatures at the South Pole in mid-summer as a function of time. In this series of simulations, the cooling "saturates" after India completely separates and moves out of an influential distance. The vertical arrows indicate a gap in temperature decrease that should lead to the continental glaciation. DeConto and Pollard (2003) suggest that this gap is closed by the decrease in concentration of CO_2. (Crowley and North (1988). Reproduced with permission of AAAS.)

a jump-over to a larger, more-stable ice cap, see Figure 7.4 and the discussion in Section 7.4.3.

The EBM-motivated theory uses the "maritimization" of the continent, combined with the seasonal small ice cap instability (Mengel et al., 1988), enabling hot summers in the interior to be mitigated to the freezing point and resulting in "rapid" spread to a stable ice sheet as the main underlying causes of the glacial inception. There are other theories proposing explanations of the Antarctic glacial inception. One of the most cited is that of DeConto and Pollard (2003); they use a simplified GCM called the *GENESIS model* to simulate the conditions for the glacial inception. They argue in favor of a CO_2 decrease from 4 times to about 2 times the present partial pressure p_{CO_2} over a 10 My period. While our argument is crude compared to the detail of these authors, the decline of CO_2 is roughly what our model requires to initiate glaciation. The argument we advance is that our solution is essentially the same as that of DeConto and Pollard. In a later paper, Pollard and DeConto (2005) discuss the small ice cap instability and hysteresis (see Section 7.5) to interpret their low-resolution GCM results. They do not discuss the effect of the recession of the Indian subcontinent and Australia away from Antarctica, as their model simulation is set and conducted after these fixed rearrangements following the separations. The paper by Crowley and North (1988) emphasizing the importance of the seasonality was published well before the recognition of the importance of CO_2 came to the attention of paleoclimatologists. In fact, both changes in seasonality and the change in CO_2 are necessary. Carbon cycle models combined with measurements give useful information about CO_2 levels over geological time (references can be found in Zhuang et al., 2014).

The $\delta^{18}O$ index actually indicates a combination of local temperature (probably at sea bottom) and total ice volume on the land surfaces. Another index taken from the skeletal material ($CaCO_3$) is based on the content of the stable isotope ^{13}C relative to its normal and much more abundant ^{12}C and its incremental change of $\delta^{13}C$ index (a measure of the isotopic deviation from normal and far more abundant ^{12}C). This index is a measure of the CO_2 concentration in the deep ocean. Bender (2013) cites the work of Coxall et al. (2005) who noted that in the neighborhood of the Eocene–Oligocene boundary (34 Ma), the changes in ^{18}O and ^{13}C are in step, indicating that as the temperature fell, so did the amount of dissolved CO_2 in the oceans. Some back-of-the-envelope estimates in Bender (2013) suggest that a 54 m drop in sea level and a fall of 3–4 °C are consistent with the isotope data. These results seem to be in line with the EBM-based theory (seasonal change due to maritimization combined with CO_2 decrease), which also agrees with the GCM results of DeConto and Pollard (2003).

Earlier theories of the glacial inception of Antarctica invoke changes in ocean circulation as the Drake Passage is opened. The argument goes that this effect leads to the circumpolar circulation of the Southern Ocean and thus leads to the isolation of Antarctica and therefore its cooling. We believe that the arguments and results presented in Chapter 5, especially those in Section 5.9 and illustrated in Figure 5.10 are pertinent here. In that discussion, the two-mode approximation provides a near-perfect fit to the poleward transport of heat, without reference to any circumpolar current. Simplicity alone suggests that the three components of heat transport (atmospheric sensible heat, latent heat, and oceanic heat) all combine to give a simple form. Just take the NH as a hypothesis and the SH as a confirmation. In light of this, it seems difficult to imagine that the presence or even the intensity of the Circumpolar Current would make a significant difference.

12.5 Glacial Inception on Greenland

The glaciations of Greenland and Antarctica present interesting problems for paleoclimatology and EBMs can contribute to understanding these glaciations (North and Crowley, 1985; Crowley et al., 1986; Crowley and North, 1990). In these studies, covering the times between 80 and 20 My BP, the authors used the rule that summertime maximum temperatures are the key to initiating glaciation. This concept dates back many years (e.g., Milankovitch, 1941). As continents drift over geological time, the maximum value of summer temperature changes depending on latitude and land–sea distributions near the poles. As an example, consider the two configurations shown in Figure 12.10.

Figures 12.10a and 12.10b show the continental configurations for 60 and 40 Ma, respectively. The solid lines depict the contours of the mid-July temperatures as computed from the EBM (North et al., 1983). In both (a) and (b), poleward of 60°N, North et al. (1983) used the value of sea ice over the ocean-covered areas ($C(\hat{r})$) is taken to be part way between land and ocean values, to take into account puddling and leads). Figure 12.11a shows the simulated mid-July temperature changes in Greenland, in which we see a definite cooling in mid-summer in the interior of Greenland over the 20 My change. Figure 12.11b shows the seasonal cycle of the mid-Greenland surface temperature for various times in the geological past, 100 Ma to the present. Two factors clearly cause the mid-summer temperatures to decline: (i) Greenland's movement toward the poles causes the mean annual temperature to fall; (ii) the seasonal cycle amplitude is diminished because Greenland has become more maritime.

Figure 12.11b suggests that at about 15 My, BP the mid-summer temperatures fall below freezing (see also Figure 12.12). If we combine this finding with the small ice-cap instability argument, we can envision that at this point, a small ice cap is not possible; hence, there must be a transition to a finite-sized ice cap. Such an ice cap would cover all of Greenland. The theory is very simple, and compelling because it gets the timing right and it calls for a continent-sized ice sheet.

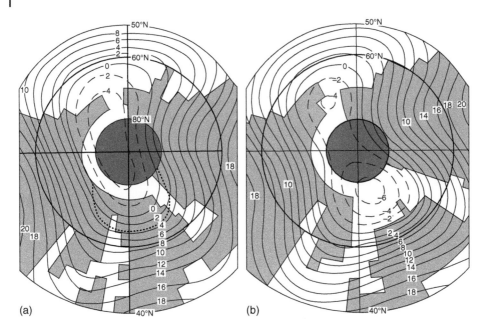

Figure 12.10 (a) Polar view of the NH continental configuration 60 My BP. Solid line contours indicate mid-July temperatures in degrees Celsius as simulated by the two-dimensional seasonal EBM (model from North et al., 1983). Opening of the land bridge connecting Greenland and Europe results in a distortion of the freezing line (0 °C) as shown by the dotted line. (b) Same as (a), except this is for 40 My BP. In the interim, Greenland shifts poleward and it becomes more maritime. Poleward of 60°N, the heat capacity of the model's surface is that of sea ice. (Crowley and North (1988). Reproduced with permission of AAAS.)

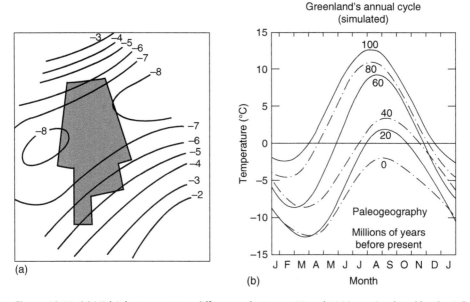

Figure 12.11 (a) Mid-July temperature differences between 60 and 40 Ma as simulated by the 2-D seasonal EBM, North et al. (1983). (b) The seasonal cycle of mid-Greenland for various times 100 My to the present. Two features are apparent: (i) the mean annual temperature is lowered by the shift, (ii) the seasonal cycle is cut by nearly a factor of 2. Both effects tend to lower mid-summer temperatures. (Crowley and North (1988). Reproduced with permission of AAAS.)

Figure 12.12 Mid-summer temperatures as simulated with the 2-D seasonal EBM as a function of time (in Ma). (Crowley and North (1988). Reproduced with permission of AAAS.)

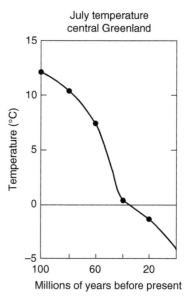

12.6 Pleistocene Glaciations and Milankovitch

Lisiecki and Raymo (2005) used a "stack" of 57 cores well distributed over the ocean floor to examine ^{18}O isotope variations in benthic[2] Foraminifera[3] to study global changes in ice volume and temperature over the last few million years. They found that roughly 5 million years ago, the ice volume on the planet began making irregular oscillations with a period close to 41 ky (see Figure 12.13). By 2 million years ago, the oscillations became more regular at this same period. Figure 12.13 shows the results of their study which combined 57 globally distributed seafloor cores of data based on ^{18}O from bottom-dwelling microspecies (benthic foraminifera). The signal is particularly strong given that so many samples were averaged together. Figure 12.14 shows the spectral density of the record of ice volume–temperature.[4] The record of ice volume being inferred from such data has a long history. There are now many independent corroborating pieces of evidence such as that from ice cores,[5] blown dust (loess[6]) deposits whose timings match well with deep-sea core data (Bender, 2013; Bradley, 2015).

---

2 *Benthic* refers to bottom dwelling. The term *Planktonic* refers to dwellers in the surface waters.
3 Foraminifera are members of a phylum or class of amoeboid protists. They have streamers of ectoplasm that can be used for catching food. Their shell or "test" is made of $CaCO_3$ and is a common sediment on the sea floor.
4 The ^{18}O record indicates a combination of both deep-water temperature and ice volume. The signal here is probably dominated by the volume of ice on land.
5 Cores taken from Greenland, Antarctica, and other ice fields provide information over the last 800 000 years about CO_2 concentrations, ^{18}O deviations and many other indicators.
6 Loess is a fine mineral dust that is picked up in the wind and deposited elsewhere. During the large glaciations, winds were stronger owing to greater gradients of surface temperature and the land more arid, leading to the transport of these materials downwind, where they are finally deposited.

336 | 12 Applications of EBMs: Paleoclimate

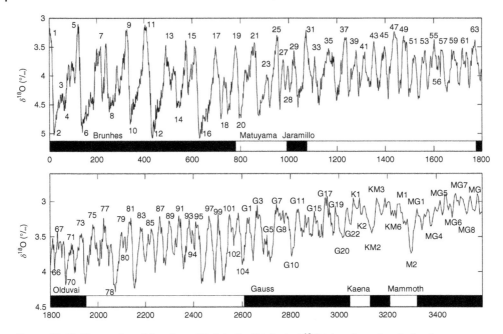

Figure 12.13 Time series of data from 57 globally distributed ^{18}O taken from the shells of bottom-dwelling microspecies (benthic foraminifera). The numbers above and below the peaks and valleys indicate the stage-name of the local extreme event. (Lisiecki and Raymo (2005). Reproduced with permission of Wiley.)

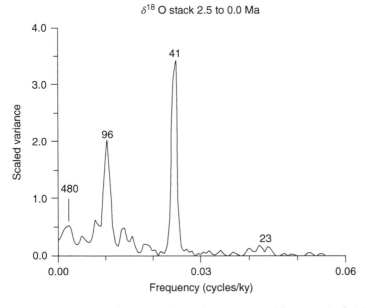

Figure 12.14 Spectra from a "stack" (similar to an ensemble average) of globally distributed seafloor cores of data on ^{18}O. The data in this figure come from 2.5 Ma to the present. Note the strong peaks at 41 ky (obliquity), 23 ky (precession) as well as 480 and 96 ky (eccentricity). (Hartmann (1994). Reproduced with permission of Academic Press.)

12.6 Pleistocene Glaciations and Milankovitch

The Milankovitch[7] theory of the ice ages (Milankovitch, 1998)[8] links the nearly periodic changes in the orbital parameters of the Earth's motion around the Sun to the periodic glaciations. It received little attention until the sea cores began to reveal the clear indication of periodicity of the glaciations, first from Emiliani (1958), culminating with the Hayes et al. (1976) paper that directly compared the sea core data with the orbital element variations based on celestial mechanics, essentially confirming the Milankovitch mechanism as the "pacemaker" of the ice ages. Figure 12.15 shows the temporal variation of the orbital elements (eccentricity, obliquity, and precession of the equinoxes) based on calculations conducted by Berger (1978). Figure 12.15 shows spectra of the insolation at different latitudes and times of year, based upon a time series of calculated insolation values over the last 468 000 years. Figure 12.15a shows the spectra of the quantity $\Delta e \sin \Pi$, where e is eccentricity and Π is the phase of perihelion. Here, the contribution from the obliquity (peaked at period 41 ky) is represented by the solid line, while that of the precession (periods of 23 and 19 ky) are in dotted lines.

Figure 12.16 shows the spectra from actual data. Figure 12.16b shows the percentage of *Cycladophora davisiana* over all other radiolaria[9] in the time-interval samples. The presence of species relative to other forms of radiolaria is known to indicate the temperature near Antarctica.

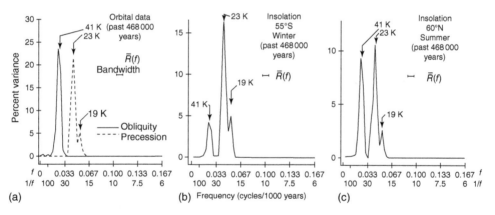

Figure 12.15 Solar insolation at different latitudes and seasons calculated over the last 468 000 years. (a) The parameter $\Delta e \sin \Pi$, where e is eccentricity, Π is the phase of perihelion. (b) Insolation at 55°S in winter. (c) The insolation at 60°N in summer. The peak frequencies are at period 41 ky (obliquity) and 23 ky together with 19 ky for precession. Figure from Hayes et al. (1976). (©Amer. Assoc. Advance. Sci., with permission.)

---

7 Milankovitch (1879–1958), educated as a civil engineer, became chair of applied mathematics at the University of Belgrade where his main work was started before WWI. He was permitted to work during the war and afterward returned to his position in Belgrade. He devoted many years to studying the celestial mechanics of the Earth's orbital parameters and the possible linkage of these to the great glaciations of the Pleistocene. He collaborated with Köopen and Wegener.

8 See the volume of papers written in honor of the 125th anniversary of Milankovitch's birth, Berger et al. (2005).

9 Radiolaria are single-celled microbes with mineral (mostly silica) outer shells. They take their name from the spiny shapes of their shells. They have been abundant for over 600 Ma. Many images are available online.

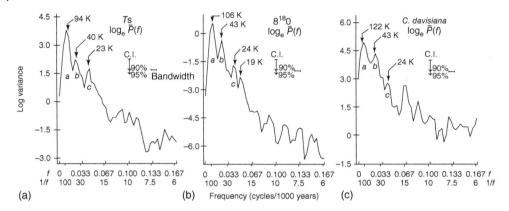

Figure 12.16 Log spectra from two subantarctic deep-sea cores. (a) The surface temperature estimated from radiolarian assemblages. (b) ^{18}O. (c) The percentage of C. davisiana relative to all other radiolaria. (Hayes et al. (1976). Reproduced with permission of AAAS.)

12.6.1 EBMs in the Pleistocene: Short's Filter

The study of EBMs leads to some interesting results for the Pleistocene. We start with the 2-D seasonal model of North et al. (1983) (see also North and Crowley, 1985). The model was run with the present TSI, then small percentage changes in the TSI were used. Ice cover was prescribed when the summer temperature fell below freezing. The model was iterated until a steady state was found. Figure 12.17 shows that a bifurcation occurs at a value between −1 and −2 °C. Below this value, the "ice sheet" extends completely across the Arctic Ocean and includes Alaska and much of western Canada, even some of Siberia. This "glaciation" is asymmetric between the Eastern and Western hemispheres—there is more ice cover in the Western hemisphere. But the big ice sheet does not resemble the Laurentide Ice Sheet much. There could be some problems with the model perhaps due to its poor spatial resolution. Experience with previous nonlinear, one-dimensional models with ice caps (Chapter 7 and especially Figures 7.10 and 7.12, both taken from Mengel et al., 1988) suggest that the transition to extensive ice cover might be rather rapid.

A comprehensive chapter by Short et al. (1991) shows a number of linear responses to orbital element changes by the 2-D seasonal model (North et al., 1983) discussed in Chapter 8. In this section, we present results from this paper. As referred to in the title of the paper, the responses are the result of filtering the changes in orbital element forcing through the model being thought of as a "filter." It is useful to note that the model is solved for its mean annual solution, its annual harmonic solution, and its semiannual solution. Then these Fourier components are composed into the complete solution for the seasonal cycle of the temperature field. But it is interesting to examine the effect on the Fourier components directly both in the insolation and in the responses.

First consider the latitude dependence of the forcing (perturbation of the insolation components). Figure 12.18 shows how the insolation as function of latitude changes for extreme values of the obliquity (a) and for the precession varying from perihelion to aphelion (b). It is interesting that the insolation in both (a) and (b) show that some harmonics of the insolation dip to negative values near the equator.

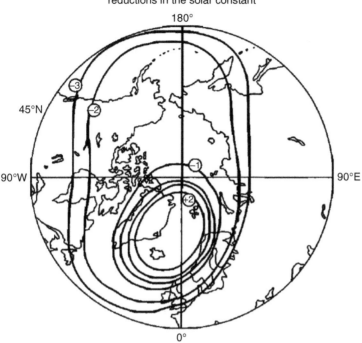

Figure 12.17 Morphology of the mid-July 0 °C isotherm for various percentage changes in the total solar irradiance. Contours are for every 1% from +2 to −3. (North et al. (1983). © American Meteorological Society. Used with permission.)

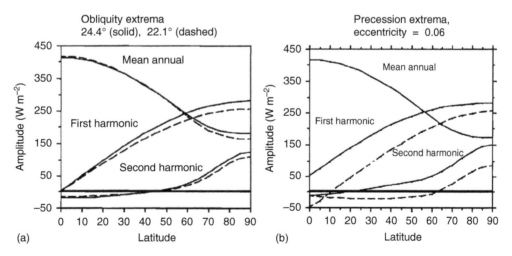

Figure 12.18 (a) The latitude dependence of the insolation for obliquity = 24.4° to 22.1° (dashed), for the Fourier components, first harmonic, and second harmonic. The eccentricity is set to zero. (b) The same but for precession extrema for eccentricity = 0.06. (Short et al. (1991). Reproduced with permission of Elsevier.)

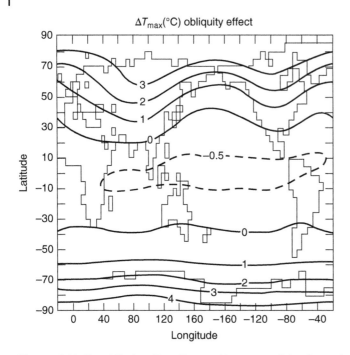

Figure 12.19 The obliquity effect. Geographic pattern of the change in maximum summer temperature as the obliquity changes from 22.1° (solid) to 24.4°. (Short et al. (1991). Reproduced with permission of Elsevier.)

Next consider the geographical distribution of the response in maximum NH summer temperatures (∘C) due to a change in obliquity from its known extremes 22.1° (solid) to 24.4° (see Figure 12.19). In this case, the eccentricity is set to zero. The continents are shown with blocklike edges to emphasize the course resolution of the model. The tropical response in this figure shows a negative response (180° out of phase with the extratropical response). Short et al. (1991) point out that this is in agreement with results of core RC24-30 taken in the tropical Atlantic (Imbrie et al., 1992).

Figure 12.20 shows the response pattern of northern summer temperatures as the summer solstice moves from aphelion to perihelion. Figure 12.21 combines the difference of extremes for obliquity and maximum eccentricity (0.06). The Southern Hemisphere is also depicted at summer solstice by combining data from the two hemispheres at six-month intervals into one map.

Next are considerations of how the changes in insolation are filtered into the thermal responses especially over oceans where deep-sea cores are collected. We are always interested in the maximum summer temperature in this series of simulations. While the summer temperature field is a linear response to the forcing, the *maximum* summer temperature is not a linear function of the forcing because finding the maximum is the result of finding the root of the equation $\partial T/\partial t = 0$ after we have found $T(\mu, t)$, where μ is the sin(latitude) and t is the time after winter solstice in the repeating seasonal cycle. In the center, the great continent of Asia where the relaxation timescale is much less than a year, the response (a) mirrors that of the forcing (b) at each of the orbital signals (see Figure 12.22).

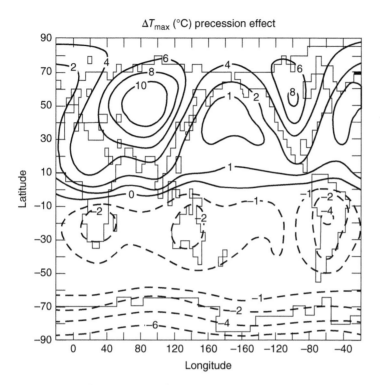

Figure 12.20 The precession effect. Geographic pattern of the change in maximum summer temperature as summer solstice moves from aphelion to perihelion with eccentricity at its maximum value, 0.06 and the obliquity is the present value, 23.25°. (Short et al. (1991). Reproduced with permission of Elsevier.)

As we look at different points in the Atlantic, the response as filtered from the forcing (and the maximum found) presents different results from points in the center of a large continental landmass. Figures 12.22 and 12.23 focused on the north Atlantic shows the time series and amplitude spectra of the maximum summer temperature and the insolation. Here we see strong peaks at the obliquity and precession periods, but also several low-frequency peaks. These are related to the eccentricity and its effect on the other orbital elements.

In equatorial Atlantic (0°N, 30°E), as seen in Figure 12.24, the obliquity signal is weak, but the eccentricity frequencies corresponding to periods of 400 and 100 ky are very strong. Note also a first harmonic of the two precession frequencies (periods of ~12 and 10 ky). These harmonics are explained in Figure 12.25 (see also Short and Mengel, 1986). Figure 12.25 shows the response as function of lag (days) after the winter solstice. At 86 ky, the temperature curve (indicated by +s) is symmetric about the summer solstice as the Sun passes over the equator twice during the year. There are two maxima of equal magnitude at this time. Then a quarter of the precession cycle later at 90 ky (indicated by ×s), there is one absolute maximum. Next consider 94 ky (indicated by ∘s), when there are two (nearly) equal maxima. Finally, in the cycle at 98 ky (indicated by △s), there is a single global maximum. This sequence shows that there are two global maxima occurring at a separation of about one half of the precession period. Finding

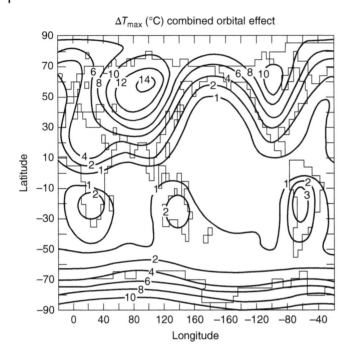

Figure 12.21 Combined effect. Geographic pattern of the change in maximum summer temperature as northern summer solstice moves from aphelion to perihelion with eccentricity at its maximum value, 0.06 and the obliquity is the present value, 24.4°. In this figure, both hemispheres are shown in summer by combining data from runs of each. (Short et al. (1991). Reproduced with permission of Elsevier.)

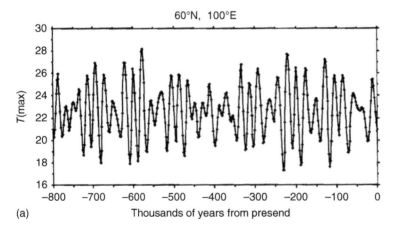

Figure 12.22 (a) Time series of the maximum summer temperature response at 60°N, 100°E (north central Asia) as simulated by the model after running for the last 800 000 years. From Short et al. (1991). (b) The amplitude spectrum of T_{max}, the time series shown in panel (a). (c) The amplitude forcing for summer at 60°N. Major peaks are labeled with the periodicity in thousands of years.

12.6 Pleistocene Glaciations and Milankovitch | 343

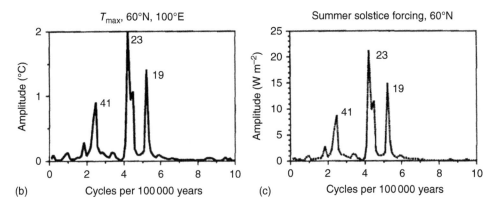

Figure 12.22 (Continued)

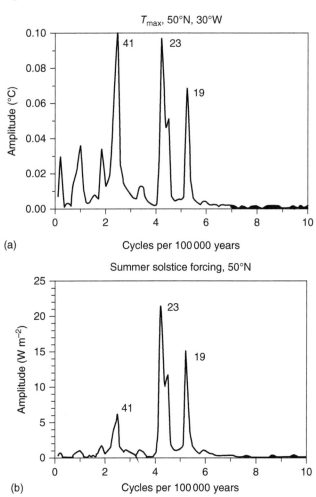

Figure 12.23 (a) Fourier amplitude spectrum of the time series of maximum summer temperature response at 50°N, 30°W (north Atlantic), as simulated by the model after running for the last 800 000 years. (b) The amplitude spectrum of orbital forcing for summer at 50°N. Major peaks are labeled with the periodicity in thousands of years. Figure from Short et al. (1991).

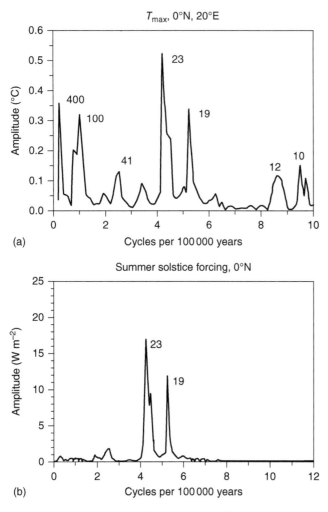

Figure 12.24 (a) Fourier amplitude spectrum of the time series of maximum summer temperature response at 0°N, 20°E (equatorial Atlantic) as simulated by the model after running for the last 800 000 years. (b) The amplitude spectrum of orbital forcing for summer at 60°N. Major peaks are labeled with the periodicity in thousands of years. Figure from Short et al. (1991).

the absolute maxima as a function lag in the seasonal cycle is a nonlinear procedure from the direct forcing amplitude: hence, the first harmonic of the forcing.

Before moving away from spectra, we draw attention to the latitude dependence of observed spectra. Figure 12.26 shows two spectral densities of temperatures estimated from microfossils along the Mid-Atlantic Ridge. The right-most spectrum from a latitude of 41° shows a prominent peak at 23 ky and the leftmost shows a similar spectrum but taken at 54°N, where the 23 ky peak has disappeared and a peak at 41 ky appears prominently. This is close to what the 2-D model would suggest. As we leave Short's work, we see how the linear model is able to produce spectra comparable to our expectation and even some data in the middle and lower latitudes. Moreover, the nonlinearity of the maximum summer temperatures as driven by linear responses of the temperature

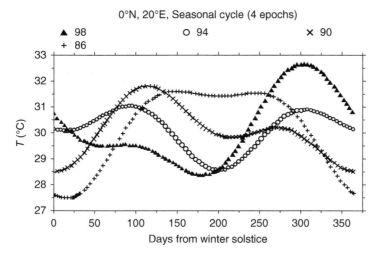

Figure 12.25 The seasonal cycle of sea surface temperature simulated at 0°N, 20°E for epoch 98, 94, 90, 86 ka. The abscissa is the number days after winter solstice. The figure shows the simulated temperature at the equator in the Atlantic Ocean. There are two local maxima at 86 ky (crosses) as the Sun passes overhead at the equator at both vernal and autumnal equinox. As time passes through the precession cycle, the two maxima become only one, either in the spring or in the autumn. This effect leads to two maximum temperatures over the 22-year cycle. The result is a peak at twice the frequency or half the period of the forcing (∼12 ky in the previous figure). Figure from Short et al. (1991).

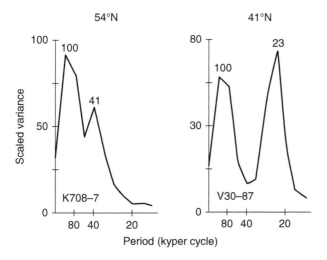

Figure 12.26 The spectral variance densities of the mid-Atlantic sea cores: K708-7 at 54°N and V30-97 at 41°N. Note that the more tropical core shows a strong peak at period 23 ky and the more polar core has its peak at 41 ky, both having a strong variance at 100 ky. Modified from Ruddiman and McIntyre (1984), Geol. Soc. Amer. Bull. 95, 381–396. (© Geological Society of America, with permission).

field invites the possibility of under- and overtones appear in the spectra along with evidence of the eccentricity cycles.

An interesting set of paleoclimatological experiments with the linear seasonal 2-D EBM were recently conducted by Zhuang (private communication, 2015). He employed the same physical model as was used in North *et al.* (1983), but as modified by Stevens who used a finite-difference numerical scheme based on multigrid relaxation as discussed in Bowman and Huang (1991), Huang and Bowman (1992), Stevens and North (1996). No parameter changes were employed in his work. Among his experiments were the following cases shown in Figure 12.27. Panel (a) shows a simulation of the linear (no ice feedback) model wherein the full Last Glacial Maximum (LGM) were employed, including the orbital parameters, the concentration of CO_2, and the placement of ice in the Laurentide Ice Sheet, the Greenland Ice Sheet, and the Fenno-Scandian Ice Sheet. As usual, no topography and no dynamical ice volume model were introduced in the EBM. It can be seen that the summertime 5 °C line hugs the equatorward edge of the ice sheet on both continents. While he did not iterate the solution to include the ice–albedo feedback, it appears that this would be a solution if the ice line were to be close to the 5 °C line. Figure 12.27b shows the simulated temperature (linear model) when the ice albedo of the Laurentide Ice Sheet is removed and replaced by bare land, while the concentration of CO_2 and the orbital elements were unaltered from their LGM settings. As the figure shows, the 5 °C line jumps to leave only Greenland and northwestern Eurasia with ice cover. It appears that the ice albedo is dominant over the other conditions, which seem to play a rather minor role. This is a rather puzzling result. It suggests that there are equilibrium solutions for the big ice sheet and for a small one. An attractive possibility is that if the ice is placed there it will stay. If remove it will remain removed. This may be a kind of neutral stability. Perhaps a large fluctuation could make it jump, aided by the orbital forcing. Another result that has been known for some time is that the orbital perturbation alone is not enough to start an ice sheet in North America. Might a large fluctuation do the trick? We leave this to the imagination of our readers.

12.6.2 Last Interglacial

Figure 12.28 shows a time series going back to 420 ka from the ice core data derived from the Vostok site in Antarctica (Petit *et al.*, 2001). Estimates of temperature departures from a modern long-term average are plotted versus time, increasing from right to left. The last interglacial is the peak between roughly 110 and 140 ka. Other information from this core including other proxies, CO_2 concentrations, and spectral estimates can be found in Petit *et al.* (1999).

Crowley and Kim (1994) employed the two-dimensional, seasonal EBM (described in Chapter 8) to study temperatures and inferred estimates of sea level during the last interglacial (LIG). They found model indications of 3–4 °C July (mid-summer) increases at high latitudes from 140 to 130 ka. In these simulations, they kept CO_2 fixed but included the forcing from obliquity and precession changes over that period. The three sites are (75°N, 120°W), Northern Canada, (75°N, 30°E) Arctic Ocean north of Scandinavia, and (45°N, 90°W), US Canadian border near Lake Superior. They note the following:

> Although it is at the same latitude as the Barents Sea site, peak warming at the Canadian Arctic site occurred 4000 years later. This response reflects the greater importance of the precession-controlled continentality effect on the Canadian site. However, all sites were warmer than they are now by 130 000 years ago.

They suggest that the Milankovitch forcing might be responsible for the decaying ice sheet (see Figure 12.29).

12.6 Pleistocene Glaciations and Milankovitch | 347

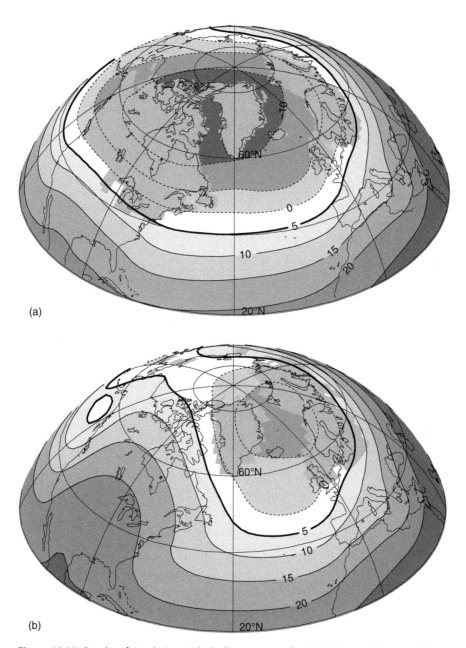

Figure 12.27 Results of simulation with the linear seasonal 2-D EBM (provided privately by Zhuang). (a) The summer average surface temperature with prescribed orbital configuration, CO_2 concentration, and land-based ice in North America (Laurentide), Greenland, and Europe (Fenno-Scandian) at the Last Glacial Maximum (LGM) 20 000 year BP. (b) The same except that the prescribed ice sheet in North America is removed. No other changes from LGM conditions were made. (Courtesy Dr. Kelin Zhuang.)

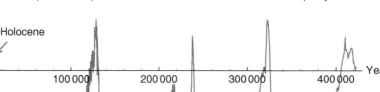
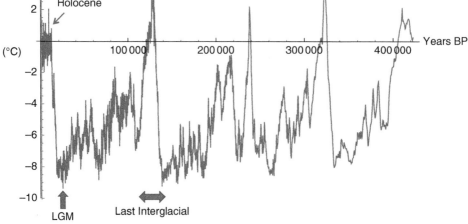

Figure 12.28 Time series of temperature estimates (departures from recent time average °C) from the Vostok ice core based on ^{2}H isotope departures. The Holocene, Last Glacial Maximum (LGM), and the Last Interglacial are indicated. The Last Interglacial is nominally the period between 130 and 110 ka. The three sites are (75°N, 120°W) Northern Canada, (75°N, 30°E) Arctic Ocean north of Scandinavia, and (45°N, 90°W) US Canadian border near Lake Superior. (NASA.)

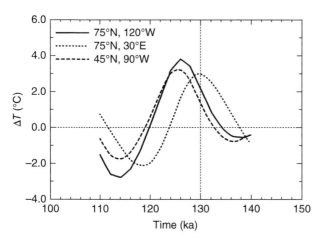

Figure 12.29 Time series of model simulations of departures of July temperatures from present at three locations over the period 140 to 110 ka. Figure from Crowley and Kim (1994). (©Amer. Assoc. Advance. Sci., with permission.)

The problem of temperature and sea level (ice volume) during the LIG has recently received attention especially because of its similarity to the present interglacial. Some studies suggest that sea level might have been much higher during the LIG.

12.6.3 EBMs and Ice Volume

Since the two-dimensional-seasonal EBM (North et al., 1983) simulates the seasonal cycle surface temperature well to first order, it was natural that Peltier and colleagues

12.6 Pleistocene Glaciations and Milankovitch

use it in their modeling of the glacial–interglacial cycles of the Pleistocene (DeBlonde and Peltier, 1991; Hyde and Peltier, 1987). Also, Pollard (1982) has tried including an ice model to an EBM comparable to the beachball model (see Section 8.1). Others, including Paillard (2001), tried combining ice models to EBMs. As noted by Huybers and Wunsch (1994), this combination always leads to a number of new tunable parameters, as there are only eight or so glaciations at the 100 ky timescale. The problem of tuning comes up immediately, as one tries to model glacial growth by inserting precipitation in the form of snow. But this is only the one of the problems encountered as one contemplates the flow and decay of ice sheets. This is not to discourage experimentation wherein much might be learned, but it is to recognize that this is a very big step fraught with potential errors of overfitting.

Crowley and Hyde (2008) re-plotted the Lisiecki and Raymo data in a way that emphasizes the increase of amplitude and the lower frequencies of the fluctuations as a function of time. They note the striking resemblance between Figure 2.10 and Figure 12.30 that have been reproduced from Crowley and North (1988). Crowley and Hyde (2008) argue that a series of threshold crossings are in progress. First is a small ice sheet in North America, then a transition to a larger one. Next is a smaller ice cover in northeastern Europe. Finally, to come is a large ice sheet in Eurasia. Their model goes beyond the simple EBMs of this book in employing a dynamical ice sheet.

An unsolved puzzle is the transition from the so-called 41 ky world to the 100 ky world. This is clearly shown in Figures 12.13 and 12.30 wherein the peaks in global ice volume suddenly change from a dominant obliquity forcing to that associated with the less well understood 100 ky forcing. Interesting attempts at explaining the transition are reviewed

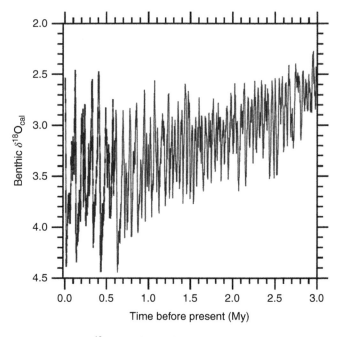

Figure 12.30 The ^{18}O record as replotted by Crowley and Hyde (2008). These authors argue that as the planet cools during the last 3.0 Ma the amplitude of the fluctuations increases. Each new oscillation seems to have a larger amplitude and a lower frequency—perhaps an indicator of another bifurcation coming up. (Crowley and Hyde (2008). Reproduced with permission of Nature.)

in Raymo and Huybers (2008). One mystery is: Why are not the 21 ky peaks evident in the "41 ky world?" Raymo notes that the precession forcing is antisymmetric between the hemispheres and she posits the ice volume forcing may be canceling. Huybers has been interested in the possibility of integrating the insolation over the entire summer season, whose length changes with the precession phase and is modulated by the eccentricity.

12.6.4 What Can Be Done without Ice Volume

Our use of EBMs in paleoclimatology has been to partially answer questions about the inception of glaciation as in Sections 12.4 and 12.5. EBMs might be able to address these issues without any new adjustable parameters. This is as opposed to trying to map the detailed time series in order to compare with a corresponding time series of data. We are guided by the results shown in Figure 7.10 where a latitude-only seasonal EBM was run with a disklike continent centered at the pole. This simulation shows how as a threshold of some control parameter is crossed there is a qualitatively different steady state. Figure 7.12 shows that as the threshold is crossed, it may take tens if not hundreds of years to adjust to the new steady state. The long adjustment time is a result of the nonlinear snow-albedo effect. This time as illustrated in that figure is sensitively dependent on the initial conditions in the problem. This is done without any noise forcing in the problem. One wonders how this bizarre behavior would be affected by background noise as in Chapter 9. Figure 7.14 shows another case where a zonal band of land replaces the pole-centered disk. In this band-of-land geography there are two thresholds. This kind of simulation is tricky because one must be very careful to use high latitudinal resolution. One cannot help but wonder what happens if there is longitudinal dependence in this nonlinear seasonal EBM. And what if noise is included? Might there be quasi-periodic jumping? Some seasonal simulations of two dimensional, seasonal EBMs including nonlinear snow feedback and with realistic and idealized two-dimensional geography (in their case, the late Permian geography) have been conducted by Baum and Crowley (1991). The two-dimensional seasonal simulations seem to confirm the one-dimensional works of Mengel *et al.* (1988) and Lin and North (1990), and Huang and Bowman (1992) that bifurcations exist when (not-too-large, or appropriately fragmented) landmasses lie near or at the poles when forced by a seasonal cycle with an interactive snowline. Then Crowley *et al.* (1994) used to Genesis GCM to investigate the effect. They showed a comparison between the EBM and the GCM results. The GCM seems to confirm the bifurcation and the corresponding catastrophic change in ice cover as the TSI Q/Q_0 is varied.

Notes for Further Reading

Bradley (2015) has written the most comprehensive book on the climate of the last few million years. It is especially good on the observational evidence including its promise and its limitations. Bender (2013) has written an nice book that covers a lot of ground from the Precambrian to the Holocene and written by an expert especially on the isotope chemistry. Another recent well written book with lots of references is that of Summerhayes (2015), a marine geologist. Bender's book in the Princeton University Press series is well written and up to date. The field is changing rapidly and sadly the book by Crowley and North (1988) is out of print and mostly out of date.

There is not much direct evidence of CO_2 changes over the long timescale spanning the Phanerozoic (last 540 Ma), hence, models have to be employed to make inferences about the record of this greenhouse gas. The book by Berner (2004) explains the many considerations that go into such a model with some details about the processes involved. This book also discusses the evolution of O_2. An excellent recent book on paleoclimatology including many historical notes written by a marine geologist is by Summerhayes (2015).

As this manuscript was being submitted for publication, a special issue of the periodical *Past Global Changes Magazine*, **24**, No. 1, August 2016, arrived. This happens to be a special issue on *tipping points*, which is of interest to readers of Chapter 12 as well as several others that cover bifurcations in the solutions of EBMs.

Exercises

12.1 For this exercise, use the program called milank.f, which computes the orbital parameters (obliquity, eccentricity, perihelion angle, precession index) based on the Milankovitch theory. (This program is available at the author's (KYK) website.)
   (a) Calculate the orbital parameters for the past 1000 ky at intervals of 1 ky. Plot the four orbital parameters.
   (b) On the basis of spectral analysis, determine the major periodicities in the time series of obliquity, eccentricity, and precession.

12.2 The magnitude of solar irradiance varies according to the distance between the Sun and the Earth. The amount of solar irradiance is given by

$$Q = Q_0 \left(\frac{1 + \varepsilon \cos \phi^*}{1 - \varepsilon^2} \right)^1,$$

where Q_0 is the mean solar irradiance and ϕ^* is the true anomaly.[10]
   (a) Calculate the solar irradiance as a function of Q_0 for three different values of eccentricity: $\varepsilon = 0.0, 0.02$, and 0.04.
   (b) What would be the percentage change in the solar constant between the perihelion and aphelion when the eccentricity is 0.04?

12.3 For this exercise, use "insol.f" which computes insolation distribution function for given orbital parameters. (This program can be found at the author's website.)
   (a) Run the program to calculate the insolation distribution function for the present orbital configuration and that at 15 ky BP (before present). *Hint*: The insolation program requires orbital parameters, which can be calculated by executing milank.f program.
   (b) Describe the difference in insolation between the two time periods. Explain why. Describe what you would expect in the NH and SH polar regions.

---

[10] True anomaly is the angle of the Earth measured from the Sun with respect to the perihelion. Thus, true anomaly varies by 360° when the Earth makes a single revolution along its orbit around the Sun.

References

Abramowitz, M. and Stegun, I.A. (1964) *Handbook of Mathematical Functions*, US Department of Commerce, 1046p, http://people.math.sfu.ca/ cbm/aands/intro.htm (accessed 7 March 2017).

Ames, W.F. (1992) *Numerical Methods for Partial Differential Equations*, 3rd edn, Academic Press, New York, 451p.

Andrews, D.G., Holton, J.R., and Leovy, C.B. (1987) *Middle Atmosphere Dynamics*, Academic Press, San Diego, CA, 489p.

Andronova, N.G. and Schlesinger, M.E. (2001) *J. Geophys. Res.*, **106**, 22605.

Angell, J.K. and Korshover, J. (1983) Global temperature variations in the troposphere and stratosphere, 1958-1982. *Mon. Weather Rev.*, **111**, 901–921.

Archer, D.A. and Pierrehumbert, R.T. (eds) (2010) *The Warming Papers*, Wiley/Blackwell, 432p.

Arfken, G.B. and Weber, H.J. (2005) *Mathematical Methods for Physicists*, 6th edn, Elsevier Publishing, San Diego, CA, 1182p.

Baum, S.K. and Crowley, T.J. (1991) Seasonal snowline instability in a climate model with realistic geography: application to carboniferous (~300 Ma) glaciation. *Geophys. Res. Lett.*, **18**, 1719–1722.

Bender, M.L. (2013) *Paleoclimate*, Princeton University Press, Princeton, NJ, 306p.

Berger, A. (1978a) Long-term variations of daily insolation and Quaternary climatic changes. *J. Atmos. Sci.*, **35**, 2362–2367.

Berger, A. (1978b) A simple algorithm to compute long term variations of daily or monthly insolation, Contribution No. 18, Universite Catholique de Louvain, Institut d'Astronomie et de Geophysique, G. Lemaitre, Louvain-la-Neuve, B-1348 Belgique.

Berger, A.L. and Loutre, M.F. (1992) Astronomical solutions for paleoclimate studies over the last 3 million years. *Earth Planet. Sci. Lett.*, **111**, 369–382, doi: 10.1016/0012-821X(92)90190-7.

Berger, A., Ercegovac, M., and Mesinger, F. (Eds) (2005) Paleoclimate & the Earth Climate System. *Serb. Acad. of Sci. & Arts*, Vol CX, Book 4. 190.

Berner, R.A. (2004) *The Phanerozoic Carbon Cycle: CO_2 and O_2*, Oxford University Press, Oxford, 158p.

Bond, T.C. *et al.* (2013) Bounding the role of black carbon in the climate system: a scientific assessment. *J. Geophys. Res.*, **118**, (11), 5380–5552, doi: 10.1002/jgrd.50171.

Bowman, K.P. and Huang, J. (1991) A multi-grid solver for the Helmholtz equation on a semi-regular grid on the sphere. *Mon. Weather Rev.*, **119**, 769–775.

Bradley, R.S. (2015) *Paleoclimatology, Reconstructing Climates of the Quaternary*, Academic Press, Amsterdam, 675p.

Briggs, W.L., Henson, V.E., and McCormick, S.F. (2000) *A Multigrid Tutorial*, SIAM, Philadelphia, PA, 193p.

Budyko, M.I. (1968) On the origin of glacial epochs. *Meteorol. Gidrol.*, **2**, 3–8.

Budyko, M.I. (1969) The effect of solar radiation variations on the climate of the earth. *Tellus*, **21**, 611–619.

Budyko, M.I. (1972) The future climate. *Eos Trans. AGU*, **53**, 868–870.

Budyko, M.I. (1977) On present-day climatic changes. *Tellus*, **29**, 193–204.

Bulmer, M.G. (1979) *Principles of Statistics*, Dover Publications, New York, 252p, cdiac.ornl.gov/trends/co2/graphics/lawdome.gif (accessed 7 March 2017).

Byron, F.W. and Fuller, R.W. (1992) *Mathematics of Classical and Quantum Physics*, Dover Publications. 672p.

Cahalan, R.F. and North, G.R. (1979) A stability theorem for energy-balance climate models. *J. Atmos. Sci.*, **36**, 1205–1216.

Callen, H.B. (1985) *Thermodynamics and an Introduction to Thermostatistics*, 2nd edn, John Wiley & Sons, Inc., New York, 493p.

Cess, R.D. et al. (1990) Intercomparison and interpretation of climate feedback processes in 19 atmospheric general circulation models. *J. Geophys. Res.*, **95**, 16601–16615.

Chandrasekhar, S. (1960) *Radiative Transfer*, Dover Publications, New York, 393p.

Chýlek, P. and Coakley, J.A. (1975) Analytical analysis of a Budyko-type climate model. *J. Atmos. Sci.*, **32**, 675–679.

Coakley, J.A. and Yang, P. (2014) *Atmospheric Radiation – A Primer with Illustrative Solutions*, Wiley-VCH Verlag GmbH, Weinheim, 239p.

Coddington, O., Lean, J., Pilewskie, P., Snow, M., and Lindholm, D. (2016) A solar irradiance climate data record. *Bull. Am. Meteorol. Soc.*, **97**, 1265–1282.

Conrath, B.J., Hanel, R.A., Kunde, V.G., and Prabhakara, C. (1970) The infrared interferometer experiment on Nimbus 3. *J. Geophys. Res.*, **75**, 5831–5857.

Cover, T.M. and Thomas, J.A. (1991) *Elements of Information Theory*, John Wiley & Sons, Inc., New York, 542p.

Coxall, H.K., Wilson, P.A., Palike, H., Lear, C.H., and Backman, J. (2005) Rapid stepwise onset of Antarctic glaciation and deeper calcite compensation in the Pacific Ocean. *Nature*, **433**, 53–57.

Cramér, H. (1999) *Mathematical Methods of Statistics*, Princeton University Press, Princeton, NJ, 575p.

Cramér, H. and Leadbetter, M.R. (1995) *Stationary and Related Stochastic Processes*, Dover, Mineola, NY, 348p.

Crowley, T.J. and Baum, S.K. (1993) Effect of decreased solar luminosity on late Precambrian ice extent. *J. Geophys. Res.*, **98**, 16723–16732.

Crowley, T.J. and Hyde, W.T. (2008) Transient nature of late Pleistocene climate variability. *Nature*, **456**, 226–230, doi: 10.1038/nature07365.

Crowley, T.J. and Kim, K.Y. (1994) Milankovitch forcing of the last interglacial sea sevel. *Sciene*, **265**, 1566–1568.

Crowley, T.J. and North, G.R. (1988) Abrupt climate change and extinction events in earth history. *Science*, **240**, 996–1002.

Crowley, T.J. and North, G.R. (1990) Modeling the onset of glaciation. *Ann. Glaciol.*, **14**, 39–42.

Crowley, T.J., Short, D.A., Mengel, J.G., and North, G.R. (1986) Role of seasonality in the evolution of climate during the last 100 million years. *Science*, **231**, 579–584.

Crowley, T.J., Yip, K.-J., and Baum, S.K. (1994) Snowline instability in a general circulation model: application to Carboniferous glaciation. *Clim. Dyn.*, **10**, 363–376.

DeBlonde, G. and Peltier, W.R. (1991) *A One-Dimensional Model of Continental Ice Volume Fluctuations through the Pleistocene: Implications for the Origin of the Mid-Pleistocene Climate Transition*, doi: 10.1175/1520-0442.

Deblonde, G., Peltier, W.R., and Hyde, W.T. (1992) Simulations of continental ice sheet growth over the last glacial-interglacial cycle: experiments with a one level seasonal energy balance model including seasonal ice albedo feedback. *Palaeogeogr. Palaeoclimatol. Palaeoecol.*, **98**, 37–55.

DeConto, R.M. and Pollard, D. (2003) Rapid Cenozoic glaciation of Antarctica induced by declining atmospheric CO_2. *Nature*, **21**, 245–249.

Donner, L., Schubert, W., and Somerville, R. (eds) (2011) *The Development of Atmospheric General Circulation Models*, Cambridge University Press, Cambridge, 255p.

Donohoe, A., Armour, D.C., Pendergrass, A.G., and Battisti, D.S. (2015) Shortwave and longwave radiative contributions to global warming under increasing CO_2. *Proc. Nat. Acad. Sci.*, 16700–16705, doi: 10.1073/pnas.1412190111.

Drazin, P.G. (1992) *Nonlinear Systems*, Cambridge University Press, Cambridge, 308p.

Drazin, P.G. and Griffel, D.H. (1977) On the branching structure of diffusive climatological models. *J. Atmos. Sci.*, **35**, 1858–1867.

Eddy, J.A. (1976) The Maunder Minimum. *Science*, New Series, **192**, 1189–1202.

Emiliani, C. (1958) Palaeotemperature analysis of core 280 and Pleistocene correlations. *J. Geol.*, **66**, 264–275.

Erdelyi, E. (ed.) (1953) *Higher Transcendental Functions*, vol. 1, McGraw-Hill, New York, 303p.

Evans, D.A.D. (2000) Stratigraphic, geochronological, & paleomagnetic constraints upon the neo-proterozoic climatic paradox. *Am. J. Sci.*, **300**, 347–433.

Feulner, G. (2012) The faint young sun problem. *Rev. Geophys.*, **50**, doi: 8755-1209/12/2011RG000375.

Fletcher, C.A.J. (1991) *Computational Techniques for Fluid Dynamics*, vol. I, 2nd edn, Springer-Verlag, Berlin, 401p.

Frank, T.D., Birgenheier, L.P., Montanez, I.P., Fielding, C.R., and Rygel, M.C. (2008) Late Paleozoic climate dynamics revealed by comparison of ice-proximal stratigraphic and ice-distal isotopic records. *Geol. Soc. Am. Spec. Pap.*, **441**, 331–342.

Gal-Chen, T. and Schneider, S.H. (1976) Energy balance climate modeling: comparison of radiative and dynamic feedback mechanisms. *Tellus*, **28**, 108–121.

Gardiner, C.W. (1985) *Handbook of Stochastic Methods for Physics, Chemistry and the Natural Sciences*, 2nd edn, Springer-Verlag, Berlin, 442p.

Gardner, W.A. (1989) *Introduction to Random Processes*, 2nd edn, McGraw-Hill, New York, p. 546.

Gasquet, C. and Witomski, P. (Translated by R. Ryan) (1991) *Fourier Analysis and Applications*, Springer, New York, 442p.

Ghil, M. (1976) Climate stability for a Sellers-type model. *J. Atmos. Sci.*, **33**, 3–20.

Golitsyn, G. and Mokhov, I. (1978) Stability and extremal properties of climate models. *Izv. Acad. Sci. USSR Atmos. Oceanic Phys. Engl. Transl.*, **14**, 271–277.

Goody, R.M. and Yung, Y.L. (1989) *Atmospheric Radiation – Theoritical Basis*, 2nd edn, Oxford University Press, New York, 517p.

Gough, D.O. (1981) Solar interior structure and luminosity variations. *Sol. Phys.*, **74**, 21–34.

Graves, C.E., Lee, W.-H., and North, G.R. (1993) New parameterizations and sensitivities for simple climate models. *J. Geophys. Res.*, **98**, 5025–5036.

Gray, L.J., Beer, J., Geller, M., Haigh, J.D., Lockwood, M., Matthes, K., Cubasch, U., Fleitmann, D., Harrison, G., Hood, L., Luterbacher, J., Meehl, G.A., Shindell, D., van Geel, B., and White, W. (2010) Solar influence on climate. *Rev. Geophys.*, **48**, 1–53, doi: 10.1029/2009 RG000282.

Gregory, J. (2000) Vertical heat transports in the ocean and their effect on time-dependent climate change. *Clim. Dynam.*, **16**, 501. doi: 10.1007/s003820000059.

Hackbusch, W. (1980) *Multi-Grid Methods and Applications*, Springer-Verlag, Berlin, 377p.

Haigh, J.D. (2010) Solar variability and the stratosphere, in *The Stratosphere: Dynamics, Transport, and Chemistry* (eds L.M. Polvani, A.H. Sobel, and D.W. Waugh), American Geophysical Union, Washington, DC, pp. 173–187, ISBN: 9780875904795.

Haigh, J.D. and Cargill, P. (2015) *The Sun's Influence on Climate*, Princeton University Press, Princeton, NJ, 207p.

Hannan, E.J. (1970) *Multiple Time Series*, John Wiley & Sons, Inc., New York, 536p.

Hansen, J., Lacis, A., Rind, D., Russell, G., Stone, P., Fung, I., Ruedy, R., and Lerner, J. (1984) Climate sensitivity: analysis of feedback mechanisms, in *Climate Processes and Climate Sensitivity* (eds. J.E. Hansen and T. Takahashi), AGU Geophysical Monograph 29, Maurice Ewing Vol. **5**. American Geophysical Union, pp. 130–163.

Hansen, J. and Lebedeff, S. (1987) Global trends of measured surface air temperature. *J. Geophys. Res.*, **92**, 13345–13372.

Hart, M.H. (1978) Evolution of atmosphere of earth. *Icarus*, **37**, 23–39.

Hart, M.H. (1979) Habitable zones about main sequence stars. *Icarus*, **33**, 351–357.

Hartmann, D.L. (1994) *Global Physical Climatology*, 1st edn, Academic Press, San Diego, CA, 411p.

Hartmann, D.L. (2016) *Global Physical Climatology*, 2nd edn, Academic Press, San Diego, CA, 411p.

Harvey, L.D.D. and Schneider, S.H. (1985) Transient climate response to external forcing on 100–104 year time scales part 1: Experiments with globally averaged, coupled, atmosphere and ocean energy balance models. *J. Geophys. Res.*, **90**, doi: 10.1029/JD090iD01p02191. ISSN: 0148–0227.

Hasselmann, K. (1976) Stochastic climate models, Part I. Theory. *Tellus*, **6**, 473–484.

Hasselmann, K. (1979) On the signal-to-noise problem in atmospheric response studies, in *Meteorology Over the Tropical Oceans* (ed. D.B. Shaw), Royal Meteorological Society, pp. 251–259.

Hasselmann, K. (1993) Optimal fingerprints for the detection of time-dependent climate change. *J. Clim.*, **6**, 1957–1971.

Hasselmann, K. (1997) Multi-pattern fingerprint method for detection and attribution of climate change. *Clim. Dyn.*, **13**, 601–611.

Hayes, J.D., Imbrie, J., and Shackleton, N.J. (1976) Variations in the earth's orbit: pacemaker of the ice ages. *Science*, **194**, 1121–1132.

Hegerl, G.C. and North, G.R. (1997) Comparison of statistically optimal approaches to detecting climate change. *J. Clim.*, **10**, 1125–1133.

Held, I.M. (2005) The gap between simulation and understanding in climate modeling. *Bull. Am. Meteorol. Soc.*, **86** (11), 1609–1614.

Held, I.M. and Soden, B.J. (2000) Water vapor feedback and global warming. *Annu. Rev. Energy Env.*, **25**, 441–475.

Held, I.M. and Soden, B.J. (2006) Robust responses of the hydrological cycle to global warming. *J. Clim.*, **19**, 5686–5699.

Held, I.M. and Suarez, M. (1974) Simple albedo feedback models of the icecaps. *Tellus*, **36**, 613–629.

Held, I.M., Winton, M., Takahashi, K., Delworth, T., Zeng, F., and Vallis, G.K. (2010) Probing the fast and slow components of global warming by returning zbruptly to preindustrial forcing. *J. Clim.*, **23**, 2418–2427, doi: 10.1175/2009JCLI3466.1.

Hoffert, M., Callegari, A.J., and Ching-Tzong Hsie (1980) The role of deep sea heat storage in the secular response to climatic forcing. *J. Geophs. Res.*, **85**, 6667–6679.

Hoffert, M.I. and Flannery, B.P. (1985) Model projections of the time-dependent response to increasing carbon dioxide, in *Projecting the Climatic Effects of Increasing Carbon Dioxide*, Report ER-0237 (eds M.C. MacCracken and F.M. Luther), US Department of Energy, Washington, DC, pp. 150–190.

Hoffman, P.F. and Schrag, D.P. (2002) The snowball earth hypothesis: testing the limits of global change. *Terra Nova*, **14**, 129–155.

Horn, R.A. and Johnson, C.R. (1985) *Matrix Analysis*, Cambridge University Press, 561p.

Houghton, J.T. (1986) *The Physics of Atmospheres*, 2nd edn, Cambridge University Press, Cambridge, 271p.

Huang, R.X. (2009) *Ocean Circulation: Wind-Driven and Thermohaline Processes*. Cambridge University Press, Cambridge, UK, 806p.

Huang, J. and Bowman, K.P. (1992) The small ice cap instability in seasonal energy balance models. *Clim. Dyn.*, **7**, 205–215.

Huang, Y. and Shahabadi, M.B. (2014) Why logarithmic? A note on the dependence of radiative forcing on gas concentration. *J. Geophys. Res. Atmos.*, **119**, 13683–13689, doi: 10.1002/2014JD022466.

Huybers, P. and Wunsch, C. (1994) Obliquity pacing of the late Pleistocene glacial terminations. *Nature*, **434**, 491–494. doi: 10.1038/nature03401.

Hyde, W.T., Crowley, T.J., Baum, S.K., and Peltier, W.R. (2000) Neoproterozoic 'snowball earth' simulations with a coupled climate/ice-sheet model. *Nature*, **405**, 425–429.

Hyde, W.T., Crowley, T.J., Kim, K.-Y., and North, G.R. (1989) A comparison of GCM and energy balance model simulations of seasonal temperature changes over the past 18,000 years. *J. Clim.*, **2**, 864–887.

Hyde, W.T., Kim, K.Y., and Crowley, T.J. (1990) On the relation between polar continentality and climate: studies with a nonlinear seasonal energy balance model. *J. Geophys. Res.*, **95**, 18653–18668.

Hyde, W.T. and Peltier, W.R. (1987) Sensitivity experiments with a model of the ice-age cycle - the response to Malankovitch forcing. *J. Atmos. Sci.*, **44** (10), 1351–1374, doi: 10.1175/1520-0469(1987)0441351:SEWAMO2.0.CO;2.

Imbrie, J. *et al.* (1992) On the structure and origin of major glaciation cycles 1. Linear responses to Milankovitch forcing. *Paleoceanography*, **7**, 701–738.

Ingersoll, A.P. (1969) The runaway greenhouse: a history of water on Venus. *J. Atmos. Sci.*, **26**, 1191–1198.

Ingersoll, A.P. (2015) *Planetary Climates*, Princeton University Press, Princeton, NJ, p. 278.

Intergovernmental Panel on Climate Change (IPCC) (2007, 2013) www.ipcc.ch (accessed 9 March 2017).

Irving, D. (2016) A minimum standard for publishing computational results in the weather and climate sciences. *Bull. Amer. Meteorol. Soc.*, doi: O.I175/BAMS-D-15-00010.1.

James, P.B. and North, G.R. (1982) The seasonal CO2 cycle on mars: an application of an energy balance climate model. *J. Geophys. Res.*, **87**, 10271–10283.

Jackson, J.D. (1962) *Classical Electrodynamics*, John Wiley & Sons, NY, 641p.

Kamke, E. (1959) *Differentialgleichungen, Lösungenmethoden und Lösungen*, vol. 1, 3rd edn, Chelsea Pull., Co., New York, 666p.

Kasting, J.F. (1988) Runaway and moist greenhouse atmospheres and the evolution of earth and Venus. *Icarus*, **74**, 472–494.

Kasting, J.F. (2010) Faint young Sun redux. *Nature*, **464**, 687–688.

Kasting, J.F. (2014) *How to Find a Habitable Planet*, Princeton University Press, Princeton, NJ, 352p.

Kasting, J.F., Pollack, J.B., and Ackerman, T.P. (1984) Response of earth's atmosphere to increases in solar flux and implications for loss of water from venus. *Icarus*, **57**, 335–355.

Kelly, J.J. (2006) *Graduate Mathematical Physics, With MATHEMATICA Supplements*, Wiley-VCH, 482p.

Kim, K.-Y. and North, G.R. (1991) Surface temperature fluctuations in a stochastic model. *J. Geophys. Res.*, **96**, 18573–18580.

Kim, K.-Y. and North, G.R. (1992) Seasonal cycle and second-moment statistics of a simple coupled climate system. *J. Geophys. Res.*, **97**, 20437–20448.

Kim, K.-Y. and North, G.R. (1997) EOFs of harmonizable cyclostationary processes. *J. Atmos. Sci.*, **54**, 2416–2427.

Kim, K.-Y., North, G.R., and Hegerl, G.C. (1996) Comparisons of the second-moment statistics of climate models. *J. Clim.*, **9**, 2204–2221.

Kim, K.-Y., North, G.R., and Huang, J. (1992) On the transient response of a simple coupled climate system. *J. Geophys. Res.*, **97**, 10069–10081.

Kirschivink, J.L. (1992) Late proterozoic low-latitude global glaciation: the snowball earth, in *The Proterozoic Biosphere: A Multidisciplinary Study* (eds J.W. Schopf and C. Klein), Cambridge University Press.

Kleeman, R. (2011) Information theory and dynamical system predictability. *Entropy*, **13**, 612–649, doi: 10.3390/e13030612.

Kopp, G. (2017) Greg Kopp's TSI Page, http://spot.colorado.edu/ koppg/TSI/ (accessed 3 March 2017).

Kopp, G. and Lean, J.L. (2011) A new, lower value of total solar irradiance: evidence and climate significance. *Geophys. Res. Lett.*, **38** (1), doi: 10.1029/2010GL045777.

Körner, T.W. (1989) *Fourier Analysis*, Cambridge University Press, 591p.

Kullback, S. (1968) *Information Theory and Statistics*, Dover Publications, Mineola, NY, 399p.

Lebedeff, S.A. (1988) Analytic solution of the box diffusion model for a global ocean. *J. Geophys. Res.*, **93** (D11), 14243–14255.

Lee, W.-H. and North, G.R. (1995) Small icecap instability in the presence of fluctuations. *Clim. Dyn.*, **11**, 242–246.

Leith, C.E. (1975) Climate response and fluctuation-dissipation. *J. Atmos. Sci.*, **32**, 2022–2026.

Leung, L.-Y. and North, G.R. (1990) Information theory & climate prediction. *J. Clim.*, **3**, 5–14.

Leung, L.-Y. and North, G.R. (1991) Atmospheric variability on a zonally symmetric land planet. *J. Clim.*, **4**, 753–765.

Levitus, S. (1982) *Climatological Atlas of the World Ocean*, NOAA/ERL GFDL Professional Paper 13, Princeton, NJ, pp. 173 (NTIS PB83-184093).

Lin, R.-Q. and North, G.R. (1990) A study of abrupt climate change in a simple non-linear climate model. *Clim. Dyn.*, **4**, 253–262.

Lindzen, R.S. (1994) Climate dynamics and global change. *Annu. Rev. Fluid Mech.*, **26**, 353–378.

Lindzen, R.S. and Giannitsis, C. (1998) On the climatic implications of volcanic cooling. *J. Geophys. Res.*, **103**, 5929–5941.

Lindzen, R.S., Hou, A.Y., and Farrell, B.F. (1982) The role of convict model choice in calculating the climate impact of doubling CO_2. *J. Atmos. Sci.*, **39**, 1189–1205.

Lindzen, R.S. and Farrell, B.F. (1977) Some realistic modification of simple climate models. *J. Atmos. Sci.*, **34**, 1487–1501.

Liou, K.N. (1992) *Radiation and Cloud Processes in the Atmosphere*, Oxford University Press, New York, 487p.

Liou, K.N. (2002) *An Introduction to Atmospheric Radiation*, 2nd edn, Academic Press, San Diego, CA, 583p.

Lisiecki, L.W. and Raymo, N.E. (2005) A Pliocene-Pleistocene stack of 57 globally distributed benthic $\delta^{18}O$ records. *Paleoceanography*, **20** (1), doi: 10.1029/2004PA001071.

Loeb, N.G. and Wielicki, B.A. (2014) *Encyclopedia of the Atmospheric Sciences*, vol. 4, 2nd edn (eds G.R. North, F. Zhang, and J.A. Pyle), Elsevier, New York, p. 67.

Loeb, N.G., Wielicki, B.A., Doelling, D.R., Smith, G.L., Keyes, D.F., Kato, S., Manalo-Smith, N., and Wong, T. (2009) Toward optimal closure of the earth's top-of- atmosphere radiation budget. *J. Clim.*, **22**, 748–766.

London, J. (1980) Radiative energy sources & sinks in the stratosphere and mesosphere, in *Proceedings of the NATO Advanced Study Institute on Atmospheric Ozone: Its Variation and Human Influences: Aldeia das Acoteias, Abi2rve, Portugal, October 1-13, 1979* (ed. A.C. Aiken), Federal Aviation Administration, US Department of Transportation, Washington, DC, pp. 703–721.

Lorenz, E. (1975) *Climatic Predictability*, GARP Publications Series, pp. 132–136, http://eaps4.mit.edu/research/Lorenz/publications.htm (accessed 3 March 2017).

Lüthi, D., Le Floch, M., Bereiter, B., Blunier, T., Barnola, J.-M., Siegenthaler, U., Raynaud, D., Jouzel, J., Fischer, H., Kawamura, K., and Stocker, T.F. (2008) High-resolution carbon dioxide concentration record 650,00-800,000 years before present. *Nature*, **453**, 379–382, doi: 10.1038/nature06949.

Manabe, S., Stouffer, R.J., Spelman, M.J., and Bryan, K. (1991) Transient responses of a coupled ocean – atmosphere model to gradual changes of atmospheric CO_2. Part I: Annual mean response. *J. Clim.*, **4**, 785–818.

Manabe, S. and Strickler, R.F. (1964) Thermal equilibrium of the atmosphere with a convective adjustment. *J. Atmos. Sci.*, **21**, 361–385.

Manabe, S. and Wetherald, R.T. (1967) Thermal equilibrium of the atmosphere with a given distribution of relative humidity. *J. Atmos. Sci.*, **24**, 241–259.

Manabe, S. and Wetherald, R.T. (1975) The effects of doubling the CO_2 concentration on the climate of a general circulation model. *J. Amos. Sci.*, **32**, 3–15.

Melezhik, V.A. (2006) Multiple causes of earth's earliest global glaciation. *Terra Nova*, **18**, 130–137.

Mengel, J.G., Short, D.A., and North, G.R. (1988) Seasonal snowline instability in an energy balance model. *Clim. Dyn.*, **2**, 127–131.

Milankovitch, M. (1941) Canon of insolation and the ice age problem (in Serbian), K. Serb. Acad. Beogr. Spec. Publ., 132 (English translation, 482ppll, Israel Program for Scientific Translations, Jerusalem, 1969).

Milankovitch, M. (1998) *Canon of Insolation and the Ice-Age Problem*, Belgrade.

Montgomery, D.C., Peck, E.A., and Vining, G.G. (2001) *Introduction to Linear Regression Analysis*, John Wiley & Sons, Inc., New York, 641p.

Morantine, M.C. and Watts, R.G. (1994) Time scales in energy balance climate models 2: the intermediate time solutions. *J. Geophys. Res.*, **99**, 3643–3653.

Myhre, G., Highwood, E.J., Shine, K.P., and Stordal, F. (1998) New estimates of radiative forcing due to well mixed greenhouse gases. *Geophys. Res. Lett.*, **25**, 2715–2718.

Nakajima, S., Hayashi, Y.-Y., and Abe, Y. (1992) A study on the "runaway greenhouse effect" with a one-dimensional radiative-convective equilibrium model. *J. Atmos. Sci.*, **49**, 2256–2266.

NASA Goddard Institute for Space Studies (2017) January 2017 was Third-Warmest January on Record, www.giss.nasa.gov (accessed 7 March 2017).

Neelin, J.D. (2011) *Climate Change and Climate Modeling*, Cambridge University Press, Cambridge, 282p.

North, G.R. (1975a) Analytical solution to a simple climate model with diffusive heat transport. *J. Atmos. Sci.*, **32**, 1301–1307.

North, G.R. (1975b) Theory of energy-balance climate models. *J. Atmos. Sci.*, **32**, 2033–2043.

North, G.R. (1984) The small ice cap instability in diffusive climate models. *J. Atmos. Sci.*, **41**, 3390–3395.

North, G.R., Bell, R.E., and Hardin, J.W. (1993) Fluctuation dissipation in a general circulation model. *Clim. Dyn.*, **8**, 259–264.

North, G.R. and Cahalan, R.F. (1981) Predictability in a solvable stochastic climate model. *J. Atmos. Sci.*, **38**, 504–513.

North, G.R., Cahalan, R.F., and Coakley, J.A. Jr. (1981) Energy balance climate models. *Rev. Geophys. Space Phys.*, **19**, 91–121.

North, G.R. and Coakley, J.A. (1979) Differences between seasonal and mean annual energy balance model calculations of climate and climate change. *J. Atmos. Sci.*, **36**, 1189–1204.

North, G.R. and Crowley, T.J. (1985) Application of a seasonal climate model to Cenozoic glaciation. *J. Geol. Soc. (London, U.K.)*, **142**, 475–482.

North, G.R., Howard, L., Pollard, D., and Wielicki, B. (1979) Variational formulation of the Budyko-Sellers models. *J. Atmos. Sci.*, **36**, 255–259.

North, G.R. and Kim, K.-Y. (1995) Detection of forced climate signals. Part II: Simulation results. *J. Clim.*, **8**, 409–417.

North, G.R., Kim, K.-Y., Shen, S.S.P., and Hardin, J.W. (1995) Detection of forced climate signals. Part I: Filter theory. *J. Clim.*, **8**, 401–408.

North, G.R., Mengel, J.G., and Short, D.A. (1983) A simple energy balance model resolving the seasons and the continents: application to the Milankovitch theory of the ice ages. *J. Geophys. Res.*, **88**, 6576–6586.

North, G.R. and Stevens, M.J. (1998) Detecting climate signals in the surface temperature field. *J. Clim.*, **11**, 563–577.

North, G.R., Wang, J., and Genton, M. (2011) Correlation models for temperature fields. *J. Clim.*, **24**, 5850–5862.

North, G.R. and Wu, Q. (2001) Detecting climate signals using space-time EOFs. *J. Clim.*, **14**, 1839–1862.

North, G.R., Yip, K.-J.J., Chervin, R.M., and Leung, L.-Y. (1992a) Forced and free variations of the surface temperature field in a general circulation model. *J. Clim.*, **5**, 227–239.

North, G.R., Yip, K.-J., Leung, L.-Y., and Chervin, R.M. (1992b) Forced and free variations of the surface temperature field in a general circulation model. *J. Clim.*, **5**, 227–239.

NRC (2006) *Surface Temperature Reconstructions for the Last 2,000 Years*, National Academies Press, 145p.

Nye, J.F. (1959) The motion of ice sheets and glaciers. *J. Glacial.*, **3**, 493–507.

Obukhov, A.M. (1947) Statistically homogeneous fields on a sphere. *Usp. Mat. Navk.*, **2**, 196–198.

Opik, E.J. (1965) Climatic changes in cosmic perspective. *Icarus*, **4**, 289–307.

Owen, T., Cess, R.D., and Ramanathan, V. (1979) Enhanced CO_2 greenhouse to compensate for reduced solar luminosity on early Earth. *Nature*, **277**, 640–642.

Paillard, D. (2001) Glacial cycles: Toward a new paradigm. Rev. Geophys., **39**, 325–346. doi: 10.1029/2000RG000091.

Paltridge, G.W. (1975) Global dynamics and climate change – a system of minimum entropy exchange. *Q. J. R. Meteorol. Soc.*, **101**, 475–484.

Paltridge, G.W. (1978) The steady-state format of global climate. *Q. J. R. Meteorol. Soc.*, **104**, 927–945.

Papoulis, A. (1984) *Probability, Random Variables, and Stochastic Processes*, 2nd edn, McGraw-Hill, New York, 576p.

Pedlosky, J. (2003) *Waves in the Ocean and the Atmosphere, Introduction to Wave Dynamics*, Springer-Verlag, Berlin, 260p.

Percival, D.B. and Walden, A.T. (1993) *Spectral Analysis for Physical Applications*, Cambridge University Press, 583p.

Petit, J.R. et al. (2001) *Vostok Ice Core Data for 420,000 Years*, IGBP PAGES/World Data Center for Paleoclimatology Data Contribution Series #2001-076, NOAA/NGDC Paleoclimatology Program, Boulder, CO.

Petit, J.R., Jouzel, J., Raynaud, D., Barkov, N.I., Barnola, J.M., Basile, I., Bender, M., Chappellaz, J., Davis, J., Delaygue, G., Delmotte, M., Kotlyakov, V.M., Legrand, M., Lipenkov, V., Lorius, C., Pépin, L., Ritz, C., Saltzman, E., and Stievenard, M. (1999) Climate and atmospheric history of the past 420,000 years from the vostok ice core, Antarctica. *Nature*, **399**, 429–436.

Petty, G.W. (2006) *A First Course in Atmospheric Radiation*, Sundog Publishing, Madison, WI, 458p.

Phillips, N.A. (1956) The general circulation of the atmosphere: a numerical experiment. *Q. J. R. Meteorol. Soc.*, **82**, 123–164.

Picard, G.L. and Emery, W.J. (1990) *Descriptive Physical Oceanography – An Introduction*, Pergamon Press, Oxford, 320p.

Pierrehumbert, R.T. (2011) *Principles of Planetary Climate*, Cambridge University Press, 680p.

Pollard, D. (1982) A simple ice sheet model yields realistic 100 K glacial cycles. *Nature*, **296**, 334–338.

Pollack, H.N. and Huang, S.P. (2000) Climate reconstruction from subsurface temperatures. *Annu. Rev Earth Planet. Sci.*, **28**, 339–369.

Prigogine, I. (1968) *Introduction to Thermodynamics of Irreversible Processes*, 3rd edn, Wiley-Interscience, New York, 147p.

Pujol, T. and North, G.R. (2003) Analytical investigation of the atmospheric radiation limits in semigray atmospheres in radiative equilibrium. *Tellus*, **55A**, 328–337.

Ramanathan, V. and Coakley, J.A. Jr. (1978) Climate modeling through radiative-convective models. *Rev. Geophys. Space Phys.*, **16**, 465–489.

Raymo, M.E. and Huybers, P. (2008) Unlocking the mysteries of the ice ages. *Nature*, **17**, 284–285.

Richtmyer, F.K., Kennard, E.H., and Lauritsen, T. (1955) *Introduction to Modern Physics*, 5th edn, McGraw-Hill Book Co., New York, 666p.

Robinson, A., and H. Stommel (1959) The Oceanic Thermocline and the Associated Thermohaline Circulation, *Tellus*, **11**, 295–308. doi: 10.1111/j.2153-3490.1959.tb00035.

Roe, G. (2009) Feedbacks, timescales, and seeing Red. *Annu. Rev. Earth Sci.*, **37**, 93–115.

Roe, G.H. and Baker, M. (2007) Why is climate sensitivity so unpredictable? *Science*, **318**, 629–632.

Roe, G.H., Feldl, N., Armour, K.C. et al. (2015) The remote impacts of climate feedbacks on regional climate predictability. *Nat. Geosci.*, **8**, 135–139.

Rosing, M.T., Bird, D.K., Sleep, N.H., and Bjerrum, C.J. (2010) No climate paradox under the faint early sun. *Nature*, **464** (7289), 744–747.

Roulston, M.S. and Smith, L.A. (2002) Evaluating probabilistic forecasts using information theory. *Mon. Weather Rev.*, **130**, 1653–1660.

Ruddiman, W.F. and McIntyre, A. (1984) Ice-age thermal response and climatic role of the surface North Atlantic Ocean, 40 to 63°N. *Geol. Soc. Am. Bull.*, **95**, 381–396.

Russell, G.L. and Rind, D. (1999) Response to CO_2 transient increase in the GISS coupled model: regional coolings in a warming climate. *J. Clim.*, **12**, 531–539.

Rygel, M.C., Fielding, C.R., Frank, T.D., and Birgenheier, L.P. (2008) The magnitude of late glacioeustatic fluctuations: a synthesis. *J. Sediment. Res.*, **78**, 500–511.

Sagan, C. and Mullen, G. (1972) Earth and mars: evolution of atmospheres and surface temperatures. *Science*, **177**, 52–56.

Salmun, H., Cahalan, R.F., and North, G.R. (1980) Latitude-dependent sensitivity to stationary perturbations in simple climate models. *J. Atmos. Sci.*, **37**, 1874–1879.

Saltzmann, B., Hansen, A.G., and Maasch, K.A. (1984) The late quaternary glaciers as the response of a three-component feedback system to earth orbital forcing. *J. Atmos. Sci.*, **41**, 3380–3389.

Schlesinger, M.E. (1986) Equilibrium and transient climatic warming induced by increased atmospheric CO_2. *Clim. Dyn.*, **1**, 35–51.

Schmittner, A., Urban, N.M., Shakun, J.D., Mahowald, N.M., Clark, P.U., Bartlein, P.J., Mix, A.C., and Rosell-Mele, A. (2011) Climate sensitivity estimated from temperature reconstructions of the last glacial maximum. *Science*, **334**, 1385–1388.

Schneider, S.H. and Dickinson, R.E. (1974) Climate modeling. *Rev. Geophys. Space Phys.*, **12**, 447–493.

Schwarzschild, K. (1906) On the equilibrium of the sun's atmosphere. *Nach. K. Gesell. Wiss. Götingen, Math.-Phys. Klasse*, **195**, 41–53.

Sellers, W.D. (1969) A climate model based on the energy balance of the earth-atmosphere system. *J. Appl. Meteorol.*, **8**, 392–400.

Shen, S.S.P. and North, G.R. (1999) A simple proof of the slope stability theorem for energy balance climate model. *Can. Appl. Math. Q.*, **7**, 201–215.

Shen, S.S., North, G.R., and Kim, K.-Y. (1994) Spectral approach to optimal estimation of the global average temperature. *J. Clim.*, **7**, 1999–2007.

Short, D.A. and Mengel, J.G. (1986) Tropical climatic phase lags and Earth's precession cycle. *Nature*, **323**, 48–50. doi: 10.1038/323048a0.

Short, D.A., Mengel, J.G., Crowley, T.J., Hyde, W.T., and North, G.R. (1991) Filtering of Milankovitch cycles by earth's geography. *Quat. Res.*, **35**, 157–173.

Simpson, G.C. (1927a) Some studies in terrestrial erudition. *Mem. R. Meteorol. Soc. II*, **16**, 69–95.

Simpson, G.C. (1927b) Further studies in terrestrial erudition. *Mem. R. Meteorol. Soc. III*, **21**, 69–95.

Smoluchowski, M. (1906) Zur kinetischen Theorie der Brownschen Molekularbewegung und der Suspensionen. *Ann. d. Phys.* (Leipzig) **21**, 756–780 (in German).

Solomon, S. and Dahe, Q. (2007) Historical overview of climate change science (What is the greenhouse effect?), in *Climate Change 2007: The Physical Science Basis, Working Group 1 Contribution to the Fourth Assessment Report of the Intergovernmental Panel on Climate Change*, Chapter 1 (eds S. Solomon and Q. Dahe), Cambridge University Press, Cambridge, pp. 115–116.

Stevens, M.J. (1997) Optimal estimation of the surface temperature response to natural and anthropogenic climate forcings over the past century. PhD dissertation. Texas A&M University, 157p.

Stevens, B. (2015) Good Scientific Practice, http://www.mpimet.mpg.de/en/science/publications/good-scientific-practice/ (accessed 9 March 2017).

Stevens, M.J. and North, G.R. (1996) Detection of the climate response to the solar cycle. *J. Atmos. Sci.*, **53**, 2594–2608.

Stewart, R. Introduction to Physical Oceanography, http://www.colorado.edu/oclab/sites/default/files/attached-files/stewart.textbook.pdf (accessed 9 March 2017).

Stoer, J., Bulirsch, R., Gautschi, W. (Translator), and Witzgall, C. (Translator) (2002) *Introduction to Numerical Analysis*. Springer, 746p.

Stone, P.H., & J.S. Risbey (1990) On the limitations of general circulation climate models. *Geophys. Res. Lett.*, **17**, 2173–2176.

Summerhayes, C.P. (2015) *Earth's Climate Evolution*, Wiley-Blackwell, Singapore, 394p.

Tarasov, L. and Peltier, W.R. (1997) Terminating the 100 kyr ice age cycle. *J. Geophys. Res.*, **102** (D18), 21665–21693.

Thiébaux, H.J. (1994) *Statistical Data Analysis for Ocean and Atmospheric Sciences*, Academic Press, 247p.

Trenberth, K.E., and Caron, J.M. (2001) Estimates of meridional atmosphere and ocean heat transports. *Bull. Amer. Meteorol. Soc.*, doi: 10.1175/1520-0442.

Trenberth, K.E., Fasullo, J.T., and Kiehl, J. (2009) Earth's global energy budget. *Bull. Am. Meteorol. Soc.*, **90**, 311–323.

van den Dool, H. (2007) *Empirical Methods in Short-Term Climate Prediction*, Oxford University Press, Oxford, 240p.

Walter H. Munk, (1966) Abyssal recipes. *Deep-Sea Research*, **13**, 707–730.

Ward, P.D. and Brownlee, D. (2003) *Rare Earth: Why Complex Life is Uncommon in the Universe*, Copernicus, 338p.

Ward, P.D. and Kirschivink, J. (2015) *A New History of Life: The Radical New Discoveries about the Origins and Evolution of Life on Earth*, Bloomsbury Press, New York, 400p.

Warren, S.G., Brandt, R.E., Grenfell, T.C., and McKay, C.P. (2002) Snowball earth: ice thickness on the tropical ocean. *J. Geophys. Res.*, **107** (C10), 3167, doi: 10:.1029/2001JC001123.

Washington, W.M. and Parkinson, C.L. (2005) *An Introduction to Three-Dimensional Climate Modeling*, 2nd edn, University Science Books, Sausalito, CA, 353p.

Watts, R.G., Morantine, M.C., and Rao, K.A. (1994) Timescales in energy balance climate models 1. The limiting case solutions. *J. Geophys. Res.*, **99**, 3631–3641.

Weart, S.R. (2008) *The Discovery of Global Warming: Revised and Expanded Edition (New Histories of Science, Technology)*, Revised edn, Harvard University Press, 240p.

Weaver, C.P. and Ramanathan, V. (1995) Deductions from a simple climate model: factors governing surface temperature and atmospheric thermal structure. *J. Geophys. Res.*, **100**, 11585–11591.

Wendisch, M. and Yang, P. (2012) *Theory of Atmospheric Radiation*, Wiley-VCH Verlag GmbH, Weinheim, 321p.

Whittaker, E.T. and Watson, G.N. (1962) *A Course of Modern Analysis*, 4th edn, Cambridge University Press, Cambridge, 608p.

Wills, A.P. (1958) *Vector Analysis with an Introduction to Tensor Analysis*, Dover Publications, New York, 285p.

Wigley, T.M. and Schlesinger, M.E. (1985) Analytical solution for the effect of increasing CO_2 on global mean temperature. *Nature*, **315**, 649–652. doi: 10.1038/315649a0.

Wu, W. and North, G.R. (2007) Thermal Decay Modes in Simple Climate Models. *Tellus*, **59**, 618–626.

Zachos, J.C., Pagani, M., Sloan, L.C., Thomas, E., and Billups, K. (2001) Trends, rhythms and aberrations in global climate 65 Ma to present. *Science*, **292**, 686–693.

Zhuang, K., North, G.R., and Giardino, J.R. (2014) Hysteresis of glaciations in the Permo-Carboniferous. *J. Geophys. Res. Atmos.*, **119**, 2147–2155, doi: 10.1002/2013JD020524.

Index

a

absorption   64
absorption spectrum   85
absorption strength   88
absorptivity   13, 63
addition theorem for spherical harmonics   232
adjustable parameters   19
advection term   229
aerosol particles   82, 229
albedo   13, 28
all-land planet   152
Andronova, N.   112, 116
Angell, J. K.   294
angular brackets   4
annual average   6
anomalies   5, 147, 229
Arakawa, A.   16
Archer, D.   93
Archer, J.   93
Arfken, G. B.   292
atmospheric scattering   64
atmospheric vertical structure   7
attractor   257
attractor basin   187
autocorrelation function   5, 158
autocorrelation time   5
autocovariance function   234

b

Baker, M.   116
B and its relation to sensitivity   31
Baum, S. K.   326
beach ball model   203
Bender, M.   326
Bessel functions   156, 180
Bessel's integral   156
bifurcation   21
bivariate pdf   5
black body   12, 28
Boltzmann constant   62
boundary conditions   67
boundary layer   8
Bowman, K. P.   195, 224
box-diffusion model   273
brightness temperature   69
Budkyo radiation rule   27
Budyko, M. I.   14, 119, 122, 198
Budyko radiation rule   29
Budyko's IR rule   122

c

Cahalan, R. F.   37, 180
centered moment   5
Cess, R. D.   108
Chandrasekhar   83
Chapman level   101
chemical composition   11
Chicago website   92
chlorofluorocarbons (CFCs)   86
circular orbit   151
Clausius/Clapeyron equation   108
climate feedbacks   105
climate sensitivity   30, 109
cloud tops   14
Coakley, J. A.   58, 83, 122, 203
coalbedo   28, 83
cold start   254
commitment   259
completeness relation   292

complex random numbers   230
Conrath, B. J.   91
convection   69
convective adjustment   15
convective-adjustment models   78
cooling rate   73
coordinate $\mu = \vartheta$ 120
CO_2 forcing   101
CO_2 spectrum   89
covariance   5, 230
covariance of errors   288
Crowley, T. J.   326

d
damped diffusion   154
damped diffusion dynamics   254
damped diffusion filter   231
decay modes   146, 233
decay times   147
degeneracy   207
degree variance   233
deterministic signals   288
detrending   5
diffusive heat transport   123
dipole moment   86
Dirac delta function   235, 254
discrete time index   6
distribution of solar flux   12
divergence of a flux   57
Doppler shift   87
dry adiabatic lapse rate   70
dust veil   253
dynamical normal modes   233

e
eccentricity   149
Eddington approximation   61, 66
eddy diffusion   75
effective heat capacity   253
eigenfunctions   205
eigenmode amplitudes   263
eigen patterns   263
eigenvalues   182, 205, 233
emission temperature   27
emissivity   63
empirical orthogonal functions (EOFs)   232, 290

empirical parameters   19
energy levels   85
ensemble   3
ensemble average   4, 28
ensemble mean   229
ensemble members   28
EOFs *see* empirical orthogonal functions (EOFs)
equation of radiation energy transfer   61
equinox distribution of insolation   121
ergodic system   5, 6
estimator   287
Evans, D. A. D.   322
exoplanets   103
extinction coefficient   61
extreme heating distributions   122

f
facula   11
far infrared   88
Farrell, B. F.   132
feedback formalism   107
Feulner, G.   325
flux density   57, 58, 65
flux at the top of the atmosphere   14
Fourier series, complex   151
Fourier transform   234
frequency dependence of length scale   191
fudge factors   28
functional   184

g
Galerkin method   163
general circulation models (GCMs)   16
generalized Stürm–Liouville system   244
Ghil, M.   180
Goody, R.   65, 86, 101, 102
Goody, R. M.   83
Gough, D. O.   12, 325
Graves, C. E.   122, 130
gray atmosphere   57, 63, 66, 85
gray spectrum   80
Green's function   253
greenhouse effect   20, 85
greenhouse gas (GHG)   85, 92

h

habitable zone   103
Hansen, J.   293
Hart, M. H.   103
Hartmann, D. L.   129
Hasselmann, K.   298
heat conduction, vertical   75
heat storage term   234
heating components   129
heating components: $H_0 = 0.70$, $H_2 = -0.53$, $H_4 = 0.061$ 130
Heaviside step function   253, 256
Held, I.   100, 109
Hermite polynomials   180
Hermitian   244
Hoffman, P. F.   322
homogeneous planet   148
homogeneous solution   257
Huang, J.   195, 224
Huang, Y.   101
Hyde, W. T.   199, 224, 329
hygroscopic gases   255

i

ice cores   92
ice feedback   109
impulse/response function   253
incoming solar radiation   121
independent estimators   288
industrial revolution   92
infrared radiation   14
Ingersoll, A.   116
initial conditions   3
initial distribution   147
insolation   121
insolation distribution   148
International Panel on Climate Change (IPCC)   306
irregular Legendre functions   128

k

Karhunen–Loève functions   232, 292
Kasahara   16
Kasting, J. F.   105
Kasting, R. F.   325
Keynes, John Maynard   287
Kidrschvink, J. L.   321
Kim, K. Y.   224
Kirchoff's laws   62, 63
Kirschvink, J.   116
Korshover, J.   294
Kronecker δ_{mn} 127

l

lagged autocorrelation   5
Langevin equation   51
Laplace series   215, 230
Laplace transform   270
lapse rate feedback   108
latitude belt   121
Law Dome   92
Lebedeff, S.   293
Legendre modes   120
Legendre polynomials   120, 145
Leith, C. E.   16
length scale   191
Lin, R.-Q.   195, 197, 199
linear stability   180
linear stability analysis   181
line broadening   87
line density   87
Lindzen, R. S.   116, 132
Liou, K. N.   83
long wave radiation   14
low-pass filters   234, 254

m

Manabe, S.   16, 58, 78
Maunder Minimum   12
mean   5
Melezhik, V. A.   322
Mengel, J. G.   193, 198
middle atmosphere   112
mid-latitude weather systems   229
mixed-layer   9, 253
mode amplitudes   152
MODTRAN   81, 92, 94, 99
moments   4
Mount Pinatubo   114
Mullen, G.   325
Myhre, G.   99, 102

n

Nakajima, S.   116
natural frequencies   85

neoproterozoic glaciations   322
network of stations   291
Neumann boundary condition   160
normally distributed   5
North, G. R.   37, 104, 122, 180, 190, 193, 195, 197, 199, 203, 224
numerical methods   159
Nye, J. F.   327

o

ocean mixed layer   9
optical depth   64
optical path length   63
orbital elements   9
orthogonality   179
orthogonality relation   292

p

Papoulis, A.   111
parsimony   19, 28
particular solution   257
Pedlosky, J.   176
Peltier, W. R.   327
pencil beam   59
perihelion   10
periodic forcing   277
Permafrost   261
Petty, G. W.   83
phenomenological parameter   19
photon energy   86
Pierrehumbert, R. T.   83, 86, 101
Planck constant   62, 86
Planck function   104
Planck radiation function   62
planet(s)   27
planetary albedo   13
polewards heat transport   122
polyatomic molecules   85
potential function   184
pressure broadening   87
probability density function (pdf)   5
pseudo-spectral method   164
Pujol, T.   104

r

radiance   58
radiant energy flux   58
radiation temperature   27, 30
radiation transfer programs   90
radiative convective model   58, 77
radiative equilibrium   58
radiative equilibrium profile   74
radiatively active molecules   62
radiative transfer   57
Ramanathan, V.   58, 104
ramp forcing   274
random field   232
randomized initial conditions   28
random number(s)   4
random number generator   231
random variable   287
random walk   153
random winds   157, 229
realizations   3, 231
recalcitrant climate effect   264
red noise   281
regression line   5
relaxation times   147
reservoir, thermal   288
residuals   5
resonant frequencies   85
rhomboidal truncation   212
Roe, G.   100, 112, 116
root mean square error (RMS)   287, 288
Rosing, M. T.   325
rotational invariance   232
runaway greenhouse   102

s

Sagan, C.   325
Schlesinger, M. E.   107, 112, 116
Schrag, D. P.   322
seasonal cycle   253
seasonal insolation   148
Sellers, W. D.   119, 123
semi-implicit method   162
Shahbadi, B.   101
Shen, S. S. P.   290, 294
Simpson, G. C.   116
single-slab ocean   254
skin temperature   58
slope-stability theorem   180
Smagorinsky, J.   16
small ice cap instability   193

Soden, B.   100, 109
soil   261
solar cycle   253
solar energy absorbed   28
solar luminosity   11
solid angle integrals   60
source function   61
spatially white noise   230
Spectral Calc.com   87, 88
spectral density   280
spectral method   163
spectral radiance   59
spectrum of variances   234
speed of light   62
spherical coordinates   120
spherical harmonic(s)   211, 293
spherical harmonic degree   230
spherical harmonic series   212
Stürm–Liouville problem   125
Stürm–Liouville system   179, 292
stability   180, 181
stationary   5, 7
statistical equilibrium   15
statistically independent   6, 230
statistical mechanics   3
statistical model   287
steam engine   92
Stefan–Boltzmann constant   62
Stefan–Boltzmann Law   27
Stevens, B.   119
Stevens, M. J.   224
stratified atmosphere   60
stratosphere   112
Strickler, R. F.   58, 78
surface albedo   64

t
Tarasov, L.   327
temporal response   253
thermal conductivity   260
thermal decay modes   244
Thiébaux, H. J.   290
time constant: $\tau_0 = C/B$ 145
tipping points   21
top of the atmosphere (TOA)   14, 67
total solar irradiance (TSI)   11
"toy" Earths   4

transient climate sensitivity   254
triangular truncation   212
truncation   212
turbulence   7
two-mode approximation   131
two-slab model   262
two-stream approximation   61

u
US Standard Atmosphere   7
UD model   272, 282
unbiased estimator   287
uncertainty principle   86
uniform slab   73
univariate   7
unstable profile   75, 77
upwelling   281
upwelling-diffusion (UD) ocean   271

v
variance   5
variate   6
vibrational and rotational transitions   85
volcanic eruption   255

w
Ward, P. D.   116, 321
Warren, S. G.   322
Washington, W.   16
water vapor feedback   108
water vapor spectrum   81
water vapor window   81
Watt, James   92
wavenumber ($\tilde{v}$)   85
weather forecast   16
weather noise forcing   229
Weaver, C. P.   104
Weber, H. J.   292
Wendisch, M.   59, 83
Wu, Qigang   224

y
Yang, Ping   59, 83
Yung, Y.   65, 86, 101, 102

z
Zachos, J., C.   322
Zhuang, K.   224
zonal strip   121